FAT MANAGEMENT:

The

Thermogenic Factor

By
Daniel B. Mowrey, Ph.D.

Important: The information contained in this book is intended for educational purposes only. It is not provided to aid in the diagnosis or treatment of any disease, nor as a prescriptive guide. The author, publisher, printer, and distributor(s) accept no responsibility for such use. Those individuals suffereing from any disease, illness or injury should consult with their doctor. It is hoped that the education of the public about the principles of thermogenesis will lead to a more widespread acceptance of those principles by orthorox medicine with the result that a greater degree of validation and acceptance will ensue.

Victory Publications
P.O. Box 372
Lehi, Utah 84043

Victory First Edition 1994
ISBN 0-936261-07-2

Cover art and internal illustrations by Derek Hegsted
Cover typography by Rick Thayne Design

Acknowledgements

This book has been long in the making. Being the first popular account of thermogenesis, it has required painstaking research, assimilation and editing. My thanks to all those who contributed to the effort. Of special merit were the editorial inputs of Harold Stephens and Lonnie Andersen who read and reread draft material until they knew it better than I did. Their comments and advice (on type and graphics and content and everything else) altered (for the better) almost every chapter in the book.

For unremitting encouragement and stimulation, even when it was not desired nor appreciated, I would like to thank my colleagues Evan Bybee and Dennis Gay. May our friendship and association outlast the vagaries of the daily grind.

My gratitude to Derek Hegsted for interrupting his burgeoning artistic career long enough to provide the illustrations and cover art for this mundane little book, made greater by his effort. And thanks to Rick Thayne for his always professional advice on matters of design, and for the cover typography.

Special thanks and appreciation are extended to my "book widow and orphans," my wife, Vickie, and family who wondered if the months of emotional, and at times physical, separation would ever end. Without their support, patience, and loyalty, this book would definitely not have been possible.

In spite of all the assistance rendered by friends, colleagues and family, I must take full responsibility for any factual errors. In defense of the almost certain occurrence of mistakes, I muster the experience of countless others who have ventured the first synthesis of a huge body of theory and data into a coherent whole. It's a thankless task. However, although second editions are invariably much better, possession of the first edition of such a work as this is most rewarding in terms of long evenings of enjoyable laughs among the knowledgeable. To such as these, I simply throw myself upon their mercy.

Other books by Daniel Mowrey, Ph.D.

The Scientific Validation of Herbal Medicine

Guaranteed Potency Herbs: Next Generation Herbal Medicine

Proven Herbal Blends

Herbal Tonic Therapies

This book is dedicated to
people everywhere
with excess fat.
May these principles guide you
to ever smaller girth
and
ever greater health.

CONTENTS

INTRODUCTION

Although the title of this book clearly states that our primary concern will be **FAT** Management, many will perceive it as yet another book on dieting and obesity. Hopefully, something different will emerge from the reading. But even if it doesn't, even if we can never separate dieting, obesity and weight from fat management, we will continue to require a knowledge of the most recent theoretical, experimental, clinical and practical innovations in this field until the final solution has been found. At least a quarter of the total adult American population is grossly overweight, even more could be considered clinically obese, and in some minority groups (American Indians, black, Hispanic women) the rate of obesity is nearing 50%. There is a clear correlation between obesity and diabetes, heart disease, stroke, hypertension, gallstones, cancer, osteoarthritis, gout, respiratory disorders, psychological problems, sleep apnea, asthma and other debilitating conditions.

The obese are characterized by "Syndrome X," a combination of symptoms, including insulin resistance, glucose intolerance, hyperinsulinemia, increased very low density lipoproteins, decreased high density lipoproteins, and hypertension. Syndrome X significantly predisposes a person to morbidity and early mortality, usually following serious battles with atherosclerotic and coronary disease. As a population, we should be aware of new developments in the obesity research. Fat management through the manipulation of the body's natural thermogenic processes is emerging as the most important development in decades.

This book summarizes our current knowledge about thermogenesis, a new field in the general area of weight management. It reviews the work of pharmacologists, nutritionists and other scientists. Our primary goal is to provide the layman with enough understanding of this topic to enhance his or her ability to make sound choices in the market place and evaluate the suitability of thermogenic drugs, nutrients and herbs for personal consumption. We also wish to stimulate the imagination of researchers and professionals in the area of *FAT MANAGEMENT*, who could make further contributions to this relatively new science.

Controlling body fat by manipulating thermogenesis may be the first physiologically sound, gimmick-free, approach to the subject in the history of medicine. It does not rely on fad diets or flimsy physiology, but instead addresses repair of a real metabolic process, known to be deficient in many obese people. As such, the science of thermogenesis may elevate the entire study of obesity to a plane on which it will finally receive consideration by the larger medical community as a physiological condition, a disease with non-psychological roots.[148]

Thermogenesis refers to the body's ability to produce heat from consumed and stored calories. More specifically, it refers to thermogenesis in brown adipose tissue (BAT), an organ specialized in turning calories into heat energy. Although thermogenesis occurs in all cells of the body, scientists are increasingly beginning to believe that a great many obesity problems are caused by improper thermogenic metabolism in BAT. BAT thermogenesis is meant to take up the slack in energy metabolism, to get rid of calories not used by the body in its daily struggle to keep us alive and healthy. The number of calories involved varies but is usually equal to about 10-20% of the total number of calories consumed over a 24 hour period.

Research results from such seemingly disparate fields as hibernation, cold adaptation and weight control have converged on the importance of BAT metabolism as a single unifying concept. These data have important consequences for the sciences of pharmacology, general nutrition and phytopharmacology, enabling scientists engaged in these fields to make substantial contributions to our understanding and manipulation of thermogenesis in the field of Fat Management.

Endocrinological, neurological and genetic factors form the foundation of thermogenics, and effective thermogenesis relies on appropriate input from all three sources. The endocrine system supplies substances such as adrenaline, thyroid hormone, insulin, adreno-corticotropic hormone (ACTH) and corticotrophin releasing factor (CRF), all of which affect thermogenesis in brown adipose tissue and other tissues. The nervous system, especially the sympathetic nervous system (SNS), supplies the BAT-activating neurohormone norepinephrine (NE), while the central nervous system controls the whole process, translating external thermogenic signals such as exercise, food, cold, and drugs into internal thermogenic stimuli to brown adipose tissue.

Perhaps the most profound effect on thermogenesis is provided by genetic factors. It is becoming increasingly apparent that genetic faults are the basis of faulty thermogenesis in BAT, and that faulty thermogenesis is a major contributor to clinical obesity. While much research remains to be done before we will understand all of the genetic, neural and hormonal intricacies of thermogenesis, physiologists, nutritional scientists and pharmacologists are already busily discovering ways to reignite long-extinguished BAT furnaces. Success in this area has shown that obesity can be reversed.

We are concerned that nutritionists understand the importance of proper dietary manipulation of thermogenesis in brown adipose tissue. Diet is supposed to be a major stimulant of thermogenesis. Diet-induced thermogenesis takes place in BAT; when BAT does not function properly, this form of thermogenesis fails to occur, and the calories are then stored as fat. We suggest that paying attention to the nature of the nutrients we ingest can substantially impact the efficiency of BAT, and that when pharmacological intervention is combined with nutritional intervention, the result is enhanced BAT metabolism and less fat storage.

We are even more concerned that pharmacological manipulation of BAT does not lead to the creation of drugs that might possess side effects more dangerous than the good they do. Recent attention has been drawn to the difficulties inherent in manufacturing drugs that are free from contamination by so-called "mirror image" molecules, i.e., chemicals that possess a reverse structure to the target drug but which pose life-threatening consequences to the consumer. Natural

3

materials are not subject to this kind of contamination. Pharmaceutical firms that ask why herb companies can make claims for food-grade products without spending the same millions of dollars on research that is required by the drug companies, might find an answer in such differences between synthetic materials and natural materials as the mirror image phenomenon.

Fat loss is an area in which vanity sometimes rules the day, and we have seen instances in which the Food and Drug Administration has been willing to permit questionable weight-loss drugs to be sold by pharmaceutical firms that could seriously disturb health. At this writing, for example, a major pharmaceutical firm is manipulating public opinion in favor of an experimental weight loss drug (a so-called 'fat blocker') that is still 2-3 years away from FDA approval. The idea, obviously, is to create such a public demand for the product that FDA will be pressured into approving it even though the firm readily acknowledges the possibility of serious, health-threatening side effects. We do not want to see this happen in the field of thermogenics. Several novel thermogenic drugs are currently being tested. Universally, these have as a serious side effect the tendency to disturb heart rhythms. Currently, the only viable thermogenic drug is ephedrine hydrochloride (and closely related chemicals). For this reason, the major focus in this book will be on this drug, as far as pharmacological manipulation is concerned.

Finally, we are concerned that phytopharmacological (plant-based) manipulation of thermogenesis be carried out in a safe and reasonable manner. Since ephedrine and other thermogenic agents are derived directly from the plant kingdom, it is possible to consume the plant materials instead of the drugs. Many companies specialize in the encapsulation of these exotic "foods" for the consumption of persons desiring to add them to their diet. Strictly speaking, phytopharmacology is a subdivision of nutrition, but a separate science is necessary since most nutritionists do not concern themselves with exotic foods. Phytopharmacology is distinct from orthodox pharmacology in that it deals with whole plant materials, and their common extracts, rather than with the creation of drugs based on isolated plant chemicals. Because phytopharmacology is presently not considered a mainstream science in the United States (unlike Europe and Asia where it is a science of considerable importance), it is not subject to the same kind of regulatory control as the orthodox pharmaceutical industry. Our concern is that manufacturers of phytopharmaceuticals comply with good manufacturing practices and do not foist on the public hastily created products of questionable scientific validity.

The science of thermogenics should revolutionize our way of thinking concerning how and why people accumulate fat and what should be done to defeat that problem. Currently, the best thermogenic solution to fat management involves the daily consumption of products containing ephedrine and caffeine in conjunction with aspirin, or their natural equivalents, precisely balanced for optimum synergy. This synergistic combination activates the thermogenic metabolic process in brown adipose tissue and results in the physiological incineration of dietary and stored fat. The outcome of this process is an overall significant decrease in fat stores throughout the body.

Notice that nowhere in that statement did I use the word "weight" or the term "weight loss." Part of the revolution involving thermogenesis is the necessity to rethink traditional terminology regarding the relationship between health and body composition. The old terms, such as weight, weight loss, weight management, etc., are gradually giving way to what are hopefully more accurate and useful terms such as fat, fat/lean body mass ratios, Fat Management, etc. Use of bathroom scales is being replaced by tools for the analysis of body composition, such as bioimpedance machines, infrared measurements, and skin-fold calipers.[960]

<u>Fat</u> is the target of the new technology. The reduction of fat is seen by most people as the answer to obesity. However, obesity has now been officially designated as a disease by the government (not the medical profession). This action makes diagnosis, treatment and cure of the disease obesity strictly the province of orthodox medicine. Since much of the material in this book deals with nutrients, plants and other food materials, as well as exercise items that fall

5

outside the area of orthodox medicine and pharmaceutical manipulation we are forced to direct our attention to something other than obesity. That "something" is <u>fat</u>.

Whether reduction of body fat impacts on the disease called obesity may ultimately be a decision only a medical doctor can make. Meanwhile, those of us that must deal with practical issues involved in preventing and reducing the amount of fat on our bodies, will continue to make use of knowledge about this subject, no matter what its source. Ultimately, we are only interested in what works.

<u>Thermogenesis is not a cure. Currently, thermogenesis is a simple physiological process, like digestion. This point cannot be stressed enough.</u>

<u>The application of drugs, nutrients and other materials described in this book requires that extreme attention be paid to ratios and standardization.</u> The when, why and how are the subject of this book. The pharmacological approach is currently patented and awaiting approval for drug manufacture. There are currently many manufacturers of nutritional and herbal products, doing so without regard for research findings and attendant precautions; products manufactured without these consideration, can, and do, pose a health risk to consumers. This book has been written to serve as a guide for consumers interested in exploring the application of thermogenic substances in the pursuit of lean body mass control and fat management.

PART ONE is a brief and simple explanation of thermogenesis. It is a non-technical introduction to the process that is discussed in increasingly greater detail later in the book. Many readers will find PART TWO to be a much more satisfactory explanation of thermogenesis. In fact, PART TWO is really the heart of the book for 90% of the readers. Here you will learn more about what thermogenesis is, how it works, what goes wrong with it, and most importantly, how to restore it. In addition, you will learn how it relates to exercise and dieting, information about the side benefits, how to use dietary thermogenic substances, and who should or should not use such agents.

PART THREE is written for those readers requiring an extensive understanding of the cellular and metabolic control processes involved in thermogenesis. Many lay readers will also be attracted to some chapters in this part of the book as their understanding of thermogenesis increases. Certainly, **product manufacturers** should carefully read PART THREE before embarking on the compounding and production of thermogenic agents. Also interspersed throughout PART THREE is information derived from research being conducted on plant- and nutrient-induced thermogenesis by the American Phytotherapy Research Laboratory (APRL).

Extensive discussions of the drugs, herbs, nutrients and other materials that impact on thermogenesis are also provided in PART THREE. Again, this material should be of interest to lay and technical readers, and *is required reading for manufacturers.*

In the chapters of this book should you choose to wade through them you will probably learn much more about thermogenesis than you ever cared to know. On the one hand, you may have to deal with big words and dense concepts. On the other hand, you will be learning about the most exciting concept in physiology. It may seem strange to say so now, but we believe that by the turn of the century, the term "thermogenesis" will have entered the vocabulary of almost everyone, and the concept of thermogenesis will have become as ordinary as bubble gum, and as common a word as digestion.

THE APRL SEAL OF CERTIFICATION

American Phytotherapy Research Laboratory (APRL) was formed to carry out basic, archival and clinical research on medicinal plants. Among several other projects, APRL has been involved in thermogenic research for over three years as of this writing, and during that time has established a set of criteria regarding natural, phytopharmaceutical, thermogenic products. These criteria, derived mostly from published literature on thermogenesis, partly on existing patents, and partly on APRL's own research, are used to develop

safe and **effective** products. Not just safe, and not just effective, but both safe and effective. Sadly, there are already a host of ineffective and/or unsafe so-called thermogenic products in the marketplace.

The presence of the APRL Seal of Certification on the label or bottle of thermogenic product is a sign that the product adheres to the standards set by APRL.

NOTE: There are things happening daily which impact the industry producing thermogenic products. For example, in October of 1991, Mr. Mitchell Friedlander of Atlanta, Georgia, was issued a United States Patent for a thermogenic composition consisting of ephedrine, caffeine and aspirin, or their equivalents. This patent is based on the wealth of research that has been generated on this concept (Friedlander assisted in some of that research by financing, and providing the chemicals used). Mr. Friedlander is additionally in the process of completing a New Drug Application through the U.S.F.D.A. and is seeking approval for a new weight loss drug containing these chemical ingredients. The FDA, with their commitment to safeguard the nation's health, will, in all likelihood, move with caution in reviewing the research, the clinical trials and the available endorsements of this chemical composition, before issuing its final approval. This process is necessary and may eventually result in an approved over-the-counter drug and the creation of a new sub-category of fat management agents based on thermogenics.

In the meantime, food supplement products, using standardized natural ingredients, will be used, providing valuable additional personal and clinical data to support the thermogenic concept. The value of that data may be positive or negative, based on the care and quality of the products formulated and produced. The unscrupulous will attempt to avoid the rights of the patent holder, and abuse the quidelines currently required by the FDA. Beware of products that are not solidly formulated according to known research-based standards. You have a right to ask for the identity of the formulators, their background, the research basis upon which the product is produced, the nature of the production facility where it is made. We believe that the presence of the APRL Seal of Certification is one indication that the product meets the highest standards of formulation and production. You may also await the arrival of approved products manufactured under the auspices of the patent.

ATTENTION: By the time this book is printed, circumstances may have already changed significantly. FDA is currently seeking to restrict the trade of all food supplements. Vitamins, minerals, herbs and related substances may be systematically removed from the shelves of retailers. The FDA seeks to ban weight loss claims and to even ban weight loss nutrients themselves, from the public domain. Whether you view this action as an infringement on your right to choose how to deal with your health, or are grateful for governmental regulation designed to control this aspect of your life, the end result will be the same: a tremendous increase in the price of thermogenic materials. No longer free to choose relatively inexpensive, natural based products, from the grocery store or health food store, you will be forced, once they are finally approved, to purchase synthetically produced drugs from the pharmacy, either by prescription or over-the-counter, depending on the type of safety data that is generated in current research.

PART ONE: INTRODUCTION TO THERMOGENICS

Let's set the record straight from the outset: "Thermogenesis" is not the name of a "hot rock 'n' roll band." Nor does it refer to some big-bang related "thermonuclear event of creation." Nor is it "cold weather underwear for genes." "Thermogenesis" is your God-given right to be thin. Despite its length, the word is not difficult. Its meaning is transparent: *thermo*- means heat, as in thermometer or thermostat; *genesis* means create, as in the first book of the Bible. It is a process that takes place in the bodies of many species of animals, including man. Thermo + genesis means the creation of heat, something the human body must do to maintain body temperature at 98.6 degrees F.

Thermogenesis also separates the thin from the fat! If you burn fat through thermogenesis, you are probably thin, or lean. If you don't burn fat, you probably store it, and if you store it, you and everybody else knows it, because it is fairly obvious. Wouldn't it be nice if we could reverse the tendency to store fat? What if you knew that beginning today you could reeducate your body to burn fat instead of store it? Would that be good news? There are at least two important reasons why it should be: health and vanity. We all want to feel good and look good. And there is nothing wrong with that. Reigniting thermogenesis may be the key for you. At least it is worth a try.

Chicken or the Egg

We can begin our discussion by addressing a question that has plagued researchers for many years. While it has been clear that defective thermogenesis and obesity go together, do defects in the thermogenic process actually produce obesity, or does obesity produce the defect?[1256] Just because the two are correlated doesn't mean there is a cause-effect relationship here. And if one does cause the other, which comes first? It's a chicken or the egg sort of thing. Surprisingly, it wasn't until recently that this question was addressed in a serious experimental manner.

In 1992, a group of French scientists recruited thirty-two women for participation in a remarkable experiment.[729] At 7:30 in the morning, following a 10-12 hour fast, each woman was led into a special, environmentally controlled room, was placed in a relaxed condition, and hooked up to a variety of measuring equipment. In fact, in this room, the scientists could measure even minute changes occurring in the atmosphere of the entire room. Then each woman was fed a quantity of pure glucose. The scientists measured how much of the glucose was used for basal metabolic processes, how much was stored as glycogen or fat, and how much was used in thermogenesis.

Some of the women used a large portion of the glucose in thermogenesis, and stored very little as fat. But others used very little glucose in thermogenesis. Lean women, whose weight profile fell within standardized norms, were those who reacted to the glucose load with increased thermogenesis. Overweight women were those who did not exhibit thermogenesis. Was the thermogenic defect a cause or a result of obesity? To answer this question, a third group of women were included in the study. These were individuals who had just recently reported the onset of obesity; some were not even yet overweight. Yet all of these women already exhibited the thermogenic defect. The scientists concluded, "The occurrence of this defect in the dynamic phase of obesity (onset) suggests that it is already present in subjects who do not yet have other metabolic features of long-term obesity." The magnitude of this defect was equal to that of the women exhibiting long-term obesity. Furthermore, it was determined that the defect in thermogenesis was not a result of defects in glucose storage; it was a true defect in the thermogenic process itself.

So now you must ask yourself the question, "If I am overweight, what are the chances that it was caused by a thermogenic defect." If you share our feelings, you have to admit that the chances are very good. If that is so, then what are you going to do about it? We suggest that you study this chapter and the next very closely, and then go about obtaining some of the nutrients, herbs or drugs that will help reverse faulty thermogenics. Then, as you introduce your body to the benefits of rejuvenated thermogenesis, you may read/study the remainder of the book to learn what is really happening.

The Fate of Excess Calories

The research presented above is typical in the field of thermogenics—the study of what happens to consumed caloric energy remaining after the body's energy demands have been met. Obesity results when the amount of energy consumed in the diet exceeds the amount of energy expended by the body. While most consumed energy is expended in maintaining bodily processes, numerous calories remain. How does the body dispose of those calories? This question has led researchers directly into the field of thermogenics, the body's main tool for ridding itself of excess calories. As the above study clearly shows, if the body loses its thermogenic capacity, obesity is certain to occur. It has been estimated that even a 0.1% deficit in the number of calories expended through thermogenesis could result in the accumulation of excess fat to the tune of 25% of body weight, i.e., obesity. And a 1% deficit over the 40 year period of adult life would yield a fat value of up to 40% of body weight.[426] Considering that men should be no more than 15-20% fat and women should be no more than 20-25% fat, this small deficit probably accounts for a great deal of observed fat problems.

In this book, you will learn that thermogenesis is one of the body's most vital physiological processes, like digestion. Though we still have much to learn about thermogenesis, we know that the presence or absence of this process makes a significant impact on the health of the human body. Thermogenesis is important for a number of reasons. Consider the following questions:

Why don't hibernating animals freeze? We know they save up fat for the winter. This fat feeds them so they don't die of starvation. But it also provides a source of fuel which is burned to provide heat.

In order to accumulate the fat, the thermogenic process must shut down for a time. Thus, there appears to be a genetic "trigger" for turning off thermogenesis. Do humans have a similar trigger that is activated by genetic cues that respond to climate, age, health conditions, dietary conditions, or certain behavioral traits? People may become overweight simply because their bodies are preparing to survive the hard times ahead. It is as if a switch is turned on—a fat switch—one minute you are not fat, the next minute you are! That may be precisely what is happening. The fat switch turns off thermogenesis. Is it possible to turn it back on? The answer appears to be affirmative.

Why do small children so easily tolerate the cold? Which of us doesn't know of the child in the neighborhood who seems to go all winter long without a coat? It can be freezing outside and this kid will be out there in his short sleeves, without a coat, fixing his bike. I have a neighbor just like that. As I watch him puttering around the backyard oblivious to the cold, I wonder, "Could we all do that if we had just done things differently when we were younger? When is this boy's thermogenic mechanism going to fail? Or will it ever?" Most lean people, who must have functioning thermogenics, don't tolerate the cold in their adult years like they could when younger. Just what are the limits to thermogenesis? Younger children have considerable thermogenic capacity that is slowly lost as they get older. Can they reclaim that ability later under the right circumstances? What are those circumstances? Are we dealing with the immutable forces of genetics, or is there some degree of flexibility involved in the thermogenic process?

Why does eating less not necessarily lead to weight loss? Why does dieting (weight cycling) usually lead to more fat, instead of less? Caloric restriction reduces basal metabolic processes and hence the amount of energy expended. Since thermogenesis is a metabolic process, restricted caloric intake may actually lower the rate of thermogenesis. Ingesting fewer calories may trigger the fat switch mentioned above. This could involve interference with the thermogenic process caused by the erratic intake of calories. Typically, a person going on a calorically restricted diet does not experi-

14

ence anticipated weight loss and decides to end the diet. When caloric intake is increased, however, basal metabolic rate does not return to pre-diet levels; hence, the body has *more* excess calories to deal with than before the diet! These calories are candidates for fat storage. The odds of building fat stores are increased even further where some kind of genetic defect in the metabolic processes of thermogenesis exists.

Two Types of Thermogenesis

In summary, thermogenesis is a fundamental physiological and metabolic process as important as any other body system, including digestion and respiration. When you enter a cold room, thermogenesis kicks in to keep you warm. There are two main types of thermogenesis. The first takes place in your muscles; it is called shivering thermogenesis. Shivering speeds up the metabolism in muscle cells, producing heat and helping to restore or maintain the appropriate body temperature. The second kind of thermogenesis does not involve shivering and it does not take place in muscle cells. It involves metabolic heat production and it takes place in a special organ of the body called Brown Adipose Tissue, or BAT.

Although we will discuss all forms of thermogenesis in this book, our main focus will be on thermogenesis occurring in brown adipose tissue. BAT thermogenesis is very special. The amount of heat generated in brown adipose tissue is substantial, as opposed to the relatively small amount generated by shivering. Even though the immediate response to cold is shivering, long term exposure to the cold requires heat generated metabolically. This is the role of BAT thermogenesis. It provides the greatest degree of protection when you enter a cold climate. Human infants do not shiver. They have such well-functioning brown adipose tissue that they don't need to shiver. BAT-generated metabolic heat does the total job of preventing hypothermia in infants. As children age, they are eventually struck with an involuntary bout of shivering. Most people probably can't remember when they shivered for the first time. However, if you observe children and catch them during their very first real shiver, you will see them expressing wonder and concern about this strange new feeling. This new sensation is the result of decreasing BAT stores relative to increasing body size.

Losing BAT

What happens, then, between infancy and childhood to cause the sudden appearance of shivering? Scientists believe it is the gradual disappearance or reduction in the amount of BAT. As a person loses this ability to generate heat, he or she must rely on shivering to an increasing extent. No one really knows why BAT "atrophies", or shrinks, but it is believed to involve certain genetic factors combined with certain behavioral factors. The genetic factors are discussed in PARTS TWO and THREE. The behavioral factors involve such things as dressing warmly every time you go out into the cold, turning up the thermostat of your home during the winter, and so forth. Each time you protect your body from the stimulus of cold (or colder than normal), you deprive it of the stimulus that keeps BAT functioning normally. Since the act of eating triggers BAT thermogenesis, continued caloric restriction or dieting will also tend to shut BAT down.

It has been speculated that man is basically a tropical animal, not meant to live in cold regions of the world.[1306,1307,1333,1334] BAT would probably suffice to keep the naked human comfortable during all climatic changes in tropical or near-tropical realms. But in colder climates, man quickly adopts dress, housing and dietary habits to minimize the trauma associated with changes in season.

Caloric Furnaces

By the time the human reaches adulthood, very little active BAT remains. This is unfortunate, since BAT is extremely important in one other regard: <u>BAT burns up excess calories.</u> The fuel that BAT burns to create heat comes from fat in the food we eat and from fat stored in the body. Imagine a situation where BAT is performing at a high metabolic rate, voraciously consuming as much fat-fuel as it can, shutting down only briefly when it has incinerated a large amount of fat. Suppose that is the normal situation. Now suppose that BAT becomes sluggish (due perhaps to a combination of genetic and behavioral factors). Operating at far less than peak efficiency, the BAT is not capable of consuming many of the calories taken in during a day's meals. What happens to the calories not consumed by BAT or used to meet the body's daily energy needs? Will they simply evaporate, or find some convenient place to hide? Probably not. As disappointing as the truth can sometimes, be excess calories are typically just deposited conveniently in our fat banks, drawing interest on our health. We need to find a way to withdraw these deposits and put them to work. Until then, we will just continue to acccumulate more ugly rolls of fat. Of course, we are not implying that those of us with excess fat are ugly; we are stating that excess fat can lead to ugly consequences, such as loss of vitality, susceptibility to heart disease and other diseases, and a shortened life span.

Our expanding knowledge of the role of BAT has led many scientists to believe that the accumulation of fat is less a function of eating too much, and more a function of inefficient BAT. It has been shown that BAT is capable of making a significant difference in the fate of ingested calories. As little as a 10% deficiency in BAT function can theoretically result in the accumulation of a three or four dozen pounds of fat over the course of a normal lifetime. If the BAT is in even worse shape, as it certainly is in most adult humans, compensating for the impact of this deficiency on health by dieting or exercising or any other means becomes increasingly more difficult.

BAT thermogenesis is therefore an extremely important physiological process; its proper or improper operation is fundamental to a person's fat profile. Notice that we have used the term 'fat' instead of 'weight' throughout this discussion. With the discovery of the importance of BAT, our understanding of weight management has undergone a total reevaluation. It is, frankly, misappropriate to label our fat problem as a 'weight' problem. The problem is not how much we weigh on the bathroom scale, but how much fat we have. More specifically, it is the amount of fat in relation to the amount of non-fat, or lean, body mass that is the important measure. This relationship is sometimes expressed as a ratio—the pounds of fat to the pounds of non-fat—the fat/lean ratio. Special techniques have evolved to measure this ratio. Those techniques do not include the bathroom scale.

Restoring BAT Function

The science of thermogenesis is concerned with how to restore BAT function in normal adult humans. Exposure to cold is one way, but not a very convenient or necessarily effective way. Some BAT function can certainly be restored by taking cold showers, leaving your coat at home during the winter, and otherwise re-introducing the concept of cold into your life. However, it has not been shown that <u>enough</u> BAT function is restored through this means to create sufficient thermogenesis to make a difference in the amount of fat we have. In other words, it takes only minimal BAT activity to compensate for cold, but it takes substantially more BAT activity to incinerate enough fat calories to make a dent in body fat stores.

Without being able to rely exclusively on cold exposure to help us in our quest to lower fat levels, we must seek a solution for our problem in another direction. Many decades ago it was observed that activation of BAT tissue could be accomplished through the consumption of certain substances in the diet. The most effective of these substances are a plant known as ma huang—a Chinese herb used by humans for at least 5000 years—and its active constituent ephedrine. These substances stimulate large amounts of activity in brown adipose tissue. The daily consumption of large amounts of ma huang or ephedrine results in BAT activation, the growth of new BAT, and increased thermogenic capacity. Under the influence of ma huang or ephedrine, long dormant BAT begins to come alive and all of the metabolic processes that support BAT activity begin to obtain new vitality. Where only small pockets of BAT were present, soon larger masses of this unique tissue appear and regain the normal range of function this organ was designed to exhibit.

Unfortunately, it turns out that we pay a high price for the consumption of <u>effective doses</u> of ephedrine or ma huang. When consumed in the large quantities necessary to activate BAT, these substances are almost certain to produce unwanted side-effects, especially on the cardiovascular system. This fact has prevented the wholesale application of ephedrine and ma huang for thermogenic purposes for many years. The FDA requires that ephedrine, in doses of more than 150mg, be dispensed by prescription.

Solving the Ma Huang Problem

Recently, however, the problem of ma huang safety has been solved. Scientists have discovered that combining ma huang or ephedrine with just the right amounts of caffeine and aspirin simultaneously augments the action of the ma huang or ephedrine and reduces the probability of side effects. Hence, a thermogenic response is possible with smaller amounts of the herb or the drug, and those smaller amounts should not create the negative side effects normally expected. This two-fold increase in the safety and effectiveness of ma huang and ephedrine means that you can now benefit from the consumption of ma huang-based thermogenic products without serious risk.

Research on thermogenics has primarily relied upon the synthetic materials (ephedrine hydrochloride, synthetic caffeine, and aspirin). Meanwhile, research at APRL (American Phytotherapy Research Laboratory) has attempted to achieve similar results utilizing the more natural forms of these substances where possible. Obviously, from a wholistic point of view, the use of whole ma huang is preferable to the use of the drug ephedrine hydrochloride, and the use of plants containing caffeine is more desirable than the use of synthetic caffeine. However, even here, the material that performed best in APRL's research was standardized to contain higher than normal levels of those alkaloids. Standardization has made producing thermogenics both cost-effective and reliable.

Without the use of standardized material, which is somewhat higher in price, reliability is sacrificed and the final product is rendered unfit for human consumption. Aspirin can be produced from salicin, derived from willow bark, combined with acetic acid as found in vinegar, or it can be synthesized from aniline dyes. Naturally, the former process, though more expensive, is most suitable for a food-grade product. These concepts are explained more fully in the chapter on Phytopharmacology, which is the science of plant-based medicinal compounds.

Cautions

Bringing thermogenics together has been a revolutionary process. It has required standardizing key ingredients and performing extensive analytical research. Now, for the first time, the extraordinary synergy of these simple materials has been confirmed, and clinically observed results have been obtained in humans.

Nevertheless, some people should avoid using the thermogenic[1]. Pregnant women, women planning to become pregnant, and nursing mothers should not use thermogenic as it may affect the nervous system of the fetus or infant. Men with serious benign prostatic hypertrophy may experience an increase in symptoms. Pre-teens should use it only as recommended by a doctor. Diabetics and people with serious hypertension should exercise some caution at first. Some people will experience mild constipation which can be prevented by the consumption of small amounts of a mild laxative, such as cascara sagrada or psyllium. Of course, people who know they are aspirin-sensitive, should only use these products under their doctor's supervision.

Those precautions aside, the use of a correctly formulated thermogenic appears to be free of serious risks.

Even though the preparation of an effective and safe thermogenic is fairly easy, it is surprising how many poorly manufactured products appear on store shelves. Please read or reread the Preface of this book to learn how to make an informed decision about which product to use.

In PART THREE we will discuss the various thermogenic constituents in greater detail. For now, please note only that the thermogenic process can be greatly influenced by other herbs and nutrients. These nutrients augment the ability of the thermogenic to mobilize fats from fat stores, the transportation of these nutrients to brown adipose tissue, the ability of brown adipose tissue to burn the fat, and the ability of the body to deal with a host of dietary fats more effectively.

USING THERMOGENIC: SUGGESTIONS, CAUTIONS, & OBSERVATIONS

Here are some suggestions about how to use thermogenic. Note: These suggestions a based on the assumption that the daily dose is split equally between three capsules or tablets of thermogenic.

1. **Take with a large glass of water. This will help disperse the contents of the capsule or tablet in the stomach, and help prevent irritation and nausea, especially for first-time users.**

2. **Take 1/2 hour before or after meals for maximum impact on stored fat. Thermogenesis appears to use up energy derived from food before it produces heat from calories derived from energy stores (as in fat). Allow sufficient time after meals to allow food to begin clearing from the stomach.**

3. **Take with meals for maximum impact on dietary fat (i.e., to prevent it from being stored as fat). This is a simple corollary of suggestion number**

4. **This approach is especially appropriate during holidays when we are prone to eat more than we should of exactly those kinds of foods we should avoid.**

5. **If any jitteriness or nervousness occurs, take with meals and a large glass of water. This will reduce the immediate impact on the body, smoothing out the action of the product. Nervous symptoms will usually disappear completely in a few days. Another approach is to consume a standard tablet of aspirin along with the thermogenic; however do not do this unless you are sure you are free from aspirin sensitivity.**

6. **Also, to minimize impact, start the thermogenic program by taking just one capsule (or even 1/2 capsule) per day, in the morning, and one capsule around noon. Basically, each individual should find his or her own level of consumption, and adjust it from time to time to reflect changing body composition, adaptation, stress levels and so forth. There is no magic guideline that applies to every person. If you**

experience some discomfort at first, do not take this as a sign that you can not use the product at all. Before giving up, try making some adjustments in dosage schedule, time of day, with or without meals, reduced dosage, taking it every other day, or even once a week. Feed your adrenals, clean your body and lower the amount of stress in your life.

7.   If you begin with one capsule or tablet in the morning, increase amount consumed to 2 capsules in the morning and one around noon after one to two weeks.

8.   For best results, do not use after 3 P.M. Most of the day's work is done by early evening. The thermogenic will have a tendency to keep you going later. Taking it after 3:00 may make you stay awake later than you desire at night.

9.   Use five days a week; reduce dose to one a day for two days a week, or, better still, do not use at all for two days per week. This allows the adrenals and nervous system to rebound.

10. To avoid jitters, do not use with other caffeinated beverages, such as coffee, tea and colas. The extra caffeine obtained from these drinks may upset the delicate synergistic balance of the caffeine-ephedrine-aspirin combination in the thermogenic.

11. If you are trying to stop smoking, a thermogenic can help prevent weight gain. Nicotine is itself a mild thermogenic agent. When people stop smoking, the absence of nicotine may depress thermogenesis, resulting in weight gain. The traditional gain in weight experienced by smokers who quit is normally attributed to an increased appetite for sweets. It is, more correctly, a defect in thermogenesis. Thermogenic will help restore thermogenesis in such instances.

12. Aspirin-sensitive people should exercise caution in consuming a thermogenic. Although properly formulated thermogenics do not contain enough aspirin to cause a problem for most aspirin-sensitive persons, individuals with a known serious sensitivity would do well to avoid the use of such products except as directed by their health care professional.

13. People with severe food allergies should exercise caution when consuming this product. As with all food products, there is always a chance that some people will be allergic to one of the constituents.

14. Diabetics seem to do best on one capsule in the morning and one capsule at noon, seven days a week. A reduced dosage has less impact on the adrenals and on glycogen metabolism. A seven-day-a-week program reduces the tendency for ups and downs in metabolic systems that might be upsetting to the diabetic.

15. Constipation and/or aggravation of pre-existing benign prostatic hypertrophy may occur in some people. Use of a mild laxative will help offset constipation. Use of a compound for reducing the symptoms of BPH will help reduce the tendency of thermogenic to aggravate that condition.

16. People with larger frames may be able to consume more than the recommended dose; on the other hand, persons with small frames and little fat may find it more effective to use fewer than the suggested number of capsules.

17. As mentioned previously, pregnant women, women attempting to become pregnant, and nursing mothers should not use the product as it may affect the health of the fetus or newborn infant.

18. Some authorities believe that pre-puberty children should not use thermogenic on a long term basis because their nervous systems are still in the developmental stage.

19. Persons using medications of any kind for any kind of health problem should consult with their doctor to obtain advise about possible interactions before using thermogenic. Not a great deal of work has been done on drug interactions with thermogenic. The key phrase to describe this product's mode of action is "mild sympathomimetic, buffered to produce minimal cardiovascular action."

20. The product may induce a cleansing cycle in some people, especially smokers and coffee drinkers. The cycle is short-lived. Possible symptoms include moderate headaches, skin eruptions and nausea. This is different from allergic reactions which will be long-lived and will usually create more severe symptoms such as hives and rashes.

21. Empirical data suggest that thermogenic has a mild anti-depressant action; persons currently using an anti-depressant drug should be made aware of this factor.

Following these suggestions and adapting them to your individual circumstances should maximize the effectiveness of thermogenic for you.

---

[1]Here, and throughout the remainder of this book, I have coined the term "thermogenic" or "the thermogenic" as a noun, to replace the cumbersome phrase "the thermogenic agent" or "thermogenic substances", etc.; before now, no simple noun-term or phrase had been implemented.

PART TWO: INTERMEDIATE LEVEL THERMOGENICS

THE BASICS

In this part of the book, we expand on the material presented in PART ONE (and repeat some), providing more detail and answering additional questions about how to use thermogenic products. To understand the basic and fundamental importance of thermogenesis, consider first of all that there are only a limited number of things the body can do with ingested calories (remember that food energy is measured in calories): Use them; Store them; or Waste them.

Use. Ingested calories are used primarily by the body to meet basic energy demands, including digestion, cardiac output, cellular metabolism, muscle contraction—that sort of thing. The fundamental measures of bodily energy needs are *basal metabolic rate* and *caloric burn rate*. Once those needs have been satisfied, *excess* calories must be dealt with. Excess calories are either stored as fat or turned into heat (thermogenesis) and removed from the body through convection and radiation.

Storage. Excess calories can be stored as fat, white fat, ugly fat. For many decades the field of nutrition and diet labored under the misconception that storage was the only thing that could happen to excess calories. That is why so many diet programs encouraged dieters to reduce caloric intake; it was felt that a balance between caloric intake and energy output could be achieved. Unfortunately, this concept failed to take into account many physiological mechanisms the body employs to compensate for caloric restriction. The frustrations of trying to fight fat storage through dieting may become a thing of the past as people learn about a relatively recently discovered alternate method the body employs to deal with excess calories: thermogenesis.

Waste. Contrary to popular opinion, the body has a fair amount of inefficiency built in. Many common metabolic cycles have overflow and waste cycles built in, so that the amount of energy entering the cycle is greater than the usable energy that exits the cycle; some energy is simply lost during the metabolic events that comprise the cycle.

The Organ of Caloric Waste: Brown Adipose Tissue

In addition to the body's natural ability to eliminate foods not digested and assimilated, the body has an organ whose primary job is to waste excess calories obtained from food that is digested and assimilated, and perhaps already stored away as fat. This organ, when activated by signals from the brain, begins to burn up excess calories at a very rapid rate. The process, as we have already learned, is called thermogenesis. The organ is called brown adipose tissue (BAT). When BAT is not functioning, thermogenesis fails, and the body falls back on the storage method for dealing with calories.

Because few of us take the time to determine the number of ingested calories needed to exactly match the number of calories required for maintenance, work and play energy, we must rely on BAT to burn off the excess. When BAT fails to do this, it is impossible to avoid fat accumulation through caloric restriction. Since increasing caloric burn rate and basal metabolic rate through exercise is a difficult, inexact and unattractive method for many people, the most efficient mechanism for dealing with excess calories is activating thermogenesis (but, see information on exercise in later chapters). You should also remember that just restricting caloric intake is not a magic answer since, as inefficient as our body systems may sometimes be, they quickly sense caloric deprivation and slow the metabolic rate in an effort to conserve fat stores, to assure continued ability to conduct our daily activities. It has been suggested that this physiological process was critical in early man who, when hungry and underfed, still had need to search for sustenance and escape those predators who pursued him.

The thermogenic approach has only recently been tried in man, simply because nobody before now could find a way to do it safely and effectively. In fact, until just a few years ago, scientists uniformly taught that manipulation of thermogenesis in man was impractical, if not impossible. Things have changed. Many researchers now believe that the proper stimulation of thermogenesis is possible, but that it must be approached carefully if we are to avoid problems. We believe that problems arise from improper pharmacological manipulation, and that a more wholistic approach poses fewer risks.

Thermogenesis occurs in virtually every cell of the body, but there is only one place where thermogenesis is the primary process. That place is brown adipose tissue. BAT is a different kind of fat than the fat that makes you look fat and that endangers your health. Bad fat is often called white fat, white adipose tissue, or WAT. The function of WAT is to store calories as fat. The function of BAT is to burn calories as heat. So BAT is 'good fat' and WAT is 'bad fat'. BAT is a special organ of the body whose soul purpose is to create heat. BAT is located between the shoulder blades, along the spinal column, atop the kidneys and in other discrete locations. Some experts believe that BAT may even occur intermixed with white adipose tissue. (See Plate I)

When BAT thermogenesis is working, excess calories are directed toward BAT where they are incinerated and dissipated as heat. When thermogenesis is not working, those extra calories are directed toward WAT where they are converted into stored fat molecules.

In summary, thermogenesis in brown adipose tissue serves two critical functions: One, it prevents fats consumed at meals from being stored as white fat; and two, it burns up stored calories acquired from white fat stores. These actions have far-reaching consequences on our overall health.

THE UNCOUPLING PROTEIN

One of the most important findings in thermogenic research was the discovery of a small molecule now known as the *uncoupling protein* (UCP). Uncoupling protein is one of the most unique molecules in the body, not because it possesses some remarkable biochemical structure, but because it performs a function unlike any other substance. First of all, UCP is found only in BAT cells. Secondly, it does not contribute to the normal metabolic needs of the body. In fact, it disrupts, "uncouples," or derails the train of biochemical events that other cells use to turn calories into energy for use by the cell. Such derailed energy goes nowhere and does nothing—it is simply dissipated as heat. Furthermore, while other cells are limited as to the amount of energy they can use, BAT cells, due to the presence of the uncoupling protein, continue to convert calories into heat as long as they are stimulated, and as long as you have white fat to feed them.

To repeat, heat generated in BAT metabolism of fat-derived calories is simply wasted; it radiates away from BAT into neighboring tissues. Since the flow of blood through BAT is very extensive, the heat is picked up by the blood and rapidly carried away and dissipated at sites distant from the BAT. Thus, some people may experience a sensation of heat when consuming thermogenic, while others who possess a more active cardiovascular system and/or more extensive BAT innervation of vessels, may experience little heat sensation as the heat is carried away from the BAT sites very rapidly.

In non-BAT cells of the body, calories are turned into energy through a special process called oxidative phosphorylation (about which we will have much to say in Part III). Oxidative phosphorylation is a cascade of enzyme-mediated events that is the very foundation of life. In active brown adipose tissue, the calories are diverted into a 'thermogenic cascade' that offers up heat instead of energy. Adenosine triphosphate, or ATP, is the currency of energy. In active BAT, ATP production is sacrificed in favor of heat production.

28

INCREASING BAT CAPACITY

Over time the capacity for burning off calories as heat seems to increase if BAT is stimulated on a regular basis. Stimulation of activity in BAT results in an increase in uncoupling protein, and an increase in the number of BAT cells found in the body. This means a dramatic increase in the amount of excess calories that can be burned off as heat through the thermogenic cascade. The increase in thermogenesis in turn stimulates the production of yet more brown adipose tissue and uncoupling protein.

Correcting Faulty Thermogenesis

The implications of these findings are truly astounding! Malfunctioning thermogenesis, under genetic and environmental control, can be corrected, and simultaneously imbued with an enhanced capacity for burning excess dietary calories and calories derived from fat stores. It follows, then, that the calorically challenged individual, suffering from expanding stores of accumulated white fat and diminishing portions of BAT and UCP, will benefit from successfully reversing this process through stimulating the body's innate capacity for thermogenesis. Perhaps this book holds one of the important keys to successfully reactivating thermogenic processes. As BAT activity increases, UCP will begin to proliferate, BAT capacity will increase, and the proper fat/lean body mass ratio will be restored.

Let's say, for example, that you begin consuming thermogenic drugs, nutrients or herbs, but nothing happens—no weight or fat loss. What can you do about it? It is probable that you don't have much BAT to begin with; it's been too many years since you used it. Or, perhaps you have BAT but it contains very little UCP. Hence, you currently have very little capacity for thermogenesis. However, if you continue to consume the thermogenic you will eventually be rewarded. Gradually, the brown fat begins to grow, UCP begins to accumulate. Suddenly, one day it all comes together, the thermogenic cascade kicks in at a significant rate, and you are on the way to serious fat loss. The wait can be discouraging, especially when your neighbor, blessed with tons of active brown fat, begins to shed fat as soon as she begins to take the thermogenic; however, if you stick with the program, you will eventually catch up.

SYMPATHETIC NERVOUS SYSTEM

Thermogenesis in BAT appears to be under the control of the nervous system. Research shows that if certain portions of the brain are lost to surgery, all BAT activity ceases, and obesity results. To understand how thermogenesis works, we need some degree of understanding of the nervous system. The next couple of paragraphs present a highly simplified, though accurate, explanation of the nervous system, how it works, and how it applies to thermogenesis.

Communication of sensory events requires an intact nervous system. Involved in a sensory event—such as pain in the big toe, or light hitting the retina of the eye—and a corresponding action—such as removing the toe from the hot water, or putting on the sunglasses—are as few as three, and as many as thousands of individual nerve cells (called *neurons*). These cells do not directly touch one another, but are separated by minute gaps, called *synapses*. In order for communication between neurons to take place, the information must be passed across the synapse from one neuron to the next neuron or group of neurons. Communication in the nervous system takes place on an electrochemical basis. That is, the signal that travels down a nerve and from one nerve to another, and from the nerves to the glands, muscles and organs of the body has both an electrical and a chemical component. The chemical events of neural conductance give rise to the electrical signal itself. Notice we alluded to communication between neurons and non-neural structures, such as glands, muscles and organs. Synapses separate nerves from these structures also. In order for information from the brain to reach these non-neural structures, it must cross the synapse.

Chemicals responsible for conductance across the synapse are called *neurotransmitters*, that is, they transmit data from one nerve to the next, and from the final nerve in the path to the gland, muscle, or organ to be acted upon. There are many different neurotransmitters in the nervous system. Two of the most important are *norepinephrine (NE)* and *acetylcholine (ACh)*. (See Plate III) NE plays an especially important role in thermogenesis. The reader may be familiar with the choline part of ACh as the nutrient labeled as a brain food by the health food industry in the late 1980's, and which spurred a buying binge of choline-rich lecithin. Unfortunately, ACh is just one of the important neurotransmitters in the nervous system; it is unlikely that dietary choline by itself would make much of an impact on total nervous system function.

Divisions of the Nervous System

The nervous system is generally divided into two broad sub-systems, the *central nervous system* comprising the brain and spinal cord, and the *peripheral nervous system*, which consists of cranial and spinal nerves with their associated *ganglia* (groups of cell bodies). The peripheral nervous system is further divided into two *functional* divisions, the *somatic* and *autonomic nervous systems*. The somatic controls voluntary motor functions. The autonomic plays a major role in the control of the more-or-less involuntary bodily functions, such as digestion and heart contractions and thermogenesis. The autonomic nervous system is again divided into two sub-systems: the *parasympathetic nervous system* (PNS) and the *sympathetic nervous system* (SNS). (See Plate I) Fibers from the PNS and SNS often innervate the same organs and structures of the body. When they do, they usually exert opposing actions. The SNS and PNS are distinguished by the type of neurotransmitters that are involved. The PNS is often called the *cholinergic* nervous system because the primary chemical involved in neural transmission is acetylcholine. The SNS is often called the *adrenergic* nervous system because the primary neurotransmitter is norepinephrine (NE). NE is also known as noradrenaline. Adrenergic comes from the adrenaline component of that word.

SNS Control of Thermogenesis

Activation of thermogenesis appears to be under the specific control of the SNS portion of the autonomic nervous system. Hence, inhibition of thermogenesis should be under the control of the PNS, although little is known about this process. Thermogenesis is thus activated when the SNS secretes NE in the vicinity of BAT. BAT responds to the presence of NE by instituting a complicated chain of events which ultimately results in conversion of calories to heat.

Arne Astrup of the University of Copenhagen in Denmark, a pioneering investigator of thermogenesis and the genetic causes of obesity, wrote:

> "In a number of obesity syndromes in rodents the sympathetic mediation (of thermogenesis) is defective, and this leads to extreme sensitivity to cold and to obesity . . . Likewise, the hypothesis has been advanced that a diminished thermogenesis in BAT may be the cause of some types of human obesity."
> (**Acta Endocrinology**, Suppl., 278, 1-32, 1986)

The hypothesis that obesity may be caused by diminished thermogenesis rests on the idea of a defect in the functioning of the sympathetic nervous system, and concomitant lack of NE. A significant body of research, particularly ongoing studies at Harvard University and surrounding hospitals, has been directed toward the possibility of reactivating thermogenesis in humans through the administration of certain synthetic medications and nutrients that affect output of the sympathetic nervous system. The findings of this research may be embodied in the following statement: *Reactivated Thermogenesis Prevents and Reduces Fat in Humans*. The Copenhagen and Harvard groups, as well as others, continue the search for safe and effective substances that stimulate the sympathetic nervous system in such as way as to activate the thermogenic cascade in BAT without endangering any other parts of the body. In the coming years we can expect to see a virtual explosion in this area of research.

GENETIC DEFECTS IN THERMOGENESIS

Complicating the thermogenic picture is the notion that brown fat processes appear to be under genetic control. For years scientists have been trying to discover what goes wrong with the thermogenic process, i.e., why some people stay thin by burning off excess calories, and why other people seem to lack this capacity, and hence accumulate pounds and inches of white fat. They ask, "Is there a 'thin' gene," or a "thrifty gene?" Many people do not have a problem with excess fat until somewhere in middle age when they suddenly begin to gain weight. This may reflect a genetic trigger telling the body that old age, lean times, slower metabolism, etc., are approaching, and that it needs to begin to consider the virtue of storing away some of that extra energy to draw upon later. Though the genetic trigger makes little sense in modern society, it may have had a very important survival value for primitive man. In other people, an inborn error of metabolism appears to be operative. These people are obese even as children, and will usually stay that way throughout life.[983]

At any rate, as the brown fat processes begin to slow down, white fat accumulation begins to accelerate. The genetic trigger is a normal mechanism. However, it appears that there may be genetic <u>defects</u> in thermogenesis that can even more drastically affect a person's fat metabolism. These defects may be inborn errors of metabolism that you experience from birth on, or that appear later in life.

Interacting with the genetic cues and defects are a whole host of environmental and/or behavioral factors that could act as cues and

33

triggers for genetic events and which could produce atrophy of BAT by themselves: traits such as insulating ourselves from the cold through clothes and warm homes. For all of these conditions, the science of thermogenesis holds forth real hope.

Reversing Obesity

Utilizing animal models of genetic obesity, it has been found that the proper application of a thermogenic can reverse genetic obesity, and restore these animals to a lean state. Almost all forms of genetic obesity have been traced to a defect in the thermogenic metabolic pathway (see the chapter on Genetics). Restoring activity in this pathway eliminates the problem.

Generalizing to humans data that was obtained from lower animals can be an admittedly difficult enterprise at times. However, these data strongly implicate genetic defects in faulty thermogenesis, and suggest with equal strength that the consumption of thermogenic will reverse the condition; meanwhile, human trials investigating the relationship between genetic faults and thermogenesis are almost non-existent. Furthermore, the heavy procedural constraints involved make it unlikely that adequate trials will be carried out in the near future. Nevertheless, when scientists apply the results of animal studies in clinical research settings, the final outcome is the same: reduction of fat. It can be argued that, ultimately, it doesn't matter what the underlying mechanisms are, genetic or otherwise, as long as the results are achieved with safety and effectiveness.

REACTIVATING BROWN FAT

The problem in all cases of defective thermogenesis is finding a way to reactivate the thermogenic cascade in brown fat. Under the right influence the body converts white fat back into calories that can be disposed of through thermal combustion in BAT. There are some simple ways to do this, such as exposing oneself to cold temperatures every day, as discussed in PART ONE. Unfortunately, the cold exposure routine is not convenient or comfortable, or even always possible for people living in warm climates (unless they have a walk-in freezer, or like cold showers). Therefore, science has been looking for foods and drugs that might reignite thermogenesis more effectively and more efficiently. Since animal research has shown that it is possible to reverse the genetic tendencies that stifle thermogenics, i.e., to turn the heat back up, it is reasoned that it should be possible to reverse genetic defects in humans. While it is still impossible to know who will benefit most from reactivating BAT thermogenesis, or how many such people there are, the concept is some of the most exciting news to come out of basic research in decades.

Thermogenic capacity can be restored and enlarged through the consumption of certain substances, i.e., through dietary manipulation. This is the good news. The bad news is that the only substances known to man that will activate thermogenesis to the extent necessary to make a dent in fat stores must be consumed in amounts exceeding safe levels. The safest of these substances, introduced in PART ONE, are ma huang and its major active constituent ephedrine. Ephedrine can be used very effectively to stimulate activity in BAT. Yet, for decades scientists have been reluctant to recommend it for this purpose. The reason has been that in order to experience significant fat reduction through thermogenesis, a person would have to consume roughly five times more ephedrine than would normally be viewed as safe, i.e., a dosage free of significant side effects. Before discussing the solution to this problem, some words about ephedrine itself are required.

Ma Huang: Natural Ephedrine

The only natural source of ephedrine is the Chinese/Indian plant ma huang. Ephedrine can be synthesized, in which case it is usually found in medicines in the form of salts, such as ephedrine hydrochloride. The synthetic form can be found in drugs for hay fever, colds and asthma, and has been used to stimulate thermogenesis in BAT, in both animals and humans. American Phytotherapy Research Laboratory (APRL) has been studying the effectiveness of the natural form of ephedrine as it occurs in ma huang. APRL has found that the natural substance is as effective as the synthetic drugs. Use of whole, standardized ma huang is recommended over pure ephedrine hydrochloride, natural or synthetic, for two main reasons. One, the consumption of synthetic ephedrine may pose a greater risk of side effects such as cardiovascular stress and adrenal stress, behaving every bit like the drug that it is. Two, the whole plant contains a broad spectrum of ephedrine-like molecules, the combined action of which may be tolerated better than the drug by the glands and organs of the body, reacting as much like a nutrient as possible. Herbs are, after all, very complicated substances, containing hundreds of different nutrients in the forms deemed most beneficial by Mother Nature.

Thermogenic Research at APRL

The mission of APRL is to promote the use of natural forms of materials instead of drugs whenever practical. Not that man can not, through his own wisdom, improve on what Nature offers, but as a fundamental rule, it is wise to respect the natural and use it as much as possible. Although this philosophy flies in the face of convention in the United States, it is embodied in the medical philosophy of most other civilized countries. One can only hope that the people of the U.S. will eventually regain total medical freedom. In many aspects, medical science in the United States is superior to every other country in the world; yet, in terms of overall health care, freedom of choice in health care, sophistication of the average doctor, and educational level of the average citizen concerning matters of health and disease, we are still lagging far behind many countries of the world. A further problem is that most Americans don't know it.

Over the past 18-24 months American Phytotherapy Research Laboratory has been overseeing the application of ma huang in humans who have varying degrees of fat problems. These studies have shown that, with proper controls, standardized ma huang can be effectively used as a thermogenic agent, and that it produces results at least as good as those obtained with synthetic forms of ephedrine alkaloids.

Improving Ephedrine/Ma Huang Safety

Scientists have been working on the problem of ephedrine safety for several decades. They have discovered that there are other substances that can be combined with ephedrine or ma huang to lower its toxicity. There are at least two substances that combine with ephedrine or ma huang to both lower the quantity of ephedrine required for thermogenesis, and to render it safe for human consumption in the amounts necessary for thermogenesis. These substances are *aspirin* and *caffeine*. Combining these substances with ephedrine substantially lowers the dose required for effective thermogenesis. Additionally, these substances remove some of the unwanted actions of ephedrine, such as the tendency to accelerate the heart beat, thus making it much safer for human consumption.

To be effective, the other substances must be combined with ephedrine in just the right proportions. Correctly combined, the result is an outstanding synergistic relationship that enhances the thermogenic action of small, non-toxic quantities of ephedrine. Unfortunately, most of the thermogenic agents available at the time of this writing are poorly conceived and poorly executed products that do not in any way mitigate ephedrine problems. Additionally, many companies fail to realize the critical need to use standardized natural materials. Consumers of unstandardized, poorly conceived thermogenic products can expect health problems down the road.

Unquestionably, there is a legitimate, beneficial and health-promoting use for aspirin and caffeine. This bald assertion may come as a surprise to authorities who usually recommend abstinence from these drugs. Prior to the advent of thermogenics, that statement could not have been issued, at least not by us who are dedicated to the promoting the more natural side of health care.

Natural Caffeine

Caffeine, a xanthine alkaloid, is found in many plants that can be used as the source for caffeine in the thermogenic. As a matter of fact, research at APRL has determined that natural sources of caffeine are much more effective than synthetic caffeine. One, the thermogenic effect is more pronounced. Two, the side effects are smaller. Three, the amount of energy experienced by users is greater. These results are probably the result of the presence in plant materials of a variety of synergistically acting nutrients. First they contain several different xanthine alkaloids, not just caffeine, and these can interact in ways that may impact on the thermogenic process. Second, they contain other substances that may further moderate, augment or otherwise affect the final physiological result. The amount of xanthine required is small. Plant sources of caffeine include bissey nut (also known as kola nut and gooroo nut), yerba mate (though it occurs here in the form of mateine), and guarana. Less ideal sources of caffeine are coffee and tea.

Caffeine affects the action of ephedrine by exerting an antagonistic effect on the action of certain enzymes in BAT that tend to inhibit the ability of ephedrine to stimulate the secretion of NE.

The ratio of caffeine to ephedrine is of paramount importance, demanding the use of standardized raw materials. If the ratio is off, the added caffeine does little to reduce the necessary dose of ma huang or ephedrine and, worse still, introduces the possibility of side effects due to the caffeine itself. For this reason, it is wise to eliminate other sources of caffeine from the diet while using thermogenic.

Aspirin, Not Salicin

Aspirin is not found in plant materials, but a precursor for aspirin—salicin—is found in small amounts in willow bark, wintergreen, meadowsweet and other plants. Salicin is unfortunately not entirely suitable for the purpose of rendering ma huang or ephedrine safe and effective. Salicin apparently lacks the critical thermogenic feature of aspirin—prostaglandin inhibition. This drawback to the use of willow bark and similar plants is totally ignored by most current manufacturers of herbal thermogenic compounds who incorporate willow bark (sometimes called wiederinde—German for willow bark, or wiedewinde—a misspelling of the German word that is unfortunately widespread in the health food industry), thinking it is a natural source for aspirin. It isn't.

This is a specific instance of the dangerous disregard for research that could undermine the efforts of the patent holder as he pursues obtaining his "new drug application" through the FDA, as discussed in the Preface. If his efforts are jeopardized by the presence of badly formulated products in the marketplace, we might all have to suffer the consequences. In addition, adverse consequences arising from the consumption of inadequately formulated products could give the FDA an excuse to step into the health food market with devastating sanctions.

Aspirin, or acetylsalicylic acid, is formed by reacting salicin with acetic acid, the primary component of vinegar. Aspirin, therefore, is only one step removed from the original plant material. Nevertheless, it is not technically a "natural" substance. Research at APRL has shown that thermogenic products that contain aspirin along with sources of salicin, such as willow bark, wintergreen and meadowsweet, are tolerated better than thermogenic with just aspirin. However, thermogenic without aspirin is simply not safe and effective.

To repeat, products without aspirin can not be regarded as safe thermogenic agents. The reason may be deduced from the above discussion. If the only way to make thermogenic doses of ephedrine safe for human consumption is to combine them with aspirin, then consumption of products not containing aspirin must lead to health risks if therapeutic thermogenic amounts are consumed, or, alternatively, if the product without aspirin is consumed in safe amounts, then the amount ingested will probably be too low to create an effective amount of thermogenesis. You can not have it both ways. You either include aspirin, or you sacrifice safety or efficacy or both.

Critical Synergy

The synergy between ephedrine, xanthine and aspirin forms the logical basis for the thermogenic concept. Without all three, combined in precise ratios, true thermogenesis will not take place safely and effectively. Large amounts of ma huang or ephedrine will create thermogenesis, true enough, but not without unhealthy, unsafe, side effects. Large amounts of caffeine produce almost no thermogenesis in BAT, and can be detrimental to health. Aspirin also has no direct effect on thermogenesis, but in large amounts can damage the gastric mucosa and affect blood viscosity.

However, when the right ratios of the active constituents are combined, a marvelous natural event occurs. The long dormant, genetically wasted, thermogenic process turns back on. The amount of BAT in the body begins to increase, its thermogenic capacity is expanded, and it goes to work burning up excess ingested calories and calories that have already been stored away as WAT. If it appears that we have unduly dwelt upon the need for proper formulation, it is because there is no other current threat to the health care industry as serious and imminently dangerous as this one. Manufacturers who complain about the "Big Brotherism" of the FDA should realize that government intervention is required when good formulation practices are being ignored.

A final point: In some people, the consumption of thermogenic instigates an immediate thermogenic reaction; in others, the effect may take a period of several weeks before it begins to work. How long it takes may depend on just how seriously deficient in brown adipose tissue a person is to begin with. If you initiate a thermogenics program and it doesn't start to work right away, give it time, and the BAT will eventually expand its capacity to the point where it can begin operating effectively. At that point, the fat will begin to be burned off. Do not approach this problem by consuming ever increasing amounts of thermogenic.

INCHES VERSUS POUNDS

As mentioned earlier, we are currently witnessing another kind of small revolution in America. Experts are beginning to realize that the best measure of overall health is not weight-to-height-to-sex-to-age ratios, but the ratio of body fat to body lean tissue. It is not how much you weigh, but how much fat you have, that indicates how healthy you are. The wisdom of this approach can easily be seen. People with large amounts of lean body mass, such as football players, would certainly be counted as overweight by normal standards, yet they are certainly not unhealthy human specimens. Likewise, their percent of body fat may be extremely low; their weight comes from muscle tissue, not fat tissue. On the other hand, many small-framed women are well within the orthodox weight limits and yet can be shown to be grossly over fat.

The revolution in weight and fat measurement standards has occurred at an extremely fortunate time in regards to the thermogenic revolution—because thermogenic removes fat, not necessarily weight. People may even gain weight when using thermogenic. There are two reasons for this paradox. One, lean tissue weighs much more than fat; hence, even small gains in lean tissue will offset the weight of a substantial amount of fat lost through thermogenics. Two, because the thermogenic concept has a side benefit of increasing energy levels, people will use that energy in a hundred different subtle ways throughout the day that tend to increase lean body mass throughout the body.

Lean body mass doesn't necessarily mean bigger muscles. Look at it this way: typically, muscles are a combination of lean body mass and fat. Over time, under the influence of thermogenic, the fat disappears and lean takes its place. Some areas of the body, however, are mainly composed of fatty tissue. Under the influence of thermogenic, the first and primary effect will be the loss of that tissue type, off the waist, off the thighs, under the arms and other places. The rule for accurate assessment of progress is to measure, not weigh. The bathroom scale is quickly becoming obsolete.

Bioimpedance

There is an even better method than measurement alone. From the start APRL has incorporated into its research a device that actually measures body fat, level of body hydration and other key indices

41

of body composition. This unit, a bioimpedance device, measures the resistivity of body tissues. Bioimpedance relies on the principle that lean body mass, containing large amounts of water, conducts an electrical current more readily than does fatty tissue that contains very little water. The bioimpedance device measures the difference in resistivity between these two body compartments and mathematically computes the percent of lean body mass, the percent of fat, the number of pints of water, the basal metabolic rate, and the caloric burn rate. In the hands of a skilled practitioner, these numeric indices of body composition can be used to accurately evaluate a person's overall health. Recently, other centers of thermogenic research have begun to incorporate bioimpedance measurements into their assessment and evaluation and procedures.[895]

Bioimpedance is an extremely easy technique to apply, yet results obtained correlate highly with traditional hydrostatic measurements. Other devices, such as infrared and skin caliper measurements can also be used, but are not as accurate as bioimpedance. However, these alternate methods may be much cheaper, well within the reach of persons on limited budgets. The difference in price will probably more than make up for lack of precision. To be useful, any of these techniques must be applied in a consistent manner. Manufacturer suggestions should be strictly adhered to.

The bioimpedance technique is especially valuable in terms of yielding an accurate measurement of fat/lean body mass. This ratio, as mentioned earlier, is the best index of health. Many people, when they begin consuming thermogenic, exhibit a very high ratio. Over a period of several weeks, this ratio comes down, enters the normal range and remains at that level. This is an important point: the ratio does not continue to plummet once the normal range has been achieved. The same goes for basal metabolic rate and caloric rate; i.e, once these levels reach the normal range, they stabilize.

On a practical level, persons using thermogenic would do well to obtain such a device (or the less expensive calipers) and use this, instead of the bathroom scale, to measure change in body composition. Physicians (mainly chiropractors and bariatric physicians)

42

throughout the country are also beginning to recognize the advantage of bioimpedance. Calling around, you may find one in your area who will provide bioimpedance-based body composition services for a nominal fee. We have recorded many instances in which the patient loses fat, dress sizes, etc., without losing any weight at all! This is possible because the person replaces fatty tissue with non-fat, or lean body tissue through increased activity and through protein accumulation due to the thermogenic itself.

APRL's pilot work with the bioimpedance/thermogenic program has been extremely successful. Physicians using the program have reported significant progress in dealing with over fat patients. Indeed, some doctors have developed a separate practice devoted exclusively to the treatment of obesity. In this program, patients consume the recommended daily amount of thermogenic and come to the clinic on a weekly or monthly basis for measurements and for evaluation via the bioimpedance technology. The patient receives a printout of results, a consultation, and a wellness evaluation that may include other health or chiropractic recommendations.

EXERCISE AND DIET

EXERCISE

As discussed earlier, caloric restriction (fasting, starvation) and exercise as such are not necessary requirements for successful fat loss through thermogenesis in brown adipose tissue and may actually lead to atrophy of brown adipose tissue.[47-48,1036] Because heat produced in skeletal muscle through exercise increases the core temperature of the body, it can actually interfere in the intricate biofeedback processes that control BAT thermogenesis. Exercising is obviously good for other reasons, but it can be counterproductive to the thermogenic process in BAT if engaged in immediately before or after the thermogenic is consumed. Ideally, one should take thermogenic in the morning and exercise 3-4 hours later, after the body has cooled off. One could reverse the order, but the thermogenic material might interfere with normal sleeping patterns if consumed in the evening. Exercise, on the other hand, when done in the evening, can actually help promote good sleep habits.

Thermogenesis in skeletal muscle is improved through exercise, which also raises the basal metabolic rate. Consuming thermogenic will also improve skeletal muscle thermogenesis and BMR. A combination of exercise and thermogenic would therefore promote the consumption of calories in skeletal muscle more than either one alone. But thermogenesis in BAT would be retarded. These facts suggest that several combinations of thermogenic and exercise may be possible and should yield predictable results.

Deep Breathing Exercise

Recently, the concept of dynamic—or deep—breathing has been suggested as a means for obtaining the same benefits derived from traditional aerobics, but without the detrimental side effect of raising body temperature (note: raising body temperature is not seen as a detriment if the goal is to burn up calories in the process of heating up the body; this kind of "muscular" heating, however, does not stimulate BAT thermogenics; raising body temperature is even a detriment to BAT thermogenesis).

The primary goal of aerobics is to oxygenate the tissues of the body, especially the muscle tissue. As the muscle tissue receives more oxygen, its ability to metabolize surrounding fatty tissue increases. Said another way, calories are burned through oxidation; the more oxygen available, the more calories that can be burned to meet the energy demands being placed on the muscle cells by the exercise. Aerobic exercise is only effective as long as the demands being placed on the cells do not exceed the amount of oxygen available to catalyze caloric combustion. Once that capacity has been exceeded, the cells go into anaerobic oxidation which is very inefficient, produces lactic acid and muscle soreness, and does not utilize many calories.

Aerobic exercise by its very name implies that its primary goal is to pump oxygen to the muscle tissue. Proponents of deep breathing justifiably ask why not pump that oxygen simply by using lung power.

Breath in deeply, hold the breath, forcing oxygen into the blood, and exhale forcefully to expel carbon dioxide and other soluble toxins. Modern experts in deep breathing have developed their own methods for maximizing the effectiveness of this technique. One deep breathing proponent is Greer Childers, who calls her method "Body Flex." Ms. Childers elaborates the processes involved in the breathing exercise itself, and adds a variety of stretching exercises to the method. The stretching is effective because it opens up capillaries in the area of stretch allowing the oxygenated blood to flood into the area. By alternatively stretching different muscle groups, one systematically oxygenates each area to a degree not possible with deep breathing alone. The stretch plus the extra oxygen serves to both incinerate the fatty tissue around the muscle fibers and to increase the percentage of lean body tissue in those areas.

Childer's program is singled out because it has the advantage that it produces only a minimal increase in core body temperature. In fact, if engaged in slowly and carefully, Childer's method will produce almost no net gain in core body temperature. This means that it will not interfere with BAT thermogenesis induced by consumption of the thermogenic, but will, in fact, contribute to the total thermogenic process. Thus Body Flex and thermogenics seem to be well-matched techniques for burning calories as heat in BAT, and for burning calories in muscles. More importantly, anyone can do deep breathing, no matter what their overall conditioning. We all breathe and we can all consume supplements; the secret is correct breathing and consuming the proper supplements.

It should be pointed out that deep breathing techniques are not without their detractors who maintain that such techniques can not appreciably affect the oxygenation of the blood due to the inherent limitations imposed by the physiological apparatus, and because the blood is already close to 95% oxygenated (98% at sea level) in the first place. According to opponents, forced deep breathing can increase the pressure of oxygen in the lungs only a small amount, and since hemoglobin is already almost completely saturated, and the liquid part of the blood can carry only so much oxygen, deep breathing can have only a minor impact.

Proponents of aerobic or deep breathing simply point to the success rate they have with the technique. Research at APRL hasn't shed any light on the ultimate legitimacy of the rationale behind deep breathing exercise, but our findings do strongly suggest that something positive is happening. The combination of thermogenesis, deep breathing and stretching apparently enhances fat loss, increases felt energy and increases muscle tone. In addition, it puts people on a path of sensible exercise they might otherwise avoid.

DIETING

Perhaps no other aspect of nutrition research is more controversial than that dealing with what happens when a person reduces daily caloric intake below the level required for maintenance of optimum basal metabolic requirements, that is when the attempt is made to consume less energy than is expended.

Dieting, or caloric restriction, is absolutely unnecessary for effective thermogenesis. In fact, dieting actually reduces thermogenesis. On the other hand, this statement should not be construed as an endorsement of bingeing. It has been shown that stimulation of thermogenesis through the consumption of thermogenic agents will increase the effectiveness of a diet characterized by mild caloric restriction. The diet, however, does not increase thermogenesis. Dieting, in fact, lowers basal metabolic rate, and overfeeding tends to raise BMR in normal people. Therefore, if you must reduce your caloric intake, consume thermogenic to offset adverse effects on BMR.

Also, it is important to remember that the action of the thermogenic will be directed at both the calories consumed and at the calories stored. Although no research has addressed this problem, consider the logic of the following hypothesis: The more calories consumed, the more thermogenic required to incinerate those calories and the less that is left over to burn off stored calories. If this idea has some basis in fact, it can be used to an individual's advantage. To burn off stored calories, consume thermogenic between meals; to prevent fat storage of calories ingested at meals, take the thermogenic with the meal. During periods of gluttony, such as Thanksgiving, Christmas, other holidays, weddings, etc., use thermogenic primarily to prevent fat gain. Research at APRL shows that people who continue to consume thermogenic through holidays do not gain weight, while those who do not take the product gain a couple of pounds (though typically not as much as they would have before beginning the program).

Excessive dieting disrupts almost every metabolic event involved in thermogenesis. Research shows that thermogenesis in BAT and even skeletal muscle and liver works best when the diet contains adequate quantities of carbohydrate, protein and fat. Restricting any of these major food categories will disrupt metabolism. This is especially true is the overweight condition was produced by a genetic fault in the thermogenic mechanism to begin with.

Fortunately, the advent of the science of thermogenics means that you have an excellent and sound alternative to caloric restriction. Consumption of thermogenic will help raise BMR thereby helping to burn off those calories you were trying avoid eating, and it will increase BAT thermogenesis thereby helping to compensate for the consumption of calories in excess of metabolic requirements. Hopefully, dieters will cease excessive calorie restriction and begin to consume thermogenic before long-term damage is done to the muscles and organs of the body. (For more detailed information on the effects of dieting, see the final section in the chapter on Categories of Thermogenesis.)

SIDE BENEFITS

We have already alluded to some of the side benefits of thermogenic substances. Some of the more important are listed below.

1. Anti-asthma effects. Ma huang is a bronchodilator that has been used for thousands of years for reducing the symptoms of asthma, and ephedrine can be found in many over-the-counter asthma preparations. Many users of thermogenic report better breathing, and freedom from hay fever symptoms that have plagued them for years. When natural sources of ephedrine and caffeine are used, the better-breathing action is not accompanied by drowsiness.

2. *Energy Production.* In a manner explained more fully later, thermogenic dramatically increases overall energy levels. Many people use the product just for this action. Increased energy arises primarily from an increased flow of blood to the muscles and brain, carrying with it extra oxygen and nutrients to feed the energy producing processes of the muscle and brain cells. This means increased energy and mental acuity. Lean body mass normally increases as the fat mass disappears (although weight can be lost from both body compartments), and this can also help increases energy levels.

3. *Appetite Suppression.* The more blood that flows to the muscles, the less that is available for the stomach and abdominal areas of the body. The net effect is a mild reduction in activity in that area, including reduced appetite. Over time, this reduction in appetite evolves into a greater appreciation for healthy kinds of foods, and greater selectivity on your part in your choice of foods.

4. *Steadily increasing BAT capacity.* As mentioned earlier, the longer one consumes thermogenic, the greater capacity for thermogenesis one experiences. Eventually, one could acquire enough brown adipose tissue that years may go by before the tissue had atrophied enough to require renewed implementation of a thermogenic program. Atrophy would only occur if the original defect in BAT metabolism was caused by a genetic fault; if not, BAT stores might never dissipate. A modest maintenance amount of thermogenic may be all that is required.

MOST OFTEN ASKED QUESTIONS

Since the science of thermogenics has only recently become known to the public, most people have several questions about how it will impact on their lives. Following are the questions that are most often asked, along with brief answers. These questions and the answers are drawn from material found elsewhere in this book, but are presented here as a quick ready-reference and for the sake of convenience to the reader.

Q: **Give me a quick definition of thermogenesis.**

A: **Thermogenesis is the body process that creates heat. The fuel for heat production is derived from lipids, or fats. These fats are obtained from the diet or from stored body fat. Thermogenesis is the body's way of disposing of excess calories so they are not deposited as fat. Unfortunately, the process doesn't work very well in most people. Research has shown that stimulation of the thermogenesis makes a significant impact on body fat stores. It is the foundation of the science of fat management.**

Q: **How soon should I see results?**

A: **This varies from person to person, the range being anywhere from immediately to several weeks or even months. Some people have virtually no brown adipose tissue in which thermogenesis can occur, or else the BAT they have contains too few mitochondria to make a difference on fat stores. Consumption of thermogenic by these people will begin to stimulate the growth of BAT and the proliferation of mitochondria; soon a "critical mass" of BAT will accumulate and significant fat loss will occur, all things being equal. Serious thyroid or adrenal disease may inhibit thermogenesis, as will a rigorous exercise program done during the same part of the day as when ingesting the thermogenic.**

Q: **What causes that jittery feeling during the first few days using thermogenic?**

A: **Not everyone experiences this problem, but for the few that do, it can cause undo anxiety. Jitteriness usually indicates that you are drinking other sources of caffeine, normally coffee and cola drinks. It may also be felt by persons unusually sensitive to the ephedrine. This feeling generally goes away after 5-15 days.**

Q: **How does thermogenic affect energy levels?**

A: **Thermogenic increases the amount of energy you feel, sometimes dramatically. It increases the flow of blood to muscles, thereby increasing the amount of oxygen and nutrients that are available for the muscles to use. The same goes for the brain. As you lose fat, you will also feel more energy. Thermogenic also has a mildly stimulatory affect on the central nervous system.**

Q: Why do some companies suggest I not take thermogenic on weekends?

A: To allow the adrenals, both the medulla and the cortex (parasympathetic nervous system) a chance to rebound. Although adrenal stimulation is not necessary for thermogenesis to occur, it is an unavoidable consequence of SNS stimulation. The body needs time each week to replenish adrenal stores being depleted by thermogenic. Constant SNS stimulation is not healthy. Fortunately, the healthy body has a tremendous capacity for replenishing the adrenals.

Q: Can diabetics use thermogenic agents?

A: Basic research shows that stimulation of the sympathetic nervous system by thermogenic does not appreciably influence blood sugar levels. Clinical trials involving patients with diabetes have been successful. Tolerable amounts for the diabetic are generally about half that of non-diabetics. In these reduced amounts, thermogenic may be used seven days a week. Use of a properly formulated thermogenic is absolutely mandatory! Both insulin-dependent and non-insulin-dependent diabetics may consume thermogenic.

Q: Can people with high blood pressure consume thermogenic compounds?

A: Again, the answer presupposes consumption of a properly formulated thermogenic, in which case the answer is yes. Sometimes dieting itself lowers blood pressure due to decreased sympathetic activity resulting from caloric restriction.[343,1392] Consumption of thermogenic may offset this tendency; however, much research shows that thermogenic does not raise blood pressure. Nevertheless, most labels will contain warnings about blood pressure. This is because thermogenic stimulates the SNS which typically tends to raise blood pressure.

Q: Can people who are pregnant or plan to become pregnant or nursing take thermogenic?

A: No. Fetuses, newborns, and infants have immature nervous systems. It is always best to avoid ingesting stimulants while you are pregnant or nursing. Women who are trying to become pregnant must use their own discretion about when to stop using thermogenic but product manufacturers are bound to include label warnings.

Q: How does thermogenic interact with anti-depressants?

A: Experience has shown that depressive persons react very positively to thermogenic; this fact suggests that persons using anti-depressant drugs should discuss the use of thermogenic with their doctor, and report their progress to him. Thermogenic may increase serotonin levels in the brain.

Q: What is the relationship between caffeine and xanthine?

A: Caffeine is a member of the xanthine family of alkaloids. Most plants that contain caffeine also contain other xanthines. The particular mix of xanthines in a plant dictates the manner in which that plant will affect physiology. Generally, the best mixes of xanthines for overall health are found in yerba mate, kola nut, and (for thermogenesis) in guarana. The worst mixes are found in coffee and green tea.

Q: Should I exercise when on the thermogenic program?

A: Thermogenic consumption will increase the effectiveness of most exercise programs if done properly. Since exercise that raises core body temperature will have a tendency to shut thermogenesis down, do such exercise 3-4 hours after taking the thermogenic. Alternatively, exercise first, but wait until you cool down completely before consuming the thermogenic. Many people adjust their exercise program to the evening. Still another approach is deep breathing exercise which can be done simultaneously since it does not raise core body temperature.

Q: What affect does thermogenic have on the adrenals?

A: As indicated in the previous answer, thermogenic tends to stress the adrenals. For this reason, responsible thermogenic compounds include as much nutritional adrenal support ingredients as will fit in the capsule after the appropriate ratios of thermogenic substances are included. For maximum health, thermogenic compounds will contain the minimal amount of thermogenic agents required for SNS stimulation together with the maximum amount of adrenal support possible.

Q: I get exhausted taking thermogenic. Why?

A: People who feel tired and exhausted immediately after consuming thermogenic usually have an adrenal problem. The thermogenic quickly depletes what adrenaline stores they have, leaving them exhausted. Remedy: stop taking the thermogenic and begin rebuilding the adrenals with consumption of licorice root, vitamin C, siberian ginseng, B vitamins.

Q: Can I take thermogenic after 2:00 P.M.?

A: You can, but it may keep you up all night. In the late evening, when you are not normally engaged in physical labor, the body is especially sensitive to the extra adrenaline circulating in the body that is unavoidably released when you ingest the thermogenic.

Q: What is the fat/lean ratio and how does it relate to weight loss?

A: Since obesity and weight management have entered the province of medicine, it is required that food manufacturers find other ways to talk about health. This turns out to be a fortunate circumstance, since it is rapidly being discovered that the best index of health is not absolute weight, but the amount of fat you have in relation to the amount of lean body mass you have. That's the fat/lean ratio. The lower this figure the better. Measures of this ratio are increasingly easy to obtain. Bio-resistance machines, infra-red devices, skin-fold calipers, and so forth can be found in many doctors' offices and even in some health food stores.

Q: What are the side effects?

A: People react differently to the thermogenic. Most will experience no side effects at all. Some may become constipated at first; if this happens to you, consume a mild laxative, drink more water and/or reduce the amount of thermogenic ingested, especially during the first few weeks. Other people may get nauseated; these persons should reduce the amount of thermogenic ingested, drink more water and take thermogenic with meals. Some men with benign prostatic hypertrophy may experience a worsening of those symptoms; these men should be taking 2-3 capsules of pygeum, serenoa repens, or other effective anti-BPH substance, every day. Transient jitters may occur in some hypersensitive persons; this condition was discussed above.

Q: Should I take thermogenic with meals or between meals?

A: Depends. For maximum impact on stored fat, take between meals. For maximum impact on dietary calories, take with meals. It's not an all-or-none deal, just one of emphasis.

Q: Should thermogenic compounds contain chromium?

A: Thermogenesis that occurs in brown adipose tissue is not appreciably affected by chromium. Thermogenesis that occurs in muscle cells may be facilitated by chromium, but this thermogenesis is not in question when it comes to body fat. Muscle cell thermogenesis seems to function in a healthy manner in the majority of people. Chromium is mainly required for efficient energy-producing metabolism in normal body cells. It plays only a minor role in heat-producing cycles in BAT. Chromium is very important for proper insulin-mediated carbohydrate metabolism. Therefore chromium need not be included in a thermogenic compound. Given the wide number of chromium supplements currently in the marketplace, persons desiring supplemental chromium should experience no trouble acquiring it. In fact, because chromium is now in most daily vitamin and mineral tablets being consumed by the majority of people desiring to consume thermogenic, the addition of chromium to thermogenic could imbalance the daily vitamin/mineral supplement. See Part Four for more information on this problem.

Q: Should thermogenic compounds contain carnitine?

A: Probably not. Research indicates that brown adipose tissue already contains sufficient carnitine to meet the maximum thermogenic metabolic demands it may experience. Also, the amount of carnitine required to make a significant difference in BAT metabolism (500-1000mg) would not fit in the typical dietary capsule, especially in the presence of other thermogenic nutrients. As with chromium, persons desiring additional carnitine, should acquire it in other forms. Since carnitine is a fairly expensive nutrient, its inclusion in a thermogenic would make the product unnecessarily expensive for the a majority of customers, who don't need it.

Q: How much thermogenic should I take?

A: This depends on the particular product. Ideally, a daily quantity of thermogenic will be present in two-three capsules. Larger capsules (800mg) are best suited for adding a good quantity of adrenal support nutrients. This is the avenue chosen by APRL. Smaller capsules (400mg) must either compromise on the amount of thermogenic or on the amount of adrenal support. Alternatively, a company could choose to put the thermogenic in one product, and the adrenal support in another. There may be other satisfactory variations.

People who are sensitive to any kind of stimulant should consume no more than 2 capsules per day. One capsule or even 1/2 capsule might be a reasonable standard for an extremely sensitive individual. People of great bulk have been known to consume more than the recommended 3 per day.

Q: I have seen a variety of so-called "thermogenic" products on store shelves; they never contain exactly the same nutrients. What are the major thermogenic nutrients?

A: The major thermogenic nutrients affecting thermogenesis in brown adipose tissue (the organ of thermogenesis) are ma huang, or ephedrine, caffeine, or caffeine-containing herbs like kola nut, gooroo nut, yerba mate, guarana, as well as aspirin, iodine (as from fucus) and cayenne. Minor thermogenic nutrients that can still make a strong impact on some part of the thermogenic process are pantothenic acid, essential fatty acids, vitamin B-6, vitamin C, ginger root, zinc, manganese, magnesium, niacin. Willow bark, wintergreen and meadowsweet are good sources of salicin and other salicylates but these do not play a primary role in thermogenesis; their inclusion in a thermogenic product modulates the action of aspirin but they are not a substitute for aspirin. Many so-called thermogenic components may stimulate thermogenesis in muscle cells, but do not affect metabolism in brown adipose tissue where the most important kind of thermogenesis occurs. Hence such nutrients do little to affect overall fat stores. The herb yohimbe and the alkaloid yohimbine stimulate thermogenesis but are unsafe for general human consumption and should not be included in a product meant to affect thermogenesis.

Q: Why not use willow bark instead of aspirin?

A: Aspirin inhibits the particular prostaglandins that restrict the action of norepinephrine. No component of willow bark has been shown to possess this property. Salicin, the main component of willow bark, from which aspirin can be made, behaves more like substances that do not inhibit prostaglandins than like substances that do, and so is highly unlikely to possess this quality. As of this writing, aspirin remains the substance of choice for enhancing the action of ephedrine through an inhibition of prostaglandin biosynthesis.

Q: Will I reach a weight plateau with thermogenics?

A: There appears to be an upper limit to the capacity of a given amount of BAT to oxidize fatty acids. Once this limit has been reached, a person may experience a temporary decrease in further fat loss. This plateau may be achieved very quickly, or it may take weeks or months to reach it. At any rate, continued fat loss from this point will depend on the growth of new BAT. Brown adipose tissue mass will increase with continued use of thermogenic. Patience may be required. In other words, quick fat loss during the first few weeks is mediated by current BAT stores; further fat loss will proceed according to how quickly BAT mass increases. Both processes depend on the consumption of thermogenic. Our research has also indicated that once a person plateaus at a fat/lean ratio that is normal for his or her sex and age, further fat loss may not be possible. Thermogenic responds fairly well to innate factors governing individual body composition. It will not stimulate continued loss of tissue of any kind in otherwise healthy individuals demonstrating normal body composition.

Q: How does consumption of the thermogenic affect subjective and objective feelings of heat in the body

A: Especially at first, activation of brown adipose tissue may create the sensation of heat in areas of the body where BAT is present, i.e., between the shoulder blades, throughout the abdomen, and down the back. In most cases, these sensations subside as the body adjusts to BAT stimulation by increasing blood vessel innervation of BAT that allows the heat to be dissipated more uniformly throughout the body.

To our knowledge, consumption of the thermogenic does not increase the frequency, duration or intensity of 'hot flashes' or night sweats. However, it can increase the sweating reflex in some people in the first couple of hours following ingestion. Again, this effect appears to decrease over time.

TOPICS IN ADVANCED THERMOGENICS

Thus far in this book we have presented only a cursory explanation of thermogenics, the science of energy expenditure as heat. For those readers desiring a more complete understanding of this science, this section has been provided.

PART THREE contains research-oriented discussions of many of the topics covered in PART TWO. These discussions delve into the whys and wherefores of thermogenesis. While technical language is often employed, most difficult terms will be defined as they are encountered. Although these topics are directed at the advanced student, they should be of interest even to the perspicacious lay reader seeking to expand his or her understanding of this fascinating subject. With both lay and technical readers in mind, it was often necessary to group the simple with the complex in several sections as demanded by the organization of the book. Hopefully, this organization will not intimidate the novice, nor bore the advance student of biology. Most physicians will find these chapters to be a welcome review of topics last encountered in medical school. We encourage all readers to read these chapters, for each contains material for both beginning and advanced students.

HISTORICAL

Scientists have been studying brown adipose tissue for over 300 years, and have understood basic cellular metabolism for decades. In that light the applied science of thermogenics is a surprisingly recent development. The metabolic events of the brown adipose cell differ only slightly from metabolic processes in all other cells of the body. Yet those small differences give BAT a unique status in the body. The meaning of that uniqueness became clear only after years of evidence accumulated from at least three different areas of research. ·

Apparently observed for the first time about 1670, BAT was considered a part of the thymus gland until the year 1817. From 1817 until 1863, brown fat was believed to be an endocrine gland. Certain authorities during this period taught that BAT was involved in the formation of blood. During the period 1863-1902 brown adipose tissue was considered to be a special kind of fat capable of storing food substances that would be released to the body as needed. This idea was abandoned when the apparent similarity of BAT to endocrine glands was once more popularized by medical researchers. Finally, in 1961, the modern era of research on brown adipose tissue was born with the announcement that BAT was a thermogenic organ whose purpose was heating blood that passed through it. Beginning in the mid-1970's the modern era was expanded to envelope the notion that faulty BAT thermogenics was the possible basis for genetically-induced obesity.[9]

Terms of BAT: From TIT to TAT

Other terms have been used to refer to BAT over the decades. An inspection of those names imparts some idea of the ambiguity in the minds of scientists regarding BAT: hibernation gland, brown body, multilocular adipose tissue, plurivacuolar fat tissue, glandular adipose tissue, lipoic gland, cholesterin gland, embryonal fat, immature adipose tissue, adipose tissue in formation, primitive fat. The cur-

59

rent designation, brown adipose tissue, is also a decidedly noncommittal description, omitting as it does any reference to function. A bold investigator at this point in history might choose a name such as 'thermogenic adipose tissue' (TAT), or 'the incinerator tissue' (TIT), or both (that way one could perhaps substitute TIT for TAT and thereby alleviate the boredom of repetition).

One interesting designation for BAT was THE "insulator gland." However the insulating capacity of brown adipose tissue is not very great. It has about as much insulating capacity as leather, and is a far better conductor than hair or feathers.[190] Because BAT *produces* heat, it may more accurately be compared to an electric blanket.

Converging Ideas

The science of thermogenics evolved in a most fascinating manner. Almost like magic, observations from widely disparate teams of researchers converged on the central idea of a thermogenic mechanism. First, investigators discovered the central role played by brown adipose tissue in hibernation.[1155] Other scientists discovered that BAT played a critical role in the process of cold adaptation in humans and other creatures. That BAT was the feature shared by hibernating animals and cold-adapting humans came as a mild surprise to many researchers. Equally surprising was the finding emerging from the study of the genetic basis for obesity that errors in BAT metabolism were the primary cause of genetically-induced obesity. Underscoring these avenues of investigation was work on adipose tissue metabolism and the sympathetic nervous system in general.[470, 978, 1035]

Linking these four areas of research and theory into a coherent science of thermogenesis took several decades. While it may seem obvious to us now that the body's ability to generate heat is involved in hibernation and cold-adaptation, it was not always so. For one thing, it was difficult to determine how the individual cells of the body were able to produce enough heat to accomplish the task of resisting and adapting to cold. It was not until the discovery of BAT and its function that we were able to account for the metabolic costs involved in generating significant quantities of life-sustaining heat.

The jump from hibernation to obesity was even more astounding. For generations, scientists firmly believed that the only thing that happened to excess calories was conversion to fat storage. They knew about white fat and brown fat, but what logic would lead to the conclusion that one kind of fat would burn up the other kind of fat? To reason thus required a certain paradigm shift in the fundamental way experts looked at fat. That shift occurred about the turn of the century, but lay dormant for several decades before next intoned by a lone investigator, and many more years were required before it became generally accepted.

Now that the idea of BAT metabolism is becoming firmly entrenched, the paradigm shift continues, producing in its logical force a need to rethink the science of obesity from top to bottom. We have already witnessed the beginnings of a shift in practical application away from bathroom scale weight measurements to body composition analysis. Such changes are certain to continue over the next few decades. Today's methods will seem primitive 20 years from now. The next few paragraphs will highlight some of the historical events associated with the discovery and advance of the science of thermogenics.

"Luxosconsumption"

Early observations that different healthy individuals seemed to expend the same amount of energy (i.e., exhibited a constant basal metabolic rate) even when ingesting a variable number of calories, suggested that caloric intake and body weight might be buffered by regulated changes in metabolic rate in the face of variable intake.[866,1174] These observations would give rise to the very first thermogenic conceptualization: Neumann's 'luxoskonsumption' hypothesis wherein he postulated that certain physiological mechanisms exist that permit excessive caloric intake to be dissipated as heat. Neumann's theory suggested that increases in body weight after chronic overfeeding were not directly related to the increase in energy consumed. 'Luxosconsumption' was renamed 'chemical thermogenesis' by M. Rubner and was given further impetus by the work of W.B. Cannon, in the 1920's, who was the first to suggest that a chemical, epinephrine (adrenaline), was involved in the production of body heat following exposure to cold. Cannon believed that the increase in heat production during cold exposure was caused by muscular contraction (shivering) and by chemical regulation independent of shivering.[199]

Non-Shivering Thermogenesis

The problem with many of the early theories of thermogenic mechanisms independent of mechanical muscle activity was that the amount of such non-shivering thermogenesis (NST) did not seem to exceed 25% of the basal metabolic rate and hence did not seem to be important in overall energy regulation. The discovery of enhanced NST capacity in animals exposed to prolonged periods of cold reignited interest in the area. It was soon discovered that newborn infants and neonates of several species had a strong ability to respond to cold exposure with heat production that did not involve shivering. In other words, the infant behaved like a cold-adapted adult, even though it had not been exposed to cold during the intrauterine life. The conclusion was obvious: the infant was provided with a special heat production plant through which its thermoregulatory system could be adjusted to the smaller body size.[157] Research in animals, meanwhile, demonstrated that cold-adapted rats had a much larger capacity for chemical regulation of heat production than their non cold-adapted counterparts.[254]

R.E. Smith, in 1961, finally suggested that BAT might serve as the source of NST in cold-adapted as well as infant animals. This hypothesis was eventually confirmed experimentally.[1184]

The Sympathetic Nervous System

Meanwhile, in the 1950's L.D. Carlson, from the University of Washington, extended the earlier work of Cannon, providing some experimental evidence for the role of the sympathetic nervous system in regulating the chemical events of heat production.[583] The verification of the role of the autonomic nervous system profoundly affected all subsequent research. Since the mid-70's, Lewis Landsberg, from Harvard University, and recently from Northwestern University, George Bray, and Arne Astrup, among others, have been involved in fully explicating the role of the sympathoadrenal system in the regulation of thermogenesis, obesity, hypertension and other control disorders.

The Gluttony Experiments

Derek Miller, of Harvard, resurrected the idea of luxosconsumption in the 1960's. Miller, intrigued by the idea that eating itself would stimulate a process that would burn up excess calories, postulated that overeating might actually stimulate enough thermogenesis to compensate not only for the extra calories, but also for a good deal of the calories that would constitute part of the "normal" meal. To test this hypothesis, Miller enrolled student volunteers in a gross overfeeding situation which Miller called "gluttony." Sure enough, some of the subjects were able to minimize fat deposition by dissipating extra energy via thermogenesis. Miller suggested that those individuals who gained weight in the study did so not because they ate too much, but because they had a defective thermogenic mechanism, probably because they were experiencing deficient sympathetic stimulation mediated by norepinephrine. Miller then suggested (the first person to seriously do so) that ephedrine could be used to treat obesity, since it stimulated the release of NE from sympathetic nerve terminals.[866,867] These early ideas of Miller flew directly in the face of conventional wisdom about the causes of obesity, which held firmly to the idea that 'gluttony' was the ultimate cause. It would take several years for the establishment to begin to seriously entertain Miller's notions.

The Hibernation Gland

Meanwhile, the field of thermogenics was being fertilized by the efforts of other individuals whose contributions help provide significant insights to the problem of thermogenics. For example, it appeared to many scientists over the centuries, that the strange-looking deposits of brown adipose tissue found in some species of animals increased the animals' resistance to cold. Most of these species were hibernators and this led to the speculation that brown fat was probably involved in keeping these animals from freezing to death during hibernation. Most investigators thought that BAT acted as an insulating blanket. At one point in time, brown fat was called the hibernation gland.

BAT as a Heat Generator

In 1961, two physiologists, George Cahill, Jr., from Harvard and Robert E. Smith, from the University of California, independently suggested brown fat actually generated heat.[1184,190] These scientists noted that BAT was capable of a high rate of metabolism, but could find no other purpose for it, but to generate heat. The heat must serve to keep the animals from freezing. In 1965, much of the work on brown fat was reviewed by Michael Dawkins and David Hull who offered a comprehensive theory of brown fat/white fat interaction and the production of heat in response to cold.[281] Their paper fell just short of seminal because it was marred by a lack of knowledge concerning the existence of the uncoupling protein and some of the other biochemical events that form the basis of thermogenesis. They also failed to note the resemblance of their work to the idea of luxosconsumption and hence missed the chance to tie brown fat metabolism to fat management. Nevertheless, their theory demonstrated a surprising degree of sophistication given the limited database from which they had to work.

The Uncoupling Protein

During the 60's and 70's, many investigators were intrigued by the idea that metabolic processes in brown fat differed significantly from those of other tissues. Mathematically, the amount of heat generated in this tissue was simply way too much to be accounted for by normal metabolic processes. It was as if BAT was receiving an inexhaustible supply of fuel in the form of free fatty acids, and that the typical negative feedback brakes on metabolism were missing. Cellular respiration is normally coupled to feedback mechanisms inherent in the process itself. In order for the cell to benefit from the energy being produced through respiration, the energy-containing molecules must pass from the inside of the mitochondrial membrane to the outside and then back to the inside. In the case of BAT, however, the molecules passed to the outside, but failed to return. Scientists, perplexed by this incongruity of metabolism, engaged in endless speculation about what was going on. The first clue to the solution came with the discovery that certain cellular poisons that 'uncoupled' the events of respiration in normal cells produced almost identical outcomes as those observed in BAT. Hence, the idea of a 'natural uncoupler' was born.

Many candidates for the nature of this molecule were nominated, ranging from free fatty acids themselves to a host of enzymes and other substances. At least eight different theories were advanced to explain the high rate of cellular respiration in BAT that was stimulated by NE. These theories competed with one another for many years. Most research papers published prior to the announcement of the discovery of the molecule responsible for uncoupling oxidative phosphorylation promoted one theory or another in an intelligent and convincing manner.[1004, 1005] The discovery of UCP in the late '70's by D.G. Nicholls resolved the issue; it must have created much excitement among researchers.[928]

Genetics

Other advances were being made in the study of the effects of genetics and surgery on obesity in animals. Slowly, the realization emerged that the most important, if not the major, metabolic defects giving rise to obesity involved thermogenesis. Applying these findings to humans has been difficult and is still the subject of considerable debate, speculation and research. However, the concept of a faulty thermogenic-induced model of obesity is certainly no more equivocal than any other current model of obesity, including models based on defects in appetite control. Animal research being currently more ethical than human research, we expect greater strides from genetic animal research than from human engineering for the foreseeable future.

Pharmacology

Another line of research has sought to explicate the role of catecholamine (norephedrine, adrenaline, serotonin and synthetics) in the regulation of thermogenesis. This research has been very productive, and while it may not provide answers to the ultimate questions concerning the cause of obesity, it has unequivocally demonstrated that the proper application of catecholamine can enhance thermogenesis which it turn significantly impacts the amount of body fat.

One of the most important applications for catecholamines, given the current proclivity for dieting, is to offset the decrease in metabolic rate that invariably accompanies reducing caloric intake, dieting, or fasting. By helping to maintain normal metabolic rate, the "thermogenic" would also help prevent the onset of obesity-related diseases (e.g., diabetes and hypertension). Awareness of these practical advantages to the consumption of a thermogenic is what provides impetus to much of the ongoing research at universities and hospitals around the world.

The most significant recent event in the history of the science of thermogenesis was an International Symposium held in Geneva, Switzerland in 1992, entitled "Ephedrine, Xanthines, Aspirin & Other Thermogenic Drugs to Assist the Dietary Management of Obesity." Papers presented at this gathering covered a multitude of subjects bearing on the research, problems and clinical applications of thermogenic substances.

The discovery of BAT, UCP and their regulation by the SNS laid the foundation for over a decade of work on the manipulation of thermogenesis by genetic, dietary pharmaceutical and behavioral means. During the past two years, APRL has turned to nutritional and phytopharmaceutical (plant-based) methods for reversing thermogenic defects. The future of thermogenic research and application looks bright. Much more needs to be learned about genetic controlling and triggering mechanisms, BAT regulatory mechanisms and other dietary considerations. Likewise, the search for thermogenic substances must continue. Almost all research on BAT activation has utilized drugs. At APRL, research has focused on the effects of herbal and nutritional agents. It is our hope that the importance of natural substances will eventually be recognized by orthodoxy, for these substances appear to achieve the same results, and probably pose a much smaller risk of long-term harm.

CATEGORIES OF THERMOGENESIS

The material in this chapter is provided to help the reader place brown adipose tissue thermogenesis within a larger framework that includes various means the body utilizes to deal with the calories in the food we eat. The very act of eating forces the body into a position where it cannot simply maintain a resting state. It must somehow deal with the myriad components of the food. Calories can be especially troublesome. Obesity is actually a result of a high degree of efficiency in caloric processing. How does the body decided to store or waste excess calories? Thermogenesis is a wonderful way of getting rid of calories the body doesn't want, but the decision to waste calories is only made on a grudging basis. The body would much rather store energy than waste it; hence, the energy wasting cycles like BAT thermogenesis are often bypassed in favor of energy storing processes. In the best of all possible world, stored energy would be quickly utilized in exercise, in cold adaptation, and so forth. But in our imperfect world, stored energy may sometimes build up to the point where it becomes unhealthy and unappealing. This tendency unfortunately leads to the atrophy of brown fat; the degree to which thermogenic capacity is lost in BAT aggravates the body's already unbalanced tendency toward energy conservation. A vicious cycle ensues in which energy storage mechanisms blunt energy wasting cycles which in turn make energy storage even more efficient, which blunts thermogenesis even further, and so on until clinical obesity results, or until something happens to interrupt the cycle. Dieting is not the proper approach; BAT stimulation is.

Thermogenesis occurs not just in brown adipose tissue, but in almost every cell of the body. Our focus in this book is on BAT because BAT is the only tissue in the human body in which heat is the primary product of metabolism, and because evidence indicates that BAT thermogenesis is probably where genetic and metabolic factors lead to deficiencies.[739] However, in the process of reactivating BAT we also impact on the thermogenics of other cells; therefore, it is important that we understand how that might contribute to the action of our intervention.

The Energy Equation

The easiest way to conceptualize the relationship between the energy we ingest in the form of food and the results of that act on body composition, is in terms of what is known as the energy equation. The equation is not difficult:

$$\text{Energy In} = (\text{Energy Stored}) - (\text{Energy Expended})$$

Energy In is easily measured, but it is very difficult for scientists to measure Energy Expended. The best measuring technique is to measure heat losses in a whole body calorimeter. However, this procedure is not comfortable. Typically, therefore, the indirect calorimeter is now used, but it is still cumbersome and not extremely accurate.[1211] Recently, a room respirometer has successfully been used without undo inconvenience.[600] For purposes of our discussion, however, it may suffice to restrict ourselves to conceptual relationships, with an understanding that there are methods for mathematically assessing the figures that represent the concepts we are discussing.

Total energy expenditure is the sum of all metabolic events occurring throughout the body. These events are regulated by dozens of different mechanisms, and it is extremely difficult, perhaps impossible, to accurately measure the contribution of each to total energy utilization.

The total energy expended by human beings is the sum of basal metabolic activity (60-70%), dietary-induced thermogenesis (10-15%), and energy used in physical activity (20-30%).[764] Obese persons demonstrate higher total energy expenditure than lean persons, usually as a direct result of increased body mass.[772, 1024, 616] The thermogenic component is generally found to be smaller in obese people. Obese individuals with impaired glucose tolerance have smaller thermogenic capacity than obese people with normal glucose tolerance.

Almost all metabolic processes exhibit a thermogenic component. That is to say that heat is a by-product of almost every biochemical event. It has become conventional to divide these thermogenic events into two broad classifications and several sub-categories which, in some cases, parallel the metabolic processes that produce the thermogenic event itself. The two broad classes of thermogenesis are *Obligatory Thermogenesis* and *Facultative (or Adaptive) Thermogenesis.*[44]

Obligatory Thermogenesis

Obligatory thermogenesis represents the minimal heat produced by all those processes that maintain the body in a basal state, fasting, at thermoneutral temperature. Obligatory thermogenesis accompanies those metabolic processes that keep us alive and warm. These processes are primarily under the control of the thyroid gland, and occur in all organs of the body.[526] Obligatory thermogenesis is energy expended during the digesting, absorbing, processing, and storing of food. Some components of obligatory thermogenesis occur in all organs of the body. Other components occur primarily in the intestines, the liver and white adipose tissue (WAT). Obligatory thermogenesis is fairly constant from one day to the next, and changes slowly under the influence of thyroid hormones.

One part of obligatory thermogenesis, the Thermic Effect of Food (TEF), represents the heat increment of eating; it is also called postprandial thermogenesis or specific dynamic action. The most familiar component of obligatory thermogenesis is the basal metabolic rate. The BMR was originally defined as the energy expended by an individual when bodily and mentally at rest 12 to 18 hours after a meal in a thermoneutral environment. Theoretically, BMR was a sound concept, but practically speaking it was impossible to remove all effects due to physical and mental unrest associated with the test itself. Therefore, a more useful concept was introduced, known as

71

the resting metabolic rate (RMR), which is usually a larger but probably more accurate figure than BMR, and one that can more effectively be related to fat free mass, as kcal/kg FFM.[578]

RMR normally accounts for 60 to 75% of our energy expenditure. Obese patients usually have higher RMR than lean persons.[429] Early observations of RMR verified that a person's metabolic rate loosely correlated with surface area of the body. Higher RMR values for women and older subjects were also observed. Correction factors for weight, age and sex are now typically part of the equations used to generate the RMR. Presently, RMR is felt to be most directly related to fat free mass, however there may be as much as 30% variation in RMR between and within individuals.

Obligatory thermogenesis is sometimes divided into two compartments: essential thermogenesis and endothermic thermogenesis. Essential thermogenesis is the energy required to maintain the basal or resting state, including synthesis and catabolism of cellular nutrients, preservation of critical ionic balances, and maintenance of tissue and organ function at the basal rate. Endothermic thermogenesis is the increment in BMR that is responsible for maintaining stable body core temperature at a level above the ambient temperature.[441]

Other obligatory processes that have a thermic component are growth, pregnancy, and lactation.

Facultative Thermogenesis

Facultative thermogenesis represents the energy spent in excess of the obligatory requirements; it is energy expended above that required to maintain the RMR. This is energy expended that produces heat, but not work.[275] Facultative thermogenesis is controlled by the nervous system, independently of obligatory thermogenesis.

Facultative thermogenesis differs significantly from obligatory processes.[536] Because they are under the control of the nervous system rather than the slow-to-respond endocrine system, facultative processes can be switched on and off fairly rapidly, responding quickly to changes in diet, exercise, temperature and other environmental events. Facultative thermogenesis occurs primarily in two areas of the body: skeletal muscle and brown adipose tissue. These sub-categories are further divided into the Thermic Effect of Exercise (TEE), Cold-Induced Shivering Thermogenesis (CIST), both of which occur in skeletal muscle tissue, and Cold-Induced Non-shivering Thermogenesis (CINST) and Diet-Induced Thermogenesis (DIT), which occur in brown adipose tissue.

Categories of Thermogenesis

CATEGORY	SITE	COMMENTS
A. OBLIGATORY		
1. Basal Metabolic Rate (BMR)	All Organs	Heat generated in processes that maintain life.
2. Thermic Effect of Food (BMR)	Liver, WAT, Intestines	Heat generated during digestion, absorption, processing, storing of food.
3. Other		Growth, pregnancy, lactation, cancer.
B. FACULTATIVE		
1. Thermic Effect of Exercise (TEF)	Skeletal Muscle	Heat generated by muscle movement
2. Cold-Induced Shivering (CIST)	Skeletal Muscle	Heat generated by shivering to stary warm.
3. Cold-Induced Non-Shiver- (CINST)	Brown Adipose Tissue	Heat generated to adapt to cold in a general sense.
4. Diet-Induced (DIT)	Brown Adipose Tissue	Heat generated in excess of TEF.

Another part of facultative thermogenesis are the "futile cycles" that occur in skeletal muscle when the ATP stored in the cell is degraded without actually providing any body building energy (i.e., creates heat without work).[925] Futile cycles lower metabolic efficiency, but act as built-in buffers to cellular metabolic processes involved in maintaining thermal and energy balance in the body.[62,685,926,1173]

The Thermic Effect of Exercise

The thermic effect of exercise (TEE), sometimes called Exercise-Induced Thermogenesis (EIT), is thermogenesis that accompanies energy expended in skeletal muscles due to exercise, work, or simply by moving. TEE is the heat increment that occurs as a direct result of muscular exertion above basal levels. TEE accounts for a great deal of the thermogenesis produced by the body; unlike thermogenesis in BAT, however, TEE thermogenesis is automatic. It is an unavoidable consequence of muscular work.

Cold-Induced Shivering Thermogenesis

Heat is also produced by skeletal muscle in the form of shivering induced by exposure to the cold (CIST). A great deal of study has been devoted to CIST. Shivering, like salivating, is primarily a thermoregulatory reaction, that is, it occurs in an attempt to maintain the temperature of the body at a pre-determined level, set and regulated by the central nervous system. As we have discussed already, and shall presently review, the body can react to cold in a non-shivering manner also. There is some debate about which reaction is preferred. One thing is certain: the newborn child does not shiver. CIST is lacking in children until they reach a certain age. Until then, non-shivering processes are used to regulate body temperature. An important question in thermogenic research is why the body begins to switch from non-shivering processes to the shivering mechanism. Is it due to loss of brown adipose tissue, genetic triggering systems, environmental factors, or a combination of these? And how reversable is it?

Cold-Induced Non-Shivering Thermogenesis

The other major sub-category of facultative thermogenesis, sometimes called *chemical* thermogenesis, takes place in brown adipose tissue. It comprises at least two sub-categories, distinguished by the eliciting stimuli. The first sub-category of chemical thermogenesis is called cold-induced non-shivering thermogenesis (CINST) to which we referred in the previous paragraph. This is thermogenesis in BAT induced by exposure to the cold; it does not involve shivering in muscles. R. E. Smith and the team of E.G. Ball and R. L. Jungus were the first to demonstrate (in 1961) that BAT is the site of non-shivering thermogenesis. They discussed the difference between the heat insulating and fat-storing properties of WAT versus the heat-producing capacity of BAT. CINST does not appear to exceed 25% of BMR, yet plays an extremely important role in thermogenesis.

CINST is exhibited by the newborn. Over the next few years the child's thermoregulatory processes will switch from CINST to CIST. This occurs for unknown reasons, but involves the gradual reduction and subsequent atrophy of brown adipose tissue. CINST does not *totally* disappear; rather some its activity appears to be taken over by shivering in skeletal muscle. With continued exposure to the cold, however, CINST begins to increase, gradually replacing CIST.[390] More on this later.

Diet-Induced Thermogenesis

The other sub-category of chemical thermogenesis, unique to BAT, is diet-induced thermogenesis (DIT), occurring as a result of eating certain kinds of food.[384,813] DIT must be contrasted to the thermic effect of food (TEF) which is the *obligatory* component of thermogenesis that results from routine processing of food. DIT is an increment of thermogenesis beyond TEF. While thermogenic differences between lean and obese persons are small, scientists are increasingly accepting the view that defects in DIT are primarily responsible for obesity.[320,326,640,1281] Defective DIT has been the subject of an enormous amount of research. Several models of genetic

75

obesity have been created in animals in which thermogenic activity is curtailed or absent altogether.[524,534,1275] (See the chapter on Genetics.) Similarly, deficiencies in DIT, BAT metabolism and thermogenesis have been created experimentally through lesions, administration of drugs, etc. Some of these procedures involve DIT alone, while others result in loss of CINST. Atrophy of BAT is a consistent finding in these models.[749,855,1274] Changes in the structure and cellular contents of BAT are often observed. Still other changes that occur in genetic and experimentally obese animals involve pancreatic, thyroid and adrenal function.[679,855,1275] All of these animal models are thought to have human counterpoints; i.e., certain types of human obesity may exhibit these same, or very similar, genetic problems.

Defective BAT in humans and animals can also be created through less dramatic means. For example, fasting, dieting or caloric restriction may induce significant atrophy of BAT.[1275,1282] Diet <u>composition</u> may also affect thermogenesis.

Obesity

Most, but not all, studies (perhaps for procedural reasons) have found that obese persons exhibit less DIT than lean persons.[165,268,748] Both categories of chemical thermogenesis are mediated or stimulated by catecholamines whose function is regulated by the sympathetic nervous system. While the heat producing capacity of BAT has been known since the early 60's, the role of CINST and DIT did not become established until the late 70's.[385,389,391,525,1066,1068,1210,437] Recognition of these forms of facultative thermogenesis in brown adipose tissue led to the intensive investigation of them in the 80's. Emerging from that study were the ideas 1) that defective BAT thermogenesis may produce obesity characterized by low energy expenditure and high metabolic efficiency; 2) that thermogenesis in BAT is regulated to buffer variable energy intake so as to minimize changes in body composition; and 3) that defective caloric buffering by BAT may contribute to obesity.[528, 529, 531, 534, 538, 793,1077, 1082]

Facultative thermogenesis in BAT is our primary focus because it provides no net gain in energy, no increase in mechanical work, etc.

More specifically, DIT will be the main focus of our attention. DIT includes stimulation by food, thermogenic drugs, nutrients and raw plant materials. DIT in BAT is activated or suppressed directly and quickly by information transmitted from the hypothalamus of the central nervous system to the SNS.[443] Sympathetic neurons, in turn, directly innervate BAT cells and instigate the thermogenic cascade.

To review, thermogenesis in brown adipose tissue is directly affected by day-to-day changes in three classes of stimuli that increase the firing of sympathetic neural fibers in brown fat: changes in ambient temperature, being in a fed state, especially after tasting and eating pleasant foods, and/or consuming specific thermogenic agents. A serious enough decrease in ambient temperature stimulates cold-induced non-shivering thermogenesis, eating stimulates diet-induced thermogenesis, as does the consumption of an agent that stimulates the sympathetic nervous system directly. These thermogenic agents are called sympathomimetics, because their action is thought to mimic the normal activity of neurotransmitters secreted by the sympathetic division of the nervous system.

Energy Sources for Heat Production

The three main energy storage organs of the body are white adipose tissue, brown adipose tissue and the liver. After birth, the body switches from glucose to fat as the main source of energy, i.e., from the liver to white adipose tissue. The human infant has large fat stores and a low metabolic rate. It is thus equipped to feed brown adipose furnaces and keep itself warm for extended periods of time when exposed to cold air. Experiments have shown that WAT is the major source of fuel for BAT. Circulating glucose may be used as fuel in BAT,[244,246,432,1309,1310,1398] but usually plays a secondary role.[209,597,808,1090,1185,1360] Circulating free fatty acids, on the other hand, are the primary source of fuel for thermogenesis in brown adipose tissue.[812,1360] The free fatty acids, or FFA, are recruited from white adipose tissue and from fat molecules circulating in the lymph system and blood. Glucose-induced thermogenesis (GIT) is mainly an experimental curiosity that proves useful for research and diagnostic purposes.[1093,1094]

A Difference Between Diet-Induced and Cold-Induced Non-shivering Thermogenesis

DIT and CINST share in common the features that they both occur in BAT and both are mediated through the SNS.[1066,1067,1384,1396] However, there is at least one important distinction between CINST and DIT. In CINST, heat conservation is the goal, and heat produced is used to maintain core body temperature. In DIT, the goal is to dissipate heat to the environment, to prevent core body temperature from rising.[742] Thus, different control mechanisms must be in place to regulate the differing metabolic processes that result in heat conservation versus heat dissipation. Little is currently known about these regulatory differences, other than that they certainly involve hypothalamic, pituitary, and lower brain stem mediation. The difficulty in doing human research limits our understanding of thermoregulation in humans.

Thermal Versus Energy Balance

It is important to realize that obligatory and facultative thermogenics are mechanisms the body uses to maintain the temperature of the body within certain narrow limits. There are certain sensing mechanisms for body temperature that invoke negative biofeedback mechanisms, much like the thermostat of your home, to turn on or off heat producing furnaces in BAT and skeletal muscles. The concept of the 'set point' is important here. The changing of the set point is a long term proposition, probably under the control of processes involved in obligatory thermogenesis. But the cues for facultative components of the temperature adjustment are the rapidly changing circumstances of ambient temperature, insulation or clothing, and so forth. In response to cold, the central nervous system initiates dozens of complex signals that reduce blood flow to the skin, stimulate piloerection (goose bumps), increase the insulating capacity of subdermal layers of the skin, and increase the responsiveness of blood vessel muscles.[633,861,1388] All of these processes are geared toward maintaining a more-or-less constant temperature in the body.[158,636,646]

However, not only does the body attempt to maintain a thermal balance, it also tries to maintain an energy balance; that is, energy ingested must equal energy expended. Thermogenesis is BAT may help balance the energy equation. It is important to note that energy balance is not nearly as tightly regulated as the thermal balance. Variation in body temperature by even a few degrees can be fatal. Thus hormonal control of body temperature is geared to maintain a stable temperature, and neural control is geared toward making any rapid corrections required by changing ambient conditions.

Changes in the energy equation, on the other hand, do not exert a life-threatening influence and are therefore much more loosely regulated. Even here, however, both hormonal and neural inputs are strongly felt. Thyroid hormones, adrenal and pancreatic hormones are geared to maintain a satisfactory energy balance, and neural input is required to initiate rapid energy production in the event it is needed. In the face of continual energy consumption (eating), the body therefore makes certain critical decisions concerning thermal and energy balance. A certain number of calories are always kept in reserve to supply the body's requirements for thermal balance. A certain number will be utilized to fulfill current obligatory energy requirements. At this point, excess energy must be dealt with. The body's natural survival instinct will dictate that the excess energy simply be stored as fat. This inclination may be offset to a large degree by properly functioning thermogenesis in BAT.

Energy Balance and Obesity

It would be convenient if the body had sensors in white adipose tissue (our energy stores) that would sense when we had enough fat, and then, through some sort of negative feedback loop, tell the body it was time to dissipate some of the fat. Unfortunately, we do not possess such a mechanism, and the body is perfectly willing to store as much energy as fat as required to balance the energy equation.

Active BAT thermogenesis has an impact on the fate of the extra calories. In normal individuals, changes in energy intake are buffered by BAT to reduce the impact on energy stores. Eating too much results in a compensatory increase in the production of heat in BAT that helps reduce or prevent the development of obesity. However, under conditions of faulty BAT metabolism, excess dietary fat will not be consumed in BAT and will therefore be available for storage as white fat. In other words, increases in energy intake lead to fat accumulation because the buffering capacity of brown adipose tissue is inadequate. In this light, the ingestion of thermogenic agents could have a major impact on the thermal and energy balance. Consumption of sympathomimetics tend to recreate a mechanism lost to genetic defects that would permit stored energy and soon-to-be-stored energy to be dissipated through DIT pathways in BAT. The idea is that once the thermal balance has been satisfied, extra calories may be utilized in BAT thermogenesis without appreciably upsetting the thermal balance, since the heat thus created would be quickly dissipated. Thus the wasting of calories occurs without measurably affecting body core temperature.

However, it must be noted that should body core temperature itself be high, due, for example, to several minutes of intense exercise, this increased body temperature may upset the body's thermal balance enough to suppress BAT thermogenesis and produce atrophy of BAT.[46,47,48,298,1036] Lactation-induced thermogenesis, fever-induced thermogenesis, and tumor-induced thermogenesis also tend to reduce BAT thermogenesis.

MORE ON DIT, TEF, TEC, AND TEE

While thermogenesis in BAT can make a significant contribution to fat loss, it should be remembered that the other factors actually play more important roles in energy balance. For that reason, some of the more important of these are discussed below. In the attempt to reduce body fat, one should use diet, exercise and cold exposure to one's advantage as much as possible. On the one hand, consumption of thermogenic can serve as an overlay, but it can not substitute for any of these. On the other hand, regular consumption of thermogenic can make an impact on stored calories independent of any of diet, exercise or cold exposure. The point is that the more of these factors one includes in one's lifestyle, the more effective may be one's efforts in maintaining good body composition. There are right ways and wrong ways to do this. One can not expect the effects of TEF, TEC, and TEE to be additive or to be equally affected by consumption of a thermogenic compound. It has been found, for example, that injection of 250 microgram/kg of NE in rats actually decreased CINST at low temperatures, while at room temperature NE stimulated thermogenesis.[1407]

The Thermic Effect of Food (TEF) and Diet-Induced Thermogenesis (DIT).

Several studies have shown that obese or over-fat persons experience a lower-than-normal thermogenic effect from ingested food.[118,327,452,673,896,993,1133,1157,1256] Though most of these studies did not establish a firm cause and effect relationship between deficient DIT and obesity, two directly addressed this question. The first was reviewed earlier.[1256] The second discovered that post-obese patients, i.e., people who were once fat, but are now closer to normal, still had only one-third of the thermogenic response to a meal as lean control patients.[327] We have mentioned that the thermogenic response to food involves both obligatory and facultative components: the thermic effect of food and diet-induced thermogenesis, respectively. TEF and DIT combined account for 5-15% of total energy expenditure.[1173] TEF accounts for 60-70% of the total TEF plus DIT. There does not appear to be a change in TEF with weight reduction.[1359]

TEF May Suppress DIT

However, the facultative component is suppressed when other sources of heat, either internal or external, are available. Thus, TEF can reduce DIT when body temperature is normal or above. This helps to explain why diet-induced thermogenesis is generally so weak and does so little to promote fat loss or prevent fat from being stored. Stated another way, when we eat food, we experience a thermic effect from the food itself, which results in raised body core temperature. The calories remaining once the obligatory obligations have

81

been met, could be used in BAT thermogenesis (DIT), except that the increased body temperature inhibits this action. So now what happens to those calories? The door to the BAT has been closed, and the obligatory requirements of the body have been met; so there is only one meaningful avenue left open—they will be stored as fat; stored energy.[46,48,1038] The science of thermogenics is concerned with finding ways to step up DIT activity in the face of forces that would suppress it.

SNS Involvement

From a research point of view, the idea that caloric excess stimulates facultative thermogenesis (DIT) has been rather controversial. Most studies show that rats fed a 'cafeteria diet' (one containing mostly very palatable foods high in carbohydrates) experience an activation of BAT metabolism mediated by the SNS.[1067,11216,1396] Systematic elimination of the sympathetic component leads to obesity in these animals. In humans, the results of such studies are not as conclusive. However, most scientists agree that there appear to be certain individuals who possess only a limited ability to 'burn off' excess calories, a blunted thermogenic response, or low SNS activity.[68,131,132] These persons are prone to obesity and glucose intolerance.[1171,1174]

Deficient TEF and DIT

Lean people often eat a great excess of calories without gaining weight, and return to their target weight very rapidly following any transitory weight gain that results from bingeing. Over-fat people, on the other hand, gain weight even when dietary calories are restricted. This difference probably rests on an impaired DIT in the obese caused perhaps by differing genetic profiles that create either true metabolic differences or a differing ability to respond to sympathetic stimulation.[429] Whatever the mechanism may be, the result is very common: experiments show that the size of the thermic response to eating is reduced in obese individuals.[318,1129,1234] However, other studies have not reported this effect.[367,731,954] Anorectics who have previously been obese exhibit a smaller TEF than anorectics with no previous history of obesity. Incidentally, persons experiencing gross <u>loss</u> of body weight apparently lose the facultative component of thermogenesis.[414] This is what we would expect if the control centers of the central nervous system respond to caloric deficit by shutting down energy wasting cycles and up-regulating energy storage processes.

Obviously, some foods are better than others in activating DIT in addition to TEF. Essential fatty acids (EFA), for example, are known to affect several factors involved in DIT.[856,1080,1085] These acids and other nutrients that affect DIT are discussed in greater detail in a later section dealing with nutritional influences on BAT metabolism.

Before leaving this section, we should also mention that excess white fat on the body has a purely physical detrimental effect on heat dissipation.[541] It has been found, for instance, that in persons with normal weight, there is leakage of heat across the abdominal wall. This leakage is reduced or prevented by the "artificial thermal insulation of obesity." The thermic effect of food is thereby significantly reduced.[159]

The Thermic Effect of Cold (TEC, CIST, CINST).

When humans are exposed to the cold, the first thing they do is put on more clothes. Once their supply of clothes have been exhausted, the body takes over, constricting peripheral blood vessels in an effort to conserve heat. After these two processes have been maximized, humans must rely on their capacity for producing internal heat.[379,685,762] Should all three mechanisms fail to compensate for the lower ambient temperature, deep body hypothermia ensues, which can eventually be fatal. The key to enhancing the effectiveness of clothing and vasoconstriction is to keep the heat production mechanism (thermogenesis) healthy. Thermogenesis in brown adipose tissue is the major neglected source of heat production. Skeletal muscle and liver thermogenesis probably remain healthy throughout life. However, thermogenesis in brown adipose tissue decreases as the amount of BAT decreases as an individual ages. We should be as concerned with maintaining healthy brown adipose tissue as we are with other life sustaining processes. Unfortunately, until we learn more about the thermogenic process, we will probably achieve only marginal success in adjusting thermogenic processes through diet, clothing, and other manipulative actions. Consumption of thermogenic is currently the best approach.

Cold Adaptation

The SNS is absolutely critical to thermoregulation and cold adaptation,[820] and all available evidence suggests that exposure to cold significantly affects BAT thermogenesis. In animal studies, cold greatly accelerates the normal development of brown adipose tissue.[903,1354,82] Anyone who has been forced to move from a relatively warm climate to a colder one will attest to the reality of man's ability to adapt to the cold.[279,657,1008] Temperatures that were originally almost unbearable become steadily less so with the passage of time. Nobody has proven beyond a shadow of a doubt that this adaptation is a function of increased BAT metabolism, but the evidence for some involvement by BAT is compelling.[195] Finnish workers exposed to the cold exhibit more BAT than office workers. Military recruits exposed to the cold for four weeks showed an increased thermogenic response to NE, suggesting the presence of greater amounts of BAT in these men.[657] Female Korean pearl divers likewise have increased BAT. People who work outside in the cold have greater BAT capacity than do office workers.[587]

Augmenting CINST

When this author first suggested to his teenage daughters that part of their problem in dealing with the winter cold was that they started wearing coats, hats and gloves too early in the fall, and that they would better tolerate the cold by leaving these garments off for a few weeks longer, he did not anticipate how literally these teens would accept his advice. Two winters later, these daughters have yet to put on a coat, hat or glove simply because of the outside temperature. One daughter keeps a horse (or ten) and spends considerable time outside during the winter months. The other daughter takes long walks, delivers morning papers and likewise spends considerable time out of doors in the winter. This kind of forced cold exposure has done much to increase the capacity of their BAT to burn calories in the generation of heat (CINST).

Resistance to cold is then increased by exposure to the cold. Consumption of thermogenic may augment this resistance, as may the ingestion of certain nutrients, exercise and other measures.[91,724,1008,1214] Finally, research suggests that consumption of substances that increase the concentration of catecholamines (NE and adrenaline) may produce many of the physiological characteristics of cold-adaptation.[297,765,766]

Certain organs of the body, especially the heart, are preferentially supplied by heat derived from metabolic processes. Related to this economy is the suppression of shivering by BAT thermogenesis, i.e., the suppression of CIST by CINST.[390] Shivering is a less economical way to produce heat since it may actually promote heat loss through body oscillations. CIST is therefore more of an emergency measure, occurring only when heat requirements exceed what BAT can supply. Cold adaptation involves growth of sufficient new BAT to meet the new *normal* thermal requirements. Attendant upon the growth of BAT is a simultaneous increase in the density of gap junctions,[1125] alpha-adrenergic receptors,[1006,1007] and thyroid hormone,[1166] all of which augment thermogenic capacity even further. These items are discussed in greater detail in later chapters.

Cold exposure also improves glucose tolerance, suggesting a positive influence on insulin sensitivity in peripheral tissues. Rats fed 'cafeteria diets' will often develop diabetes following a period of time during which they exhibit glucose intolerance. This tendency is reversed by prolonged stimulation of energy expenditure through exposure to cold.

Cold also increases benefits experienced through exercise, and some research suggests that cold exposure may actually be more effective in preventing diet-induced obesity than exercise. These results are all in addition to the ability of cold exposure to enhance BAT thermogenesis induced by the consumption of thermogenic.

So how does one take advantage of cold exposure? Here are some suggestions. Take cold (or cool) showers; shower before and after exercise; exercise in the cold whenever possible; don't bundle up in warm clothing when going outside—adapt; from time to time, especially during exercise, apply a cold towel to the back of the neck, between the shoulder blades and around the waist.[512]

Finally, it should be mentioned that people who lose weight quickly, through disease or excessive dieting, run a risk of upsetting the control of temperature regulation of the body. Studies show that a loss of 10% or more of body weight can change central control processes of thermoregulation to such an extent that the body fails to respond to cold appropriately. Loss of CINST will persist until weight is regained.[827]

The Thermic Effect of Exercise (TEE).

The thermic effect of exercise is obvious to anyone who has ever worked up a sweat. The amount of energy expended in exercise is directly related to the weight of the person. The same amount of exercise consumes more calories in the obese than in the lean.[154,1359] Obese persons are not any more nor less metabolically efficient when exercising, nor can they depend on changes in efficiency during caloric restriction. However, it has been shown that the thermic effect of exercise does decrease during and after weight reduction.[1339,1359] Two reasons for this have been postulated. One, decreased body weight results in less workload. Two, activity levels decline steadily throughout the dieting period.[1023] It is clear that the positive effect of exercise experienced by lean individuals is not experienced to the same degree by over-fat people.[867,1139,1142]

Exercising combined with a low calorie diet may help maintain resting metabolic rate and fat free mass.[191,265] Combined with a thermogenic compound, the tendency toward reduced RMR would be counteracted even further; the fat reducing tendency would be enhanced and the conservation of lean body mass reinforced.

It is also important to note that RMR is often more elevated in the obese than in the lean. This is because heavy individuals often have a large fat free mass in addition to a large fat mass.[149,512] Alternatively, RMR is often the same in lean and obese patients;[1140] however exercise prior to a meal significantly increases the thermic effect of food in the obese, but not in the lean.[1139] Postmeal exercise, on the other hand, enhances TEF only in the lean. These kind of findings suggest that RMR is influenced both by metabolic efficiency and by the amount of fat free mass.[1023,1141,1142]

Exercise is a positive behavior because it increases the metabolic rate and hence the amount of energy expended.[265,427,1104,1131,1172] In one animal study, exercised subjects ate less than sedentary control animals. They also experienced significantly lower gains in weight, fat and protein accumulation.[1037] Exercise probably activates the SNS and hence increases the amount of BAT and its sensitivity to NE.[485] Aerobic exercise also increases the level of circulating growth hormone, which in turn helps to reduce body fat and increase lean body mass. Exercise has been found to stimulate futile cycles in skeletal muscle, and may thus help dissipate excess energy as heat. Exercise may help maintain the efficiency of these (non-efficient) cycles and their sensitivity to circulating NE and adrenaline. Exercise has been found to increase DIT in males but not in females.[851]

While exercise itself does not consume a great number of calories, the increase in RMR caused by exercise may continue to eat up calories for hours following the exercise itself.

The type of exercise is important for two reasons. One, the nature of the exercise dictates how much energy is consumed and from where it is drawn. Two, because heat produced in skeletal muscle by exercise tends to raise core body temperature, it has a tendency to counteract the effects of stimulant-induced thermogenesis in BAT.[46-48,298,1036,1038] A form of exercise called deep breathing was explained in PART TWO. Deep breathing purportedly oxygenates muscle tissue without raising core body temperature. Isometric exercise is also strongly recommended for the same reason. Any type of outerwear that restricts evaporation should not be used. Weight loss produced by restrictive clothing, body wraps, etc., is temporary at best, and may even be dangerous to some people.[393]

It should also be noted that to be really effective in affecting body fat, exercise should last longer than 30 minutes. During the first 30 minutes, the body is primarily using calories derived from stored carbohydrates in liver and muscles. This is especially true of high-intensity, peak performance (anaerobic) exercise. Moderately intensive aerobic activities use both carbohydrate and fat during the first 30 minutes. After the first 30 minutes the body begins to use steadily more fat than carbohydrates.

While it is generally believed that exercise is the most potent 'natural' stimulus of thermogenesis in skeletal muscle and the liver, it is equally true that high intensity exercise of long duration that makes a significant impact on energy expenditure is unrealistic and dangerous.[1104]

Conclusion: to be effective in weight and fat reduction, exercise should be done at least 20-30 minutes at least 3 days per week, and should be geared to burn at least 300 kcal per session. Shorter sessions more frequently also works. In addition, exercise inhibits BAT thermogenesis least when performed 3-4 hours before or after the consumption of a thermogenic compound.

Perhaps the best solution is to develop a more active life-style, a natural consequence, by the way, of consuming herbal-based thermogenic agents. Greater activity in turn feeds the thermogenic cycle by increasing sympathetic nervous system activity and increasing effectiveness of NE in mobilizing lipid fuels for the BAT furnaces. Furthermore, exercise increases insulin sensitivity (decreases insulin resistance) and hence increases diet-induced thermogenesis.

Finally, the idea of exercising in the cold should not be discounted as a means for enhancing thermogenesis if engaged in properly. To be effective, exercise should be light and severe cold should be avoided. The reason for this is because air that is extremely cold will tend to constrict blood vessels in peripheral adipose tissue thereby reducing lipid mobilization.[309] Fewer free fatty acids means less fuel for thermogenesis. In other words, the combination of exercise and cold exposure does not automatically enhance metabolism of fats. In fact, if the temperature is too cold and the exercise is too strenuous, or not strenuous enough, fat metabolism can even be reduced. Hence, exercise and cold exposure are tricky entities to balance.

Dieting and Obesity: Some Observations.

The consequences of reducing caloric intake in the effort to lose weight are not pleasant. There have been literally thousands of diets based on the concept that the restriction of daily caloric consumption will result in loss of fat and/or weight, with a resulting improvement in health, looks and even sexual ability. The results of these kinds of diets have been uniformly disappointing.[1217,1189,651] Most diets are based on the erroneous notion that since surplus calories are all converted to stored fat, fat stores will be diminished if no surplus calories are available. Unfortunately, it hasn't worked out that way. What happens is that total body energy expenditure, as calculated by the caloric burn rate, falls to equal energy intake.[106,118,266] Caloric restriction also tends to lower basal metabolic rate in BAT. Lowered BMR more than compensates for the reduction in calories—the body is still going to store calories as fat, meaning the dieter will lose little, if any weight in the long run.

Gluttony is Better?

The seriousness of the dieting mistake is highlighted by research that shows that the opposite of dieting, namely overfeeding or gluttony, tends to increase BMR, or maintain it. Overfeeding also produces an increase in TEF and DIT, the latter through an activation of the SNS.[267,1128] Such findings are not a justification for gluttony, but they do show that in persons with normal brown adipose tissue function, consuming a great number of calories can be compensated for through thermogenesis. In persons with defective BAT, overfeeding would probably result in fat accumulation.

It seems that energy balance can be achieved at various body weights.[1352,1353] Therefore, no matter how much one restricts caloric intake, there will probably always be excess calories due to the difficulties inherent in trying to force caloric intake to exactly match caloric utilization. Under drastically reduced caloric intake, a wasting of non-adipose tissue begins. When the diet ends, and caloric intake increases, the wasted tissue is not replaced immediately, but the fat deposits are! Hence the common observation about yo-yo dieting: an increase in fat.

Obesity and Metabolic Rate

Ironically, some studies have shown that obese people demonstrate a <u>higher</u> basal metabolic rate than lean people. These data are often misconstrued to prove that BMR has nothing to do with obesity. How could BMR be higher in the obese—shouldn't this mean that the higher metabolic rate would make them lean? This misinterpretation is then used to justify caloric restriction or dieting as the best way to lose weight. Actually, the process goes more like this: A person who is mildly overweight, perhaps because of a thermogenic defect, begins to diet by restricting caloric intake; metabolic rate at first goes <u>down</u> to compensate for the lower caloric level; the person continues to gain weight; eventually the person becomes obese; now, in this state, the person has accumulated additional fat pounds <u>and</u> additional fat free pounds (muscle tissue) required to carry the fat around. As a result, the person's metabolic rate increases in proportion to the amount of extra <u>lean</u> mass. The amount of increase may slightly or greatly exceed the BMR value the person had before the diet. Combined with a certain degree of insulin resistance developed during the process, which would tend to promote glucose storage in adipose tissue, the overall result is the creation (aided and abetted through lack of exercise) of the ideal environment for promotion of continued obesity.[992]

An important question remains: Is the lower energy expenditure the result of the loss in metabolically active tissue mass and reduced physical activity or is it the result of a metabolic adaptation that results in a more efficient utilization of energy?[618,643,696] This question is partly answered through the science of thermogenics which has revealed that hundreds of excess calories can be burned away in BAT. This factor has helped clarify much of the conflicting data to come out of studies on dieting. As we learn more about the SNS regulation of BAT metabolism, the clearer it becomes that caloric restriction suppresses sympathetic activity which causes a decrease in metabolic rate and a reduction in sympathetic mediation of thermo-genesis.[151,484,659,661] It also produces a decrease in the conversion of inactive thyroid hormone, T_4, to the thermogenically active form, T_3, which further depresses metabolic rate since this thyroid hormone normally potentiates the effects of NE.[1303]

On the other hand, it has been shown that sustained overfeeding, i.e., consumption of a large number of calories, in healthy lean men, results in only a small gain in weight, and that even that small weight gain is difficult to maintain without continued overfeeding by least 50%.[1174]

In animal and human studies energy expenditure was lower in chronically underfed subjects than in controls as a result of meta-bolic adaptation to the low caloric intake.[29,439,1175] Other studies have failed to observe this effect, but have instead found that the pro-cesses that regulate energy expenditure during under-feeding are af-fected by the changes that occur in body weight and body composi-tion, that is, in the reduction of the mass of metabolically active tis-sues.[382,439,618,643,687,696,872,873,939,1342] Even in the latter studies, however, adaptive changes in energy expenditure occurred and in one study accounted for about 10% of total energy expenditure.[618]

Reconciliation of these apparently disparate findings can be achieved if we take into consideration the dynamic nature of the pro-cesses involved. The evidence at this time definitely supports the notion that most of the changes in energy expenditure induced by dieting (and overfeeding) can be accounted for on the basis of de-creased body weight and depletion of metabolically active tissue. Rapid increases and decreases in weight may be explained on this basis alone. However, we must also consider the impact on the course of obesity delivered by the remaining 10% of change that depends on true adaptive metabolic processes. When we consider that energy storage is a function of time, a 10% effect on storage efficiency pro-jected into the future five or ten years or more can make a significant impact on body weight, body composition, as well as on the overall health of the body. Reactivating the thermogenic process in in brown fat may significantly alter the fate of that 10%.

Further light has been shed on this topic by a prominent team of English scientists headed by C.A. Geissler of the University of London. In an interesting series of experiments in obese and lean women, closely matched for height, weight and age, differing only in lean body mass, the researchers found that obese women exhibited a significantly lower energy expenditure than the lean. Utilizing control procedures, they determined that the difference was due to metabolic differences in thermogenesis.[429] Basal metabolic rate accounted for 40% (15% after adjusting for differences in energy intake).[1153] The researchers admit that these results may be interpreted in two ways: 1) weight gain occurs as a result of lower metabolic rate; or 2) dieting reduces metabolic rate. Geissler cites, in support of the first interpretation, data showing that in adult American Indians and in newborn infants of obese women, a reduced energy expenditure predisposes to increased body weight.[1024,1053] Geissler's hypothesis has received support from other studies.[722,849,1288]

Diet Composition

It has been suggested that since carbohydrates and fat stimulate the SNS, and protein does so only minimally,[536] it would be advisable for persons wishing to diet to consume foods low in protein (but enough to satisfy the body's basic nitrogen requirements), but relatively high in carbohydrate (with some fat); this practice would in turn stimulate thermogenesis which would dissipate excess calories derived from the carbohydrates and fat as heat, and would thereby avoid the accumulation of fat.[642] The problem with this hypothesis is that protein is the only source of essential amino acids required to make the entire thermogenic process work. Adequate amounts of protein are also required to support the life of all cells of the body. In the final analysis, it is not the amount of protein, carbohydrate or fat that matters, it is the quality. More on this topic is found in the chapter on Nutrition. Now let us turn to one of the most controversial issues in all of obesity research.

LOW-ENERGY-ADAPTED OBESITY

Newmann: How comes it, my lord, that thou dost en-
gorge thyself and add nere a stone to thy weight, whilst
thine peer, under restrain, dost so sadly augment himself?   —Shakespeare, Henry VIII, Part IV, Act II,
Scene 1.

Along with Shakespeare's Newmann, we all know people who
must eat a great excess of food in order to gain weight, and we know
others who claim that no matter how little they eat, they still gain
weight.[1174] Another Neumann, R.O., in 1902 reported a series of incredible experiments in which he demonstrated, using himself as the
sole subject, that weight could be maintained at a uniform plateau
even when daily caloric intake varied over a wide range.[924] Neumann
referred to this phenomenon as 'luxoskonsumption.' Before
Neumann, C.Voit and M. Rubner had described a similar notion.
Rubner clearly had thermogenesis in mind when he wrote,

> "The stream of food increases, but it does not determine
> the size of consumption. Apparently the organism does.
> At first the organism builds reserves, then it deposits ad-
> ditional substance, and finally, with increasing heat pro-
> duction, it gets rid of the ample food intake, at least in
> part."

He reasoned further that this process would have survival value
in that an organism could consume an excess of foodstuff of poor
quality in order to obtain an essential nutrient without being subjected to obesity. Although it is probable that overfeeding in normal,
healthy, adult humans does not produce clearly defined 'luxosconsumption,' the concept of impaired thermogenesis is probably an
important contributing factor to obesity.

Many people who are susceptible to obesity appear to have depressed metabolic rates that simply cannot handle a normal day's
worth of calories.[163] Why is this so, and what can be done about it?

Generally, scientists speak of these people as being low-energy-adapted, i.e., they have a reduced capacity to expend energy. In terms of the energy balance equations, there is virtually no amount of energy that they can consume in the way of food that will not exceed the metabolic requirements of their body. The excess is then stored as fat. It seems that whenever they diet by decreasing caloric intake, their bodies adapt to the lower calorie value, and they simply find it impossible to make significant headway in their weight loss efforts. This condition may be caused by one or more genetic, environmental and dietary factors.[1024,1130]

In an extensive study using animals, scientists in Switzerland, under the direction of L. Girardier and A.G. Dulloo, determined that the consequences of an extended low-calorie diet were the stabilization of lean body mass percentage equal to that of normal control animals (protein = 22%, water = 72%, and residue = 6%).[439,320] This meant that the animals on the low calorie diet had adapted to the lower energy level through an adjustment of basic metabolic processes. Lower energy going in led to lower energy going out. What happens now, if the low-energy adapted animals are permitted unrestricted access to food? It turns out that "refeeding" led to a 10% reduction in the amount of energy expended by these animals. What happened to the 10%? It was deposited as fat rather than protein.![319]

Among low-energy adapted people, the end of the diet period is usually followed by a period of extremely rapid weight gain. In fact, following an initial period of moderate weight loss, many of these people show significant weight gain even while still on the restricted-calorie diet! The refeeding experiments suggest that increased caloric intake following a period of dieting accompanied by depressed metabolism would result in the expected energy sparing process, but unfortunately, the spared calories are deposited as fat not as protein. The Swiss refeeding experiments seem to support the notion that yo-yo dieting is detrimental to body composition. Thus, the end result of yo-yo dieting is the accumulation of fat stores that typically exceed original fat deposits; this is accompanied by reduced accumulation of lean body mass. Is there a relationship between the 10% of reduced energy expenditure in low-energy adapted humans and thermogenic mechanisms?

Typically, low-calorie-adapted people demonstrate an impaired thermogenic response to food. Low calorie diets lead to reduced SNS activity, reduced BAT, reduced BAT mitochondria, and reduced UCP. Thus, the 10% of consumed calories deposited as fat may very well be due to impaired BAT thermogenesis, although reduced thermogenesis in skeletal muscle may also contribute a certain amount.[16,990] This suggests that an intervention program designed for improving thermogenesis might be beneficial. An Italian study carried out a few years ago addressed the problem.[974] The Italians evaluated the effects of ephedrine on a group of women selected because they were low-energy-adapted. All of the women were over fat and had shown resistance to weight loss programs of the typical low calorie type. In a double-blind cross-over design, the women were placed on a restricted diet (1000-1400 kcal/day) and administered 50mg of ephedrine hydrochloride or placebo three times per day, just before each meal, for two months. On average, subjects using ephedrine lost significantly more weight than the placebo subjects (2.41 kg vs. 0.64 kg). Individual BMR values did not make a difference, nor did physical activity or life-style differences. Surprisingly, the 150mg dose did not produce serious side effects, with only a few women complaining of agitation, insomnia, palpitation, headache and giddiness.

A similar treatment regimen in a population of unselected obese patients did not yield significant results.[968] This suggests that the particular dose of ephedrine was not sufficient for people who were not low-energy-adapted.

Procedural aspects of the first study should be noted. One, the ephedrine was given with meals. Given between meals, it may have had a greater impact on weight and it may have caused a greater number of reported side effects. Two, by not controlling life style factors, the study provides further evidence for the importance of the thermogenic approach to fat loss. Approaches that require significant changes in lifestyle, either in terms of exercise or diet, are simply not going to work for most people. Third, it is unclear what the restricted diet used in this study may have contributed to the results. What would have happened if the patients had been allowed to consume 1800-2400 calories per day? Our data suggest that the thermogenic approach works even when no great care is being paid to caloric intake. Data from other research groups concur with our observations. For example, the Astrup group found that 60mg/day of ephedrine treatment for 12 weeks without any caloric restriction resulted in an increase in basal energy expenditure of 7-11% and an enhanced thermogenic response.[64]

Recently, L. Girardier has resurrected a model first published by A.E. Dugdale and P.R. Payne in the late 70's, to help account for the process by which the body decides how to utilize consumed and stored energy.[315,439,981] The model proposes four types of metabolic tissues: 1) storage fat; 2) structural fat, or the fat in cell walls and cytoplasm; 3) fast lean tissue, or metabolically active tissue in splanchnic (visceral) organs and bone marrow; and 4) slow lean tissue, or primarily muscle. The body attempts to keep the composition of these compartments constant. Thus, as excess (or deficient) energy is consumed, decisions of what to do with available energy depend on the relative degree of balance in each of the compartments, immediate energy demands, and the likelihood of future demands. The model nicely accounts for many experimental and clinical observations, accurately predicting the yo-yo phenomenon, or weight cycling, as well as such findings as a greater loss of muscle tissue in metabolically lean people relative to metabolically obese people when put on a low-calorie diet.[424] We hope to witness an expansion and investigation of this model in future years.

In summary, two major metabolic adaptations occur under conditions of restricted caloric intake: 1) reduction in energy production to prolong life in the face of diminished consumption of energy sources, assessed by reduction in basal metabolic rate, resting metabolic rate and caloric burn rate; and 2) increase in substrate (fat) mobilization to prevent damage of vital body organs.[45] The sympathetic system mediates both adaptations. First, suppressed SNS activity helps reduce the metabolic rate by reducing thermogenesis.

Increasing SNS activity through the consumption of thermogenic helps compensate for the decreased SNS activity due to dieting.[1156] A mildly caloric restricted diet in conjunction with thermogenic may be indicated in some cases.

Second, stimulation of the adrenal medulla increases the mobilization of fat stores.[45] However, the effect is somewhat paradoxical since the act of dieting both stimulates and *limits* this adrenergic action in such a way that only enough substrate (fat) mobilization is allowed that meets the bare minimum metabolic needs of the body.[742] (See Figure 6) Hence, as diet-induced adrenal medullary action increases, energy stores are mobilized but not to such a degree that it compensates for the lowered metabolic rate imposed by decreased SNS activity.[45,255,416,1176,1393] Apparently, the increase in adrenaline output is an emergency reaction to a perceived threat to life signalled by the presence of starvation. Incidentally, with reduced SNS activity, there is a lowered conversion of T_4 to T_3 which further depresses metabolic rate.

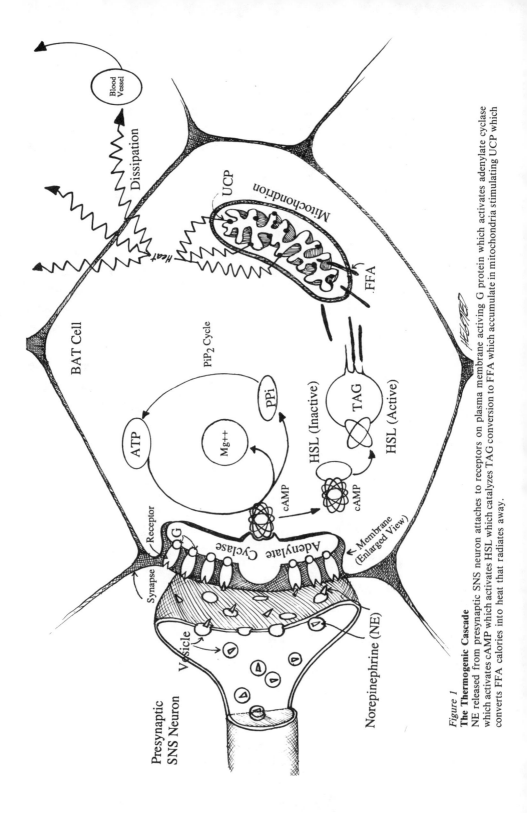

Figure 1
The Thermogenic Cascade
NE released from presynaptic SNS neuron attaches to receptors on plasma membrane activing G protein which activates adenylate cyclase which activates cAMP which activates HSL which catalyzes TAG conversion to FFA which accumulate in mitochondria stimulating UCP which converts FFA calories into heat that radiates away.

BROWN ADIPOSE TISSUE: WHAT IT IS AND HOW IT WORKS

A good portion of this book is taken up with discussions of *Brown Adipose Tissue* (BAT). BAT is found in almost all mammals,[1075,1184] and some birds.[949] It is the critical and central feature of the entire thermogenic mechanism.[539,1045] When brown adipose tissue functions correctly, most excess dietary calories are burned off, a significant dent can be made in white storage fat under normal conditions; when BAT fails, excess calories are deposited in storage as fat, and fat stores seem to hang around without ever being used up or reconverted back into utilizable energy. Thus, it has been postulated that much of human obesity is the direct result of impaired brown adipose tissue function.[524] BAT is the most unique organ of the body. Although scientists have studied BAT for many years, and excellent reviews have appeared in the scientific literature,[1048,512,531], no comprehensive treatment of this tissue has appeared in the popular press before now; therefore, both its presence and importance are generally unknown to the public and to scientists not directly involved in BAT research. It is our hope that this book may generally increase the public and the scientific community's understanding of BAT and its centrality in fat metabolism and weight management.

The Yin and the Yang

In a 1964 symposium on fat, a professor from the State University of New York delivered an address entitled, "Adipose Tissue: Yang and Yin." It was his thesis that fat, typical plain white fat, had really been dealt with in a harsh manner. There were, he contended, actually two sides to fat, the good and bad, the Yang and the Yin. Yin constituted "dark and somber Yin thoughts about obesity," while Yang considered behaviors "made possible by the distinctive biochemical capability of adipose tissue cells." In the latter category he placed the exploits of hibernating animals (more properly a brown fat achievement), migrating salmon and birds (stories of "fat mobilization and love"), and the mating male seal.

> "Yang is survival; what of Yin? Yin is obesity, susceptibility to a long list of diseases, early morbidity and death. The fat man is a poor operative risk, the fat lady trips and falls because her view of the floor is obscured; both are candidates for diabetes, early atherosclerosis, gallstones, and other disabling illnesses. But even in the dismal recital of these Yin contingencies one can discern a trace of Yang, for the adipose tissue itself has not yet been shown to be the primary villain in any of these circumstances. . . . . . .the adipocyte loyally performs its normal line-of-duty function. In fact it performs almost too well, and in the very efficiency of its performance one can detect an echo of the remote evolutionary past."

The professor's observations are true enough. But in the interim since 1964, brown adipose tissue has definitely emerged as the real embodiment of adipose Yang. White adipose tissue can be the Yin to brown fat's Yang. The two tissues, properly balanced one against the other comprise a unity of metabolism that preserves the survival tendencies of the species and ensures a minimum of health risks at the same time. All of the carefully programmed inhibitors of BAT metabolism and the relative lack of control on WAT metabolism suggest that it is the current purpose of Nature to encourage survival at the expense of optimum health. Hopefully, with an increased understanding of BAT and methods for improving its effectiveness, we can restore balance to the Yin and the Yang of energy metabolism.

Bat is Unique

BAT appears to be the major, if not the only, site of non-shivering, diet-induced, and drug/nutrient-induced thermogenesis.[174,391,1069]

BAT appears to serve no other purpose, and its metabolism is unique to the entire body. In many newborn animals (including rats and humans), BAT is the principal site of cold-induced thermogenesis. Since the newborn of many fur-bearing animals are susceptible to hypothermia, BAT-generated heat serves the critical function of protecting these infants against the cold. Why human infants have a large amount of BAT is still the subject of speculation, but it probably involves some aspect of enhanced survivability because it provides some capacity to adapt to changes in seasonal temperature, or an ability to adapt to colder areas of the globe.

With maturity, BAT activity declines in most animals and humans, and the tissue begins to take on the physical appearance of white adipose tissue as mitochondria disappear. The process can be reversed by prolonged exposure to cold and/or by the consumption of thermogenic agents.[1184]

Researchers often assume that adult humans do not have enough BAT metabolic capacity to make a significant impact on white fat stores through thermogenesis. This assumption turns out to be false. Although the amount of BAT in the human body is limited, there is enough to incinerate a considerable number of calories each day.[9,438,1073] The presence of less than 0.1% of body weight of BAT could have a dramatic impact on energy expenditure.[438,1066,1073] A low fat/lean body mass ratio is characteristic of people who work outdoors in the cold, while a high fat/lean body mass ratio is a common finding in office workers.[587]

(**Note**: Although the remainder of this chapter is technically oriented, and may be difficult to follow for the lay reader, careful study and review of these sections will reward the reader with a deeper understanding of the many factors that affect thermogenesis, fat management and body composition. Some attempt has been made to simplify and illustrate by analogy, but some of the concepts are just plain difficult to explain and hard to absorb. We suggest very strongly that every reader attempt to assimilate as much of this chapter as possible, since this material is really the heart of the book. The average reader is advised to skim the chapter first, and then study each section more careful.)

BAT MORPHOLOGY AND DISTRIBUTION

Before turning to a discussion of the basic biological processes underlying cellular metabolism in general and BAT metabolism in particular, we should probably establish a few of the distinctions between BAT and other tissues of the body, especially the closely related yellow, or as we prefer to say, white fat cell.[1278]

TAGs

Earlier chapters made repeated reference to white fat and brown fat (WAT and BAT respectively). But what are the real differences between these tissues? Both tissues are able to store relatively large quantities of lipids or fats in the form of triacylglycerol droplets (TAGs). White adipose tissue stores TAGs, and brown adipose tissue stores and burns them. In white adipose tissue, the triacylglycerol droplets are described as being 'unilocular' since they occur as a single large droplet within the cytoplasm (or cytosol) of each cell (the cytoplasm is the part of the cell that lies outside of the nucleus). In brown adipose tissue, on the other hand, the TAGs occur as multiple smaller droplets, or 'multilocular.[1278,915,1112] The implications of this distinction are not completely clear, but probably involve the relative ease by which the TAGs are metabolized. One large droplet may be much harder to convert to FFAs than several small droplets.

The fat in the TAG droplets is in the form of triglyceride molecules. A triglyceride consists of a glycerol molecule with three long-chain fatty acids attached. (See Plate IV.) Before the triglyceride can be used as fuel by the brown adipose tissue cells, it must be split into smaller, more soluble units, i.e., into glycerol and free fatty acids (FFAs). Free fatty acids are the fuel source, since the glycerol cannot be burned. Once the glyceral molecule has been separated from the three free fatty acids, it is recycled by the body to form new triglycerides.

Mitochondria

Other distinctions between BAT and WAT involve the number and appearance of mitochondria present; BAT contains many more mitochondria than WAT, with a distinctly more healthy appearance. Mitochondria are the site of thermogenesis. The more mitochondria there are, the greater the thermogenic capacity. Mitochondria are among the largest objects in the cell, and can occupy as much as 25% of the total intracellular space. (See Plate II.) Mitochondria have outer and inner membranes which differ from one another both structurally and functionally. The outer membrane is permeable to even fairly large molecules. The inner membrane in folded back on itself several times. These infoldings are called cristae and protrude into the central space, or matrix, of the mitochondrion. The inner membrane and the matrix are the sties where most of the important reactions in the mitochondria take place. The matrix contains a concentrated mixture of hundreds of different enzymes. The folds of the inner membrane dramatically increase its total surface area thereby increasing its metabolic capability, as will become clear later. (See Plate IV.) In atrophied BAT, mitochondria have a distinct "abnormal" look.[549]

Even though only about 40% of brown adipose tissue is composed of brown adipocytes, the tissue may nevertheless be distinguished from white adipose tissue by examining differences in cell size, shape, nuclear position, concentrations of certain amino acids, and distinctions among the various enzymes for the respective tissues.[809,1145]

BAT as an Organ

BAT has a lobulated appearance and is blessed with a rich network of blood capillaries and nerves. It is because of these factors that brown fat resembles an endocrine gland. Even though BAT is called a 'tissue,' several authorities have argued that it be regarded as an organ.[546,1406] In view of the specific role of BAT as a thermogenic effector, the notion of redesignating the tissue as an organ should be given some consideration. Under conditions of exposure to cold, DIT, or the influence of thermogenic drugs, nutrients and herbs, there can be observed an enlargement of BAT tissue due to an increase in both the size and number of brown adipocytes, and in the number of blood vessels which, in turn, depend upon the integrity of the nerve supply to the tissue.[169,176,543] Incidentally, it is the numerous mitochondria (containing iron cytochrome pigments), nerves, and blood vessels that give BAT its distinctive brown hue.

Sympathetic Innervation

An extremely important distinction between BAT and WAT is the richness of sympathetic innervation in BAT.[252,1396] (See Plate II.) As new brown adipose tissue is formed, nerve fibers in the region undergo a certain amount of outgrowth. These new fine fibers lack the typical synaptic contacts with brown adipocytes. Instead, all along the length of the fiber, as it winds among blood vessels and fat cells, can be found points of transmitter release. This constitutes an open system of innervation which can undergo rapid growth and may quickly extend a network of neural filaments through the brown adipose tissue. As brown adipocytes proliferate, the neural and vascular support is thus able to keep up with and even stimulate the new growth.[253]

Uncoupling Protein

The most important difference between white adipose tissue and BAT is the presence of uncoupling protein (UCP), a special molecule found only in brown adipose tissue.[189,208,755,1046] UCP, or thermogenin, is found on inner mitochondrial membrane. (See Plate IV.) Some authorities believe that in the final analysis the only reliable difference between white and brown adipose tissue is the presence of the uncoupling protein. Recently, cDNA probes have been utilized to identify the mRNA for UCP.[139,611,1050] This technique has helped prove the presence of UCP in brown fat cells. It is used to demonstrate the conversion preadipocytes to BAT during stimulation.

104

BAT Changes With Disuse

When BAT malfunctions it becomes gorged with lipid, as occurs in obesity. Then, the BAT appears very much like white adipose tissue. Alternatively, some researchers believe that, with disuse, brown adipocytes are replaced by white adipocytes.[1184] Much of the controversy surrounding the issue of whether BAT atrophies, changes to WAT, or is replaced by WAT as we get older originates in the difficulty of distinguishing the appearance of brown fat from white fat in organisms whose thermogenic processes have been shut down for a long period of time. We stated earlier that adults have very little brown fat; actually, some research seems to demonstrate that the brown fat itself is there, at least as measured by the concentration of UCP,[757] but these data also underscore the notion that thermogenesis is shut down for some reason, in spite of the presence of the organ responsible for doing it. With repeated stimulation, such tissue quickly reacquires the typical appearance of active brown adipose cells.[262,758,761,1049,1184]

Following behind Heaton's pioneering explorations of BAT distribution in humans,[509] Japanese researchers recently reported an investigation of perirenal BAT samples obtained from 215 fresh necropsy cases of both sexes (Japanese), aged from one month to 93 years.[603] They could find BAT in only 75% of the samples. It appeared that BAT in the other samples had transformed into white adipose tissue. The scientists were able to differentiate six types of BAT, differing from each other in the morphological features of their lipid droplets. According to the authors, these six states correspond to an equal number of functional thermogenic states. This suggests that perhaps one day we will have amassed enough human clinical data to be able to identify six or more different levels or types of thermogenic responders. Such information would perhaps help researchers build thermogenic programs designed for each type of thermogenic responder that would help the individual overcome some of the observed limitations in current thermogenic fat loss programs.

Researchers have reported that in some strains of genetically obese rats, the BAT mitochondria are deformed to such an extent that they are unable to function properly. These observations have not as yet been reported for other strains of obese rats or other organisms.

Preadipocytes

Undifferentiated, or "convertible", preadipocytes have been observed in brown adipose tissue.[169,173,430] It is believed by many authors that these cells are actually precursors of brown adipocytes and will convert into full blown BAT under the proper circumstances.[801] When stimulated by thermogenic agents or the cold, these cells are recruited as needed and fill in the ranks of the brown adipose cells. These preadipocytes may all be converted to white fat cells if BAT deteriorates enough.[81,1225] Thus, three types of adipocytes exist in the human body: white ('storers'), brown ('users'), and convertible ('undecided').

Distribution

Until recently, it has been difficult for researchers to determine the exact location and quantity of brown adipose tissue in living organisms.[509] A recent advance in magnetic resonance imaging (MRI) promises to usher in a new era of BAT morphotometry. MRI provides the ability to accurately measure volumes of BAT deposits and changes in BAT volume under different conditions (e.g., cold exposure, meals, drugs, etc.).[1108] In fact, it is this technique that has conclusively shown that BAT is present in adult humans.[261,755-757,760-761]

BAT is strategically located along the neck muscles (pars cervicalis) about the abdominal viscera, between the shoulder blades (interscapular), atop the kidneys, along the breast bone and about the thoracic viscera (pars thoracia) alongside large blood vessels, and especially all along the spinal cord and key bones (pars axillaris) and paravertebral ganglia.[9,509,587,756,757,760,761,915] In man, BAT occurs most prominently as lobes around the muscles and blood vessels of the neck and extensions thereof passing under the clavicles to large deposits in the axillae.[281,509,587] The large vessels entering the thoracic duct are also well covered with BAT. The abdominal masses envelope the kidneys and adrenals, pancreas, autonomic ganglia, chromaffin tissue and the aorta.[9,281,509,587] As such, BAT is extremely well positioned to radiate heat in such a way as to maintain the temperature of the spinal column and main internal organs, and export heat to distal tissues throughout the body. Warming of the blood vessels entering the heart serves the important function of insuring that blood returning to the heart is warm; cold blood could have adverse affects on the health and/or function of the heart.

How BAT Works: The Thermogenic Cascade

On the surface of BAT cells are located a variety of special receptors that mediate communication between chemicals outside the cell and chemicals inside the cell. Thermogenesis in BAT is initiated by the action of extracellular catecholamine (norepinephrine, or NE) on Beta-receptors on the surface of BAT membranes that results in membrane depolarization.[440] Immediately, a complex series of intracellular events occur that result in the uncoupling of cellular respiration (oxidative phosphorylation) in BAT mitochondria through activation of a special and very unique proton conductance pathway.[929,931,1195] Activation of this pathway requires the presence of UCP.[927,930.931]

Uncoupling of metabolism in BAT mitochondria allows the diffusion of hydrogen ions into the inner mitochondrial matrix which results in the creation of heat through the oxidation of substrates (free fatty acids) instead of the production of energy. The substrate is provided through catecholamine-induced lipolysis within brown adipose tissue that provides a source of free fatty acids from stored TAGs. We have coined the phrase "thermogenic cascade" to refer to these events. The thermogenic cascade will be discussed in detail later in this chapter. The lipid content of BAT can be depleted by exposure to cold, prolonged fasting or starvation, hibernation, the consumption of thermogenic activating agents, or by any combination of these.

Vascular Effects

The heat created through thermogenesis is quickly carried away from BAT via the blood, a process which is facilitated by a dramatic, though indirect, catecholamine-induced increase in blood flow through BAT. Blood flow is also affected by ambient temperature, increasing as temperature decreases.[109] This increase in blood flow also increases the rate of delivery of oxygen and substrate for BAT metabolism.

Suppression of BAT

We have mentioned that BAT atrophies under the normal maturation process. Other factors can suppress the growth of brown adipose tissue. Some of these we have discussed already; others will be discussed in other parts of the book. They include surgical denervation of BAT (a strictly experimental factor);[85,125,467,961] exercise;[47,48,503,1036] fasting, dieting, or caloric restriction;[295,296,1282,1283] lactation;[1273,1276,1286,1320] diabetes;[84,85,1151,1167] dressing warmly (high ambient temperature, etc.);[1076,1079] pregnancy;[1273,1274] fever;[505] and hypothalamic lesions.[963]

Before leaving this section, we should also mention that accumulation of brown adipose tissue occurs in association with several diseases, including congestive heart failure and Chagas disease (South American trypanosomiasis). In these conditions, chronic hypoxia (lack of oxygen) is suspected of being the cause of the proliferation of brown adipose tissue in affected organisms.[1188]

ORDERING THE UNIVERSE: CELLULAR RESPIRATION AND OXIDATIVE PHOSPHORYLATION

In order to fully appreciate how normal metabolic processes are disrupted in brown adipose tissue and how heat is generated as a result of that disruption, we need to understand the underlying processes of energy metabolism. To understand the beauty of the operation of uncoupling protein, we should understand what is being uncoupled, why it is coupled in the first place, and how all of these processes fit into the overall scheme of things. Thermogenesis necessarily involves the most basic metabolic processes of the cell. To appreciate and understand thermogenesis, we simply must come to terms with the difficult language and concepts underlying cellular respiration and oxidative phosphorylation.

All life, all existence, all forces of the universe and of nature exert their most basic effects at the level of the individual cell, and more specifically in the mitochondria of the cell, and even more specifically still in the inner membrane of the mitochondria. The bold assertion that even the chromosomes and genes rank behind mitochondrial processes in importance can be supported by the realization that all life-sustaining processes require energy for vitality. And all energy is provided by the mitochondria. Literally, as a cell breathes, or respires (cellular respiration), it uses oxygen (oxidative phosphorylation) to create the basic molecule of energy and life—ATP (adenosine triphosphate).

Order and Disorder

The universal importance of cellular, i.e., mitochondrial respiration can be clearly understood by an inspection of the basic laws of thermodynamics that govern all of nature. The great opposing forces of universe are order and disorder, organization and chaos. These forces are expressed in the human body in terms of catabolism (chaos) and anabolism (organization), the opposing forces of metabolism. The Second Law of Thermodynamics states that the amount of order in a system left to its own devices must always decrease. This is entropy, the process that leads to the eventual degradation of all natural systems when no external source of energy is available for use in building and maintaining order in the systems. In the human body this means that without some energy-consuming ordering mechanism, the entire structure of the body, even life itself, must eventually disintegrate and end. Death may be the result of entropic processes winning some cosmological battle against life itself.

Opposing Entropy

The body must work very hard to retard this inevitable disintegration, to build and repair, to grow, and to create order. Still, according to the laws of thermodynamics, the amount of order created in the body must be compensated for by an even greater amount of disorder in the rest of the universe. This battle for order in the face of increasing disorder takes place in the individual mitochondrion of the cell. If all goes well order replaces disorder yielding in the process a net cost, or a net loss to the universe, of a certain amount of heat. We must continually replace the energy needed by the cell, and it must be in a form other than heat. Heat is unorganized energy; it occurs in the form of random disordered commotion among molecules. Heat released from a cell increases the general confusion of the universe. There is a very exact quantitative relationship between heat and order that allows us to calculate, at least in principle, the exact amount of heat a cell must release in order to compensate for a given amount of ordering in the cell.

The final outcome of the process is a net increase in universal disorder. In accord with the laws of thermodynamics it is actually the release of heat that makes the ordering of the system possible. The ordering process is under feedback control; once sufficient order has been established, a signal is generated that inhibits further ordering events. Uncoupling the ordering process from the disordering events removes feedback control of the entire system, allowing heat to be freely generated for as long as fuel is provided.

From Sun to Cell

The energy that drives life on earth and keeps it from running down is provided by the sun. Energy from the sun can not be utilized by members of the animal kingdom directly. It must be channeled through photosynthetic organisms: the plants. Plants derive energy directly from the electromagnetic radiation of the sun. We derive energy from the sun by eating the plants—second hand energy—or by eating the animals that ate the plants—third hand energy. This

110

energy is stored in the covalent bonds of the organic molecules we eat. We use this energy and deliver heat back to the system. Energy cannot be created or destroyed in chemical reactions. The heat lost in cellular activities drives the production of order, and is returned to the universal pool of energy.

Digestion

When we eat food, the processes of digestion and assimilation begin to break the proteins, lipids and polysaccharides down into smaller, individual molecules that the body can utilize. The body utilizes enzymes to do this; without enzymes, the amount of energy needed to break food down could equal or exceed the amount of energy derived from the food and there would be no net gain. In fact, because of heat loss, there would be a net loss of energy. Enzymes combine with food molecules in such a way that they reduce the amount of energy required to break the bonds that hold molecules together.

Enzymes are proteins with special three-dimensional shapes that act as binding sites for other molecules (substrates). When substrate is bound to an enzyme, chemical reactions that the substrate can undergo are accelerated dramatically, typically by a factor of 10^6 or more. Enzymes also govern which metabolic pathway will be 'fed' by any particular molecule, i.e., they selectively lower the 'activation' energy of one of a number of possible pathways.

Catabolism and Anabolism: Twin Processes of Metabolism

The metabolic process of breaking food down is called 'catabolism.' Its single goal is to break the energy-containing bonds of the food molecules in the most efficient way possible; that is, to obtain as much energy as possible from every molecule. The catabolic forces of life begin in organs of digestion and end in the mitochondria of the cell.

Once the body has the energy, it can use it to create, to build-up and/or restore order to the system. Enzymes are also used for this purpose; without them, it would take way too much energy to accomplish all of the required building-up tasks. The metabolic process of building up is called 'anabolism.' The anabolic tasks of the body begin in the mitochondria.

Thus catabolism and anabolism, breaking apart and building up, come together in the mitochondria of the individual cell, in every cell, of the body. In fact, at the final common pathway of cellular respiration the two processes are coupled together like a zipper. The zipper is the inner membrane of the mitochondrion. The zipper ties the synthetic, or anabolic, chemical reactions that create biological order to the degradative, or catabolic, reactions that provide the energy. As we shall learn, the uncoupling protein in BAT mitochondria unzips the zipper, disrupting the ordering processes of cells, allowing catabolic chemical reactions to flow down one side of the zipper without contributing to any synthesis on the other side.[919]

Oxidation

Catabolism, or the enzymatic degradation of foodstuffs, involves three somewhat distinct (and difficult to comprehend) stages. One, the breakdown of large molecules into simple subunits (proteins into amino acids; polysaccharides into simple sugars; lipids into fatty acids and glycerol). Two, the breakdown of subunits to acetyl Coenzyme A (acetyl CoA). Three, oxidation of acetyl CoA to H_2O and CO_2 (water and carbon dioxide). The second stage yields a small amount of energy in the form of adenosine triphosphate (ATP). The third stage yields the production of a great deal of an intermediate molecule known as NADH which yields a good deal of ATP via electron transport through the citric acid cycle and oxidative phosphorylation.

To understand the above process better, and make them more comprehensible, consider that high energy organic molecules contain a great deal of carbon and hydrogen atoms. The process of splitting apart the carbon and hydrogen bonds is called oxidation. Oxidation refers to the removal of electrons; the opposite of oxidation is reduction which refers to the addition of electrons. If an atom loses an electron it has been oxidized. If it loses another electron, it has been oxidized still further. The most energetically stable form of carbon is carbon dioxide and the most stable form of hydrogen is water. Cells obtain energy from foods, therefore, by allowing carbon and hydrogen containing foodstuffs to combine with oxygen to produce CO_2 and H_2O respectively. Enzymes facilitate this process. Several enzyme-dependent metabolic (catabolic) steps may be required before a substrate (such as a protein molecule) has been totally converted to ATP plus CO_2 and H_2O. In nature, oxidation sometimes takes the form of rust, or corrosion. In the body, excess oxidation, due to free oxygen radicals, is thought to be responsible for much of the aging process. Truly it can be said, "Life is oxidation, and then you die."

Combustion of food materials in a cell converts carbon and hydrogen atoms in electron-rich (reduced) organic molecules to carbon dioxide and water, during which they give up electrons and are, therefore, highly oxidized. The whole process involves a series of highly regulated intermediary reactions that typically involve the transfer of electrons of a hydrogen atom between one molecule and another, from one side of the inner membrane (or zipper) to the other, back and forth, yielding increasingly more NADH and ATP as the process continues, with water and carbon dioxide being the final by-products. ATP acts as the carrier of chemical energy because it is highly unstable and can be easily broken apart to yield a great deal of usable energy throughout the cell. The use of ATP by the cell generates order in cells.

So far, in a discussion that was supposed to address the issues of cellular respiration and oxidative phosphorylation, we have not once mentioned oxygen which would appear to be an integral part of respiration and oxidative anything. We did discuss oxidation, but we didn't relate it to oxygen. One might well ask why the donation of an electron is called oxidation in the first place. What does it have to do with oxygen? As it turns out oxygen is the common denominator of catabolic processes. Both water and carbon dioxide are built around the oxygen molecule. The oxygen atoms required to make CO_2 (in the citric acid cycle, stage Two) are supplied by water. The oxygen molecules required for oxidative phosphorylation (stage Three) are supplied by the air we breathe. The mitochondrion is the place where carbon and hydrogen atoms of food molecules are oxidized. It is the center toward which all catabolic processes lead.

Oxidative Phosphorylation

Oxidative phosphorylation is the last step in catabolism, running down one side of the zipper. It is the final point at which the major portion of metabolic energy is released. During this process, NADH and $FADH_2$,—another important intermediary—transfer the electrons they received from the oxidation of the food molecules to oxygen, O_2, releasing an enormous amount of energy. Part of this energy is used to make ATP; the rest is liberated as heat.

In the course of oxidative phosphorylation, electrons from NADH and $FADH_2$ are passed down a chain of carrier molecules, each of which might represent one section of the zipper. At each transfer point, the electrons fall to a lower energy state until at the end they are transferred to oxygen molecules. Oxygen molecules have the highest affinity of all for electrons; hence, electrons bound to oxygen are said to be in their lowest energy state. Thus we see the importance of the oxygen molecule. It acts as a sort of chemical magnet at the bottom margin of the zipper, irresistibly drawing electrons to it down the electron gradient. All forces of nature tend toward this state.

The energy released as the electrons fall to lower energy states is harnessed, in a manner not fully understood, to generate an electrochemical proton gradient (protons are the atomic material left behind as electrons desert the various intermediate molecules in search of the waiting oxygen molecule) across the inner mitochondrial membrane (the zipper). This gradient drives a flux of protons back through the enzyme complex in the membrane that adds a phosphate group to ADP (adenosine diphosphate), generating ATP (adenosine triphosphate) inside the mitochondrion. This is the heart of phosphorylation. The new ATP molecule is then transported out to the rest of the cell where it, in turn, fuels a myriad, life-promoting, metabolic pro-

114

cesses. The creation of the ATP molecule is therefore the first step on the anabolic side of the zipper, the first step in the creation of new life, the first ordering process of the biological universe.

Taking a closer look at the inner mitochondrial membrane, we see that it is ideally suited for carrying out the tasks of oxidative phosphorylation. It contains three major types of proteins: (1) those responsible for catalyzing the oxidative reactions of the respiratory chain; (2) an enzyme complex called ATP synthetase that converts ADP into ATP; and (3) the specific proteins that regulate the transport of metabolites into and out of the matrix.

Disrupting Oxidative Phosphorylation

In the quiescent state, i.e., in the absence of external stimulation from the sympathetic nervous system, the rate of cellular respiration in BAT is closely controlled. It is the rate at which the proton-translocating enzyme complex ATP-synthetase dissipates the proton gradient across the inner mitochondrial membrane that determines how efficient cellular metabolic events will be. As protons produced during FFA metabolism or respiration leave the mitochondria and enter the cytoplasm of the cell they are immediately pumped back in by the electron transport system. It is during this process that they are instrumental in oxidizing the reduced coenzymes NADH and $FADH_2$. The rate of oxidative phosphorylation is therefore tied to the rate at which protons derived from FFA in the mitochondrial membrane enter the cell. The rate of oxidative phosphorylation is also determined by the rate at which its substrate, ADP, is provided by the use of ATP for work of various types. The multitude of enzymes and coenzymes involved in the several steps of the electron transport system are tied to each other, one to the next, in a lock-step sort of way. Dehydrogenases in the mitochondria produce the coenzymes for the electron transport system from FFA, via the tricarboxylic acid (citric acid) cycle.

However, when UCP is activated, it acts as a proton translocator that accelerates the rate of electron transport down the thermogenic cascade, speeds up the action of certain mitochondrial enzymes and speeds up substrate (FFA) oxidation. In functioning BAT a great amount of lipid is combusted; water and carbon dioxide are created, but no hydrolysis of ATP is required as part of the control mechanism. This uncouples the two transport systems and thereby disrupts oxidative phosphorylation. Subsequently, under the influence of UCP, a great deal of FFA is used up without being coupled to the events that would normally result in the synthesis of ATP. When BAT is stimulated to produce heat it uses at least two-thirds of the extra oxygen consumed by the tissue. The heat energy derived from these metabolic events is enormous. With no ordering processes in place to deal with the energy, to put it to work, its random internal molecular disorder is transferred to the blood in vessels coursing through the BAT and to surrounding tissues, warming them up a bit, but otherwise producing no net anabolic effect on the system.

Thus the introduction of UCP into the organized events of catabolism and anabolism uncouples the processes the universe has devised for creating order out of the disordered products of digestion. There is almost something cosmically (comically, cosmetically?) unsettling about this process, until we remember that obesity is really a matter of too much order, a system running at too high a rate of efficiency, unwilling to give up even a few miscellaneous calories to the universe that spawned its life. We don't believe for a minute that there is actually anyone, no matter how obsessively organized or anal retentive, that is unwilling to contribute a little net disorder to the surrounding environment if it means losing a few pounds of unhealthy white fat.

THE UNCOUPLING PROTEIN

To review the material in the previous section in terms more specific to the uncoupling protein, the primary mechanism for heat production in BAT is the mitochondrial proton conductance pathway. This pathway involves a "short circuit" across the inner mitochondrial membrane such that the proton gradient induced by oxidation of substrates is dissipated as heat rather than linked to ATP synthesis and energy production through the mitochondrial ATP synthetase as happens in all other cells of the body.[1277]

Mitochondria are the energy generators of all body cells. When nutrients, or substrates, enter cells they are transported to the mitochondria where they enter into a complex cascading metabolic pathway that begins in the cell wall of the mitochondria. The end result of the dozens of chemical reactions that occur in the mitochondria is the accumulation of energy. ATP, the currency of energy, powers all life-sustaining cellular processes. Every step in the chain of biochemical events that precedes ATP synthesis is critical. Interrupt, or uncouple, the chain at any point, and the pathway is destroyed. The pathway is sometimes called the <u>mitochondrial proton conductance pathway</u> because on an atomic level that is just what is happening: protons are being traded back and forth among different chemicals and passed along to the next stage of the reaction. The formation of ATP is simply the acceptance of a proton by the last chemical in the chain.

In BAT, and no other tissue, a natural uncoupler exists whose sole function is to disrupt the proton conductance pathway, thereby "short-circuiting" the events leading to ATP synthesis.[193,196,754] Uncoupling Protein (UCP, or Thermogenin) acts as a transmembrane proton (H+) transporter. The energy of the reaction is converted to heat (thermogenesis) and dissipated through surrounding tissues. Increased thermogenesis is accomplished by stimulating the synthesis of UCP. In animals adapted to the cold, UCP may make up 15 percent of the inner mitochondrial membrane protein.

UCP, itself, is a small protein consisting of 306 amino acids.[31] It is located in the inner mitochondrial membrane, which it completely spans, one end projecting from the outer membrane surface, and the other end projecting toward the interior.[31,341] When BAT is in a quiescent state, UCP is also inactive, thus permitting normal cellular metabolism to occur (left in a quiescent state for too long, UCP and mitochondria diminish in quantity). Normal ATP production is necessary, of course, to keep the BAT cell alive and well. Under the surge of free fatty acids created by the stimulation of alpha- and beta-receptors through the consumption of thermogenic, or, to a lesser extent, by exposure to the cold, or by other dietary substances, UCP is activated—awakened from a state of slumber—and begins to uncouple cellular respiration.[168,170,262,263] It is believed that required life-sustaining basal cellular metabolism is maintained during high UCP activity by an alternate glucose-fueled pathway.

The new pathway created by UCP may be called the thermogenic proton conductance pathway, or the thermogenic cascade. It may be helpful to think of UCP as a switchman for a railway whose job it is to throw the switch that shunts the proton train to a side track. The proton train begins its journey on the track to ATP, but is switched at a critical moment to a new track whose destination is HEAT. Because UCP is unique to (the country of) BAT, the thermogenic proton conductance pathway is also only found in BAT.

The amount of UCP in any given BAT cell depends on the number of mitochondria present in the cell. It has been postulated that brown cells become white fat cells as the number of mitochondria decrease due to lack of use. Likewise, with repeated stimulation, thermogenic capacity increases as fat cells that appear white may become brown fat cells as the number of mitochondria and UCP increase and the TAGs change from unilocular to multilocular.[532,1009] It has also been postulated, as mentioned earlier, that the body contains numerous undifferentiated fat cells that can go either way depending upon the circumstances. The fact that the amount of UCP and the number of mitochondria increase with repeated stimulation through consumption of thermogenic substances, cold exposure, and DIT,[212] is therefore one of the most important findings of thermogenic research. [50,51,124,295,481,499,516,961,964,871, 1027,1279]

To summarize, under the influence of long term administration of thermogenic substances, BAT cells adapt by increasing in number themselves, by increasing in the number of mitochondria they contain, and by increasing in the amount of UCP that is available.[1009] This translates into significantly greater capacity for thermogenesis.

NOTE: Recently, two German researchers reported the discovery of a membrane component that acts as an anion conductance channel similar to, but distinct from, that created by the uncoupling protein.[708] Future research may uncover yet other mechanisms by which BAT is able to utilize such an incredible amount of energy so quickly.

MEMBRANE EVENTS

Before the events of cellular respiration can be disrupted by UCP, this natural uncoupler must be activated. This section details the events leading up to UCP activation. Beginning with the arrival of NE at the membrane of a BAT, we can trace a series of events that leads ultimately to the production of heat.[614] These events mostly take place in the cell membrane and on both sides of it, and in the membrane of the mitochondria. Hence, we have chosen to group them together under the term 'membrane events.'

Thermogenesis does not occur in a vacuum. It is a very real event, requiring the same considerations as any other physiological process. Though the end product of BAT cell metabolism may be different from every other cell of the body, BAT cell metabolism calls for many of the same basic processes. There must be fuel to burn. The fuel must be derived from dietary sources and/or from storage areas. The fuel will typically occur in a form that cannot be used by the BAT cell and must therefore be converted into a utilizable form. The fuel will need to be mobilized and transported to BAT. There will be several feedback control mechanisms in place to stimulate BAT metabolism and to inhibit it. None of these processes will be as straightforward as we would like them to be.

Though much remains to be discovered, much is also known about the control mechanisms of BAT metabolism and its underlying biochemical events. We understand, for example, the importance of plasma membrane-bound receptors, adenylate cyclase, cAMP, free fatty acids, several lipases, adenosine, prostaglandins and phosphodiesterase. We have been able to trace the cellular metabolic events of the mitochondria of brown adipose tissue, and we know how the uncoupling protein uncouples oxidative phosphorylation.

119

The entire thermogenic process is ultimately under the control of the central nervous system via the hypothalamic-pituitary-adrenal network. The action of thermogenic substances is to activate critical relays in this network that permit effective stimulation of sympathetic or SNS neurons in the vicinity of BAT. A complex series of events then occurs as summarized below.

The Events

(1) Norepinephrine (NE), a primary adrenergic neurohormone, is secreted from the presynaptic membrane of sympathetic neurons into the synapse separating neuron from BAT cell. (Figures 1-4, Plate III) NE, along with adrenaline, are catecholamines (CA). CA are synthesized in the cytosol of cells in the adrenal medulla and the sympathetic nervous system, and are then packaged into vesicles, tiny storage structures found in sympathetic neurons, especially in the neighborhood of active junction sites.

The association of adrenal medulla cells and neurons of the SNS typifies the coordination between endocrine cells and nerve cells that occurs throughout the body and is responsible for regulating the activities of the body's billions of cells. Information transmitted along nerve cells travels very quickly. Information transmitted by hormones secreted by endocrine glands travels much more slowly. The two systems converge at the synapse, that is, where the electrical signal traveling the nerve is transformed into a chemical signal mediated via hormones. The hormone (in the form of a neurotransmitter) is released from the terminal of the nerve (the presynaptic membrane) and diffuses across the synapse to the next nerve or to the membrane of another type of cell, such as BAT, (the postsynaptic, or plasma, membrane) where it interacts with membrane-bound molecules in such a way as to recreate another electrical signal or to initiate further chemical events. The gap of the synapse is microscopic, and diffusion of the neurotransmitter substance takes place in less than a millisecond. Thus electro-chemical conductance takes place very rapidly throughout the body. Since endocrinological processes are slower, the hormones must already be in place when the nerve signal arrives at the synapse. Depletion of the neurotransmitter (hormone) would drastically curtail neural transmission.

The endocrine and nervous systems are physically and functionally linked in the hypothalamus (see chapter on the CNS). The bridging of endocrine to nervous system is mediated by cells in the hypothalamus that exhibit properties common to both nerve cells and endocrine cells. They transmit information electrically, like neurons, but they release hormone directly into the blood (instead of a synapse). In this latter regard, they behave like cells of the endocrine system. These cells of the hypothalamus are often called neurosecretory cells.

(2) NE diffuses across the synapse separating the nerve from the BAT cell. (Figures 1-4, Plate III) The synapse is not just empty space. It contains hundreds of different chemicals, each of which has the potential for modifying the action of the neurotransmitter. How effective the neurotransmitter will be in initiating activity in the postsynaptic membrane is a function of its interaction with many of these synaptic chemicals. Associated with the release of NE are high activity rates among the various enzymes responsible for the synthesis and metabolism of the catecholamine.

(3) NE attaches to the BAT cell at the *receptor* site on the BAT cell membrane. (Figures 1-2, Plate III) Before a neurotransmitter can initiate activity in the postsynaptic cell, it must attach itself to the cell membrane at some receptor site. The same is true of circulating hormones such as adrenaline. The membrane is often called the *plasma membrane* because it faces out into the extracellular environment. Some molecules of the extracellular, or *interstitial*, spaces enter cells directly by diffusion through the plasma membrane; they can do this by possessing certain chemical characteristics in common with the membrane. Other chemicals can not enter a cell directly; they require the presence of some transport system. Other substances never penetrate the plasma membrane; they simply initiate events in the plasma membrane that are conducted to the inside

121

of the cell. These latter kinds of substances, such as NE, are often called *first messengers*.

The plasma membrane transport mechanisms are also important in the thermogenic process. There are several membrane-bound enzymes, called permeases; these proteins form channels that allow certain molecules to enter or leave the cell. For instance, the uptake of potassium, sodium and calcium ions is mediated via certain of these permeases. Almost all cells contain a glucose permease protein that allows glucose, but not sugars with related structures, to cross it.

Certain hormones, such as thyroid hormone, diffuse directly into the cell due to their structural similarity to the phospholipid bilayer that constitutes the cell membrane.

From the above discussion, it can be ascertained that the plasma membrane plays several important roles. It constitutes the outer perimeter of the cell; it recognizes and reacts to certain extracellular signals, such as hormones and polypeptides; and it communicates and interacts with other cells.

In general, receptors are proteins that recognize hormones and communicate to the cell that something is knocking at the door. The cell may react to the hormone or polypeptide in a number of ways. One, it may excrete some metabolite or protein; two, it can undergo a change in metabolism; three, it can generate an electric current, as in a nerve; and four, it can generate a contraction, as in a muscle.

Water-soluble hormones interact directly with the cell surface receptors. Included in this class of hormones are the large polypeptides such as insulin, growth hormone and glucagon, and small charged compounds such as the catecholamines adrenaline and NE. All known neurotransmitters, as well as most hormones are water-soluble. These molecules usually exert an immediate, short-lived effect. The cell reacts within milliseconds or within a few seconds at the outside.

Lipid-soluble hormones, in contrast to the water-soluble substances, persist in the blood for hours or even days. Target cells are therefore exposed to signals arising from these substances for long periods of time, and their responses may last for hours or longer.

Certain steroids and thyroxine from the thyroid gland are members of this class. Release of these substances from presynaptic terminals is closely regulated. For thyroxine release, thyroid-stimulating hormone (TSH) triggers the release of the iodinated precursor protein thyroglobulin and its conversion to thyroxin. Steroid-producing cells, such as those of the adrenal cortex, may store a few hours' worth of precursors; then, when stimulated, the cells convert these precursors into the finished hormone which diffuses across the plasma membrane into the blood to circulate throughout the body in search of receptors with which to interact.

Water-soluble substances may be considered hydrophilic, while lipid-soluble molecules are considered hydrophobic. Hydrophilic substances "love water" and are repelled by lipid (fat) materials, such as cell membranes. They can not enter the cell or affect the contents of a cell without the aid of receptors and transport mechanisms. Hydrophobic substances "hate water" and love fat and can pass easily through the plasma membrane of the target cells; these hormones bind to specific receptor proteins *inside* the cell.

There are several different types of receptors. Adrenergic receptors bind NE. There are several categories of adrenergic receptors. Beta-adrenergic receptors are present on many kinds of cells. They bind NE and adrenaline. The subsequent events may differ widely depending on the type of cell. However, the binding action and the translation of the NE signal into some form of intracellular event is the same for all of these cells, as discussed below. The human BAT cell has both beta and alpha adrenergic receptors.[179,953] For more information on receptors see the chapter on the SympathoAdrenal System.

In summary, cell-surface receptors bind signaling molecules, or ligands, and convert this synaptic, extracellular event, into an intracellular signal that changes the internal processes of the cell.

(4) The receptor-NE complex activates *adenylate cyclase*, an enzyme found in almost all cell membranes. (Figures 1-2, Plate III) The beta adrenergic receptors and one type of alpha adrenergic re-

123

ceptors for catecholamine, when activated by NE in BAT and other tissue, in turn activate an enzyme in the membrane of the cell called adenylate cyclase. Both the receptor and adenylate cyclase are proteins. However, the activated receptor does not directly activate adenylate cyclase. A third protein, called G, couples the receptor and adenylate cyclase together and helps these two molecules communicate with each other. G is so-called because it binds guanosine phosphates (GTP and GDP). The G protein cycles between active (GTP) and inactive (GDP) forms.

In an unstimulated cell, most G molecules are bound to GDP. Binding of a hormone to the beta-adrenergic receptor changes the conformation of the receptor, causing it to bind to G in such a manner that GDP is displaced by GTP. This activates adenylate cyclase. The binding is only momentary, however, as the GTP is quickly reconverted to the inactive GDP, which in turn inactivates adenylate cyclase. Thus the GTP-GDP cycle is absolutely crucial to the activation and inactivation of adenylate cyclase by the conformational change in the receptor. It is also the site of major regulatory interactions.

As yet unexplored is the role that the integrity or lack of integrity of GDP binding may play in defective thermogenesis. However, at least one study has discovered errors of metabolism that involve GDP. Researchers from Florida described a diminishing of thermogenic capacity with age that apparently involves a failure of BAT cells to increase GDP binding, either because of insufficient UCP, or a failure to unmask reserve GDP binding sites on cell membranes.[1111] Similar findings from a University of California team substantiate that females exhibit a greater increase in BAT and GDP binding with increasing age than do males.[844]

Cold exposure normally stimulates initial increased GDP binding, followed by a relative decrease with prolonged exposure.[460] Increased sensitivity and capacity of BAT with special reference to UCP apparently down-regulates the role of GDP. Hence, GDP may be especially important in compensating or smoothing out the rough edges of early adaptive thermogenic mechanisms, becoming less important as BAT and UCP and begin to assume the dominant role in thermogenesis.

In summary, the activation of adenylate cyclase in the membrane of the cell requires at least three plasma membrane-bound proteins: receptor, G, and adenylate cyclase itself. Then a set sequence of events can occur, as follows: 1), the hormone (e.g., NE) binds to the receptor (e.g, beta-adrenergic receptor) which alters the shape of the receptor and allows it to bind to and activate the G protein; 2) the G protein then binds GTP and displaces GDP at its surface in such a way that the shape of the G protein is changed just enough to allow it to activate an adenylate cyclase molecule; and 3) the G protein then hydrolyzes the bound GTP to GDP, which returns the cyclase to an inactive state. This process can occur dozens of times per second.

The presence of the G protein may initially appear to unnecessarily complicate a rather straightforward process. But the protein actually increases the sensitivity and power of the reaction. Each activated receptor can collide with and activate many molecules of G protein, thereby greatly multiplying the initial extracellular signal. Other advantages will become apparent as we proceed. Binding of GDP is often used as the key element in activation of the proton conductance pathway, as it is easily measured experimentally.[481,982,1274]

(5) Adenylate cyclase in turn activates a molecule inside the cell called cyclic AMP, or cAMP. (Figures 1-4, Plate III) To this point all events have been geared toward converting the action of an extracellular messenger, NE, into expanded activity via an intracellular messenger, or second messenger. This second messenger is a protein called cyclic AMP, or cAMP. The amount of cAMP in the cell increases as it is stimulated. The amount of calcium ion in the cell also increases as a result of adenylate cyclase activity.[242,243] An increase in the concentration of either of these second messengers in cells throughout the body triggers a rapid change in the activity of several possible enzymes or nonenzymatic proteins. The intracellular level of cAMP can increase fivefold in a matter of seconds. Cyclic AMP is synthesized from ATP. The presence of ATP is therefore required in order to activate BAT metabolism and thermogenesis. Yet BAT in an active thermogenic state does not produce ATP through normal channels (there may be some produced through another pathway). This factor tends to become a self-limiting force in the regulation of thermogenesis. The uptake and use of glucose, the storage and mobilization of fat, and the synthesis and secretion of other cellular substances are among the processes stimulated by increased cAMP concentration.

Cyclic AMP is not only synthesized quickly, it is also destroyed rapidly by intracellular cAMP phosphodiesterases (PDEs). To this point, inhibitors of thermogenesis have acted on the metabolism of NE. Now, inhibition or disruption of thermogenesis occurs due to the action of materials like the PDEs that act on the second messenger, cAMP. Type 3 PDE hydrolyzes cAMP to adenosine 5'-monophosphate (5'-AMP), an inactive form of cAMP. PDE thus inhibits the mobilization of lipids from adipose tissue. One of the goals of thermogenic intervention is to inhibit the action of PDE so that cAMP functions for a longer period of time.

One of the advantages of involving cAMP in this series of metabolic events is that many cAMP molecules are activated by one adenylate cyclase molecule, thus continuing to expand the action of the single NE molecule that instigated the chain of reactions. Another advantage is that cAMP easily passes from one cell to another through openings in cell walls called gap junctions. This means that hormonal stimulation of just one or a few cells can initiate a metabolic chain reaction in many surrounding cells. However, these gap junctions tend to close as the concentration of calcium ion increases. Increases in calcium ion are often induced by neuronal or hormonal stimulation. Calcium ion binds to calmodulin, a ubiquitous binding protein. The calcium ion-calmodulin complex binds to cAMP PDE and the enzyme begins to hydrolyze cAMP. This is yet another instance of metabolic feedback control. So far, no thermogenic agent has been constructed that would selectively block the action of the calcium ion in brown adipose tissue.

(6) cAMP activates an hormone-sensitive lipase (HSL). (Figure 1, Plate IV) The effects of cAMP differ substantially from one type of cell to another. In all cells, however, the actions of cAMP are thought to be mediated in basically the same way, through the action of specific cAMP-dependent enzymes called protein kinases. In brown adipose tissue, cAMP produces an increase in a protein kinase known as hormone-sensitive lipase (HSL). In keeping with the tendency of events to spiral upward, here again the action of one molecule (cAMP) initiates activity in several molecules (protein kinases).

(7) HSL catalyzes the conversion of TAGs to free fatty acids plus glycerol. (Figure 1, Plate IV) HSL is responsible for accelerating the breakdown of stored fat molecules (TAGs). In the blood, dietary fats are broken apart by a similar enzyme called lipoprotein lipase (LPL); the breakdown products (of HSL and LPL) are then recruited for uptake by BAT mitochondria. This process is discussed in detail in the next section on lipolysis, the formal name for the process of breaking triacylglycerol droplets down into free fatty acids.

This is the final stage in the upward cascading series of enzymatic reactions that begins with the NE-receptor binding. By now, that one short-lived NE molecule has generated many thousands of effector molecules within the target cell. Some of these work in conjunction with one another; some work in opposition to keep the reaction under control. When the system gets out of balance, either too little or too much metabolic action takes place. The fundamental postulate of all BAT research is that in brown adipose tissue, the inhibitory processes have a disproportionate effect on metabolism. The uncoupling protein helps restore balance; when it isn't working, not enough thermogenesis takes place. But that's getting ahead of ourselves. Returning to the thermogenic sequence of events, we note the following.

(8) The free fatty acids (FFA) diffuse into the mitochondria. (Figure 1, Plate IV) Due to their lipophilic nature, fatty acids can freely cross the mitochondrial membrane. In normal cells, FFA are used as the primary fuel for the cell. Free fatty acids are converted into ATP during cellular respiration. When too many free fatty acids enter the cell, their increasing concentration tends to shut down respiration.

(9) In BAT the FFA accumulate in the mitochondrion until a threshold value of concentration is reached. (Figure 1, Plate IV) In brown adipose tissue, free fatty acids are used to fuel the metabolic processes of the cell just like any other cell, with just one exception: As the concentration of free fatty increases it does not shut down metabolic respiration. Instead, as the concentration reaches a certain threshold value, just about where respiratory shutdown would normally occur, something unique occurs.

(10) When the threshold concentration of FFA is reached, UCP is activated. (Figure 1, Plate IV) Not present in other cells of the body, the uncoupling protein awaits precisely the series of actions we have so far presented. Attached to the inner side of the mitochondrial membrane, UCP is perfectly positioned to interact with the events of cellular respiration taking place across that membrane. This brings us to the point at which universal biological catabolic processes are uncoupled from anabolic synthesizing events.

 (11) UCP interrupts ongoing oxidative phosphorylation, or cellular respiration and initiates the thermogenic cascade. (Figure 1) This interruption removes the brakes on FFA utilization by diverting the electro-chemical events down a pathway that has no normal feedback inhibition other than its own inherent FFA concentration-dependent activation. Energy produced during UCP-dependent metabolism is diverted down the thermogenic proton conductance pathway and dissipated into surrounding tissue as heat. This is thermogenesis.

(12) Thermogenesis continues as long as substrate (FFA) is being provided. (Figure 1) As the effects of cold, or food, or thermogenic agent begin to subside, the level of TAGs and FFA begin to decrease until they reach a point too low to activate UCP any further. At that point normal mitochondrial respiration is reinstituted, i.e., FFA oxidation is once again used mainly to produce ATP. There appears to be a limit to the amount of FFA oxidation for thermogenic purposes that any given cell can produce. Increases in total BAT thermogenesis depend on growth of new brown adipose tissue, recruited from the reserves of undifferentiated preadipocyte tissue at hand. Hence, consumption of thermogenic would probably produce an immediate amount in thermogenesis that depends on the amount of BAT present. Further increases would depend on the further accumulation of BAT.

Regulation of the Thermogenic Cascade

To emphasize one point: the several steps listed above are carefully regulated by an assortment of biochemical, primarily enzymatic, activating and inhibiting devices. The synthesis, release and degradation of all hormones are closely regulated. Under conditions of faulty BAT metabolism, the inhibitory devices exert a disproportionate amount of influence, resulting virtually in the total suppression of thermogenesis. One goal of the thermogenic is to inhibit these natural inhibitors. This process, called disinhibition, is explain in some detail elsewhere. Normally, we think of processes simply being activated. Often, in human physiology, metabolic events are struggling to occur at all times, and are virtually being held in biochemical chains until some event releases the locks, usually by destroying the chemical that is preventing the events from happening.

Adenosine

For example, at step (3) a substance known as adenosine interferes with the ability of NE to bind to the receptor. (See Figure 3.) Adenosine is produced by BAT cells as a consequence of NE stimulation;[117-119,1238] whereupon it then helps to increase the flow of blood through BAT and hence facilitates the delivery of oxygen and substrate to BAT, and accelerates the dissipation of heat away from BAT. However, as the amount of adenosine accumulates it tends to bind to the same BAT receptors as NE and hence tends to shut down the entire thermogenic process. This type of feedback control of metabolic processes is extremely common.

Prostaglandin

Another inhibitor of NE activity is the prostaglandin E2 (PGE2).[113] (See figure 4.) PGE2 is one of a group of lipid-soluble hormone-like substances produced by almost all body tissues. These 20 carbon fatty acid derivatives serve as special, highly localized, mediators; they bind to cell surface receptors just as do neurotransmitters. They exert their effects in a very limited area, generally less than a millimeter from the cell wherein they are synthesized. Prostaglandins are synthesized and degraded very rapidly, making their study extremely difficult. Nevertheless, in recent years, we have made gigantic strides toward elucidating their method of operation. Needless to say, much more remains to be understood. Unlike most signaling molecules, prostaglandins are not stored (like NE is stored in synaptic vesicles, and adrenaline is stored in the adrenal medulla), but are continuously released to the cell exterior. PGs are constantly being synthesized in cell membranes from precursors that are cleaved from the membrane by certain enzymes (the action of phospholipases on the phospholipids of the membrane).

There are at least 16 different prostaglandins in nine different classes designated PGA, PGB. . ., PGI. They modulate the responses of other hormones and may exert important and profound effects on a wide variety of cellular events.[804] Prostaglandins of the E subtype are especially important in the modulation of thermogenic events. PGE_2 is a very potent antilipolytic agent, i.e., its influence results in the inhibition of processes involved in the breakdown of triglycerides to free fatty acids. In the present case, PGE_2 is known to attach to adenylate cyclase in such a way as to disrupt the ability of norepinephrine to activate adenylate cyclase. With PGE_2 attached, adenylate cyclase molecule remains unafffected by plasma membrane activity. Hence the entire thermogenic process is slowed down or inhibited.[11,113]

NE Inhibits NE

Another mechanism that affects membrane events underlying thermogenesis involves norepinephrine itself. As already mentioned, this catecholamine stimulates thermogenesis by interacting with receptors on the BAT cell membrane. The particular receptor is the beta-receptor (actually, there are three beta adrenergic receptors involved: beta-1, beta-2, and beta-3; B-3 may be a specific thermogenic receptor; but norepinephrine indiscriminately binds to all three). However, norepinephrine also binds to alpha-adrenergic receptors on brown adipose cell membranes, some of which tend to inhibit adenylate cyclase and thermogenesis. (See Figure 3.)

Reuptake and Degradation of NE

The fate of synaptic NE is also governed by the tendency of the presynaptic membrane to recover NE from the synapses, i.e., resorb it back into the presynaptic membrane. (See Figure 2.) Finally, NE must contend with other enzymes in the synapse, such as COMT and MAO, whose job is to degrade NE to inactive forms. Signaling cells such as those of the sympathetic nervous system typically store at most one or two day's supply of hormone in vesicles just under the plasma membrane. The signaling cell is stimulated from time to time to synthesize and store more hormone in order to replenish the cell's supply. Released neurotransmitter substances, such as NE, have a lifetime in the blood of only seconds or minutes and are rapidly degraded by blood and tissue proteases as mentioned. The rate of blood flow through BAT will affect NE levels in the synapse. Increased flow rates due to increased sympathetic activity will tend to increase the rate at which NE is taken up by the presynaptic membrane.

In summary, the eventual ability of NE to transfer positive action to the second messenger cAMP is a function of the sum of the various stimulatory and inhibitory factors at play in the synapse and in the BAT cell membrane.

An important question that remains to be conclusively answered asks to what degree the thermogenic process is subject to adaptation.

While much evidence indicates that thermogenic capacity increases with continued activation via a thermogenic agent such as ephedrine,[466,468,469] other research suggests that cold stimulation may eventually lead to an uncoupling of adenylate cyclase from the adrenergic receptors, thereby decreasing responsiveness to NE.[172,917,1110,1232,1233] Obviously, this is an important question and the development of effective thermogenic agents may very well depend on obtaining a clear resolution of the apparent contradiction.

LIPOLYSIS

We have been discussing the metabolic events of BAT thermogenesis in reverse order. Backing up even further, we can now proceed to discuss the manner in which the body supplies BAT the primary fuels of oxidation. It is here, after all, that the main impact on body composition is observed. It is the removal of fat deposits, their conversion to free fatty acids (FFA), and their delivery to BAT for incineration that results in changes in the lean/fat ratio.[129] In this light, thermogenesis and fat mobilization are really manifestations of the same function. Hence, increased thermogenesis is a direct consequence of increasing levels of free fatty acids in brown adipose tissue.[881] Conversely, thermogenesis is impaired when FFA pools are low.[462] Lipolysis is instigated by the stimuli discussed in this book, mainly cold, exercise, diet and thermogenic agents.[1228]

Thus it is, that the first critical task in BAT metabolism is to obtain fuel. The primary fuel is free fatty acid.[162,461,511,532,1271,1272] Over 95 percent of the lipid contained in adipose tissue is present as triglycerides, not free fatty acid. A small, insignificant fraction is present as FFA. Enzymes called lipases are present in tissues and provide a link between the TAGs and the FFA because of their ability to hydrolyze TAG to FFA.

A secondary source of fuel in BAT is glucose.[725] Both lipids and glucose are derived from dietary fats and stored triglycerides. Most of the fat we consume is in the form of triglycerides. To absorb this fat effectively, digestion or lipolysis is required. In the intestine, this hydrolysis is catalyzed by the action of pancreatic lipase. However, considering how resistant lipids are to degradation, even this enzymatic action is retarded unless aided by some form of detergent. In this case, the detergent is supplied by conjugated bile salts from the gallbladder. Once the lipids have been digested (or 'hydrolyzed'), they are absorbed by the intestinal mucosa and reesterified (opposite of hydrolyzed) into triglycerides.

The Lipid Pool

One portion of these triglycerides are then shunted to the lymphatic system in the form of chylomicrons. Chylomicrons are particles of emulsified fat, cholesterol ester, phospholipid and protein that contain most of the long-chain fatty acids recruited from dietary fat. Another portion of the lipid products of digestion are shunted to the liver. These are released into circulation for use by adipose tissue in the form of very low density lipoproteins (VLDL). Short-chain fatty acids do not form chylomicrons and have a greater water solubility; they usually enter the portal rather than the lymphatic system. These various groups of lipids form a pool from which lipids are selected for inclusion in brown adipose mechanisms.[588] In order for these lipids to be stored in white or brown fat they must be broken down into triglycerides again and then reassembled inside the cell. This is accomplished with the aid of the enzyme lipoprotein lipase (LPL). LPL is synthesized in adipose tissue under the influence of glucocorticoids, insulin and other hormones, secreted and transported to the surfaces of endothelial cells lining blood capillaries; here it exerts its action on TAG of circulating chylomicrons and VLDL.[1056,130] Exposure to thermogenic stimuli (cold, food, herbs, drugs) produces a rapid rise in the amount of circulating LPL, mediated by beta-adrenergic receptors.[203,204,878]

Chylomicrons contain several million molecules of triglyceride, yet their residence time in the circulation is less than 10 minutes. During that time they are hydrolyzed at least 90% by LPL. Thus, in 10 minutes something on the order of 6×10^6 hydrolytic events occur. A large chylomicron may interact with 20 - 30 LPL molecules in order to complete the degradation process. A typical very low density lipoprotein, on the other hand, may contain less than 15,000 triglyceride molecules. Thus, each VLDL can be hydrolyzed by one LPL in less than 1 minute.[950] As stated, the unesterified fatty acids generated through the action of LPL are either taken up by adipocytes for storage, oxidized, or stored elsewhere.

Fuel From Fat Stores

The sympathoadrenal system (SNS plus adrenal glands) through the action of NE and adrenaline plays a critical role in the process of recruiting free fatty acids from triglyceride stores in white adipose tissue, brown adipose tissue and from circulating very low density lipoproteins.[44,194,203,204] Involved are several important enzymes, including lipase and lipoprotein lipase, and hormone-sensitive lipase. FFA are also derived from blood triacylglycerol (TAG) molecules in chylomicrons and through the de novo synthesis of TAG from glucose in blood and BAT. The relative importance of adrenaline supplied by the adrenal gland is small compared to NE, but is still a molecule of some importance. It appears, on the basis of the foregoing discussion, that the major form in which lipids cross the fat-cell membrane, both in BAT and WAT, in either direction is the free fatty acid molecule. The SNS also participates to some degree in the rapid growth of BAT that results from NE or prolonged cold exposure, and in glucose uptake.[245]

The delivery of fuels derived from white adipose tissue to BAT is aided by an increased cardiac output, accompanied by a fall in total peripheral resistance, and increased pulse rate and pulse pressure. This results not only in a more rapid delivery of substrate but also in an increased delivery of oxygen. Sympathetic stimulation of the heart helps to sustain cardiac output.

Nutritional Influences

Nutritional factors profoundly influence the outflow of FFA from white adipose tissue. When white adipose tissue is actively using carbohydrate, the release of free fatty acids is inhibited. On the other hand, when carbohydrate is missing, as in fasting or diabetes, there is an increase in circulating FFA. In the presence of glucose and/or insulin, the rate of FFA production and release by WAT is dramatically accelerated.

Hormone-Sensitive Lipase

In WAT and BAT, cAMP and the enzyme hormone-sensitive lipase (HSL) is primarily responsible for the conversion of TAG stores of lipids into FFA (plus the glycerol skeleton) that serve as the main fuel for mitochondrial oxidation.[167,168] HSL activity is promoted by lipolytic hormones, e.g., NE, glucagon and ACTH, and inhibited by insulin. It is also subject to regulation by other enzymes (protein kinases and phosphoprotein phosphotases). In white fat cells, neurohormones (NE and adrenaline) induce a cascade of events that regulate the synthesis and degradation of TAGs. Activation of the beta-adrenergic receptor site on the WAT cell triggers an increase in cAMP and activation of the cAMP-dependent protein kinase. The hormone-sensitive lipase that hydrolyzes TAGs to FFA is activated by phosphorylation by this kinase.[712,1172]

The main reaction sequence that constitutes adipose tissue lipolysis can be summarized as follows: TAG to 1,2-diacylglycerol + FFA, to 2-monacylglycerol + FFA, to glycerol + FFA. HSL catalyzes the first two steps. The last step is catalyzed by 2-monoacylglycerol lipase.[98] NE increases the activity of HSL by increasing the phosphorylation of the enzyme with cAMP-stimulated protein kinase. Insulin interferes with this action.[98]

Albumin and Carnitine

The released fatty acids are transferred either to the mitochondria of the cell itself, or to the blood, where they are bound to albumin, a major serum protein, thereby forming a lipid-protein complex, but remaining in an unesterified form. In order for this reaction to take place, an acceptor molecule must be present to solubilize the reactions products of the enzyme-mediated conversion of TAG to FFA; the FFA may then be transferred via the acceptor molecule across the interstitial space separating fat cell from blood vessel to albumin, the final acceptor molecule within the blood. The entire process can be limited by the availability of an adequate supply of albumin. The WAT-derived FFA are then carried via the blood-borne albumin to tissues throughout the body, including BAT, where they are easily incorporated into the mitochondria of individual cells, possibly via both carnitine-dependent and carnitine-independent pathways.[376] Thus, adequate levels of carnitine are required to mediate this important step in the thermogenic cascade.[142]

In the carnitine-dependent pathway, fatty acids are activated outside the mitochondria of the cell to Acyl-CoA by an enzyme called ATP-thiokinase. Then Acyl-CoA is converted into acylcarnitine, transported into the mitochondria, and there oxidized. This is known as the 'carnitine barrier' in mitochondria; it allows passage of carnitine-bound materials but is impermeable to Acyl-CoA. The carnitine-independent pathway involves the transport of FFA into the mitochondria where they are activated by ATP-dependent thiokinase, without the necessity of a carnitine carrier. The carnitine-dependent pathway has been shown to be very active in BAT mitochondria.[402,403,404] Brown fat is naturally rich in carnitine. The concentration of carnitine required for maximal lipolysis is very close to actual measured level of carnitine in BAT.[312,719] Since carnitine restores oxidative phosphorylation in isolated uncoupled mitochondria, it may be reasoned that too much carnitine might inhibit thermogenesis by some unknown amount.[544]

Summary

Thus, the release of FFA from white adipose tissue and its delivery to and uptake by BAT are controlled by nervous, hormonal and dietary factors. Blood flow may be regarded as an additional, extremely important factor in this process. Not only does it affect the rate of substrate and oxygen supply, it carries albumin, the acceptor of released fatty acids, to the tissue. Alterations in blood flow, through diet, exercise, drugs, nutrients and herbs can dramatically affect adipose tissue lipid mobilization.

Fuels From Dietary Fat

Dietary TAG is hydrolyzed (broken down to glycerol and FFA) by the enzyme lipoprotein lipase (LPL) which is found on the capillary endothelium at the surface of BAT cells, a location that markedly improves the incorporation of the FFA into the BAT cell. Dietary carbohydrates are converted to TAG in chylomicrons and low density and very low density lipoproteins in the liver (lipogenesis) which are then hydrolyzed by lipoprotein lipase at BAT. Fatty acids can also be synthesized from glucose in BAT itself.[340] In fact, it is thought that the lipogenic capacity of BAT is about the same as its thermogenic capacity.[929,931]

UCP Activation

Free fatty acids derived from these exogenous and endogenous sources eventually combine to form a pool of FFA within the brown adipose cell. When this pool reaches a certain size, the uncoupling protein is activated and uncoupling of oxidative phosphorylation occurs. The accumulated FFA serve as the fuel to drive UCP-mediated thermogenic metabolism.[166,168,170,263,400,436,705-707, 746,747,929,930] The mechanism by which FFA interact with UCP is not known. When BAT is in a quiescent state, i.e., when it is not being activated by NE, the oxidation of FFA occurs in BAT the same way it does in all other cells. This implies that thermogenesis is not a continuous process; it is switched on and off, and is therefore being controlled by circulating NE, FFA and other forces.

Research in lipolysis is complicated by the difficulty in generalizing from one species of animal to another. Some authorities feel, for example, that almost none of the hormones that stimulate lipolysis in animals works very well in man. These hormones include secretin, vasopressin, glucagon, ACTH, and parathyroid hormone. Thyroid-stimulating hormone is active during infancy, but may not be later in life. The single class of hormones all authorities agree does stimulate lipolysis in adult humans comprises the catecholamines, the chief of which is NE, and next important is adrenaline.

Lipolysis Inhibition

On the other side of the coin are numerous substances which inhibit lipolysis in BAT cells, chief of which are adenosine,[477] phosphodiesterase,[44] insulin,[15] growth factors,[695] certain prostaglandins,[1041] and the catecholamine themselves.[43,180] The pharmacological and nutritional control of thermogenesis is concerned with trying to inhibit the inhibitors and maximize the stimulatory role of catecholamines while down playing their inhibitory effects. It sounds complex because it is. Clearly, however, the inhibitory forces greatly outnumber the stimulatory forces. So great is the disparity, in fact, that we must rethink the basis upon which thermogenesis occurs. Rather than viewing thermogenesis as the result of the activation of something, as presently thought, it may be better to view thermogenesis as a process that occurs only when it has been released from inhibitory control.[680] In other words, thermogenesis is constantly on stand-by, ready to kick in when the opportunity presents itself. Pharmacological/nutritional manipulation is thus designed to inhibit the inhibitors of lipolysis.

Phosphodiesterase

In the preceding section we discussed some of the inhibitors of membrane events that prevent the signal provided by NE from being translated into the "second messenger" cAMP inside the membrane. Assuming a successful conversion of the NE-induced signal, the events of lipolysis may proceed. However, there are yet other inhibitors to contend with. One of the most important and potent class of these disrupting agents is the phosphodiesterases.[349] These enzymes are involved in several events that regulate lipolysis. Type 3 phosphodiesterase is thought to be the dominant enzyme involved in regulating the breakdown of cyclic AMP which in turn governs lipolysis.[44,351] This enzyme thus inhibits the mobilization of lipids from adipose tissue. Its activity is modulated by thyroid hormones, and may undergo substantial alteration in effect in the presence of obesity and diabetes.[352-354]

The role of the thermogenic is to inhibit phosphodiesterase, adenosine and PGE2, while stimulating the production of NE. This process will be discussed in detail in the chapter on pharmacology. Briefly, ephedrine stimulates NE production and lipolysis; caffeine inhibits phosphodiesterase and adenosine; and aspirin inhibits the synthesis of PGE2.

In conclusion, it is our hope that this chapter has presented some admittedly very technical material in such a manner that we have succeeded in shedding at least a ray of light upon the important role of brown adipose tissue metabolism in maintaining body fat at healthy levels. Whether viewed as just another physiological process or as the mediator of a unique disrupter of the body's attempt to resist the irresistible degradative forces of the universe—the forces intending to reduce all organized life to the random movement of molecules— brown adipose tissue will certainly be the subject of an increasing body of research in years to come.

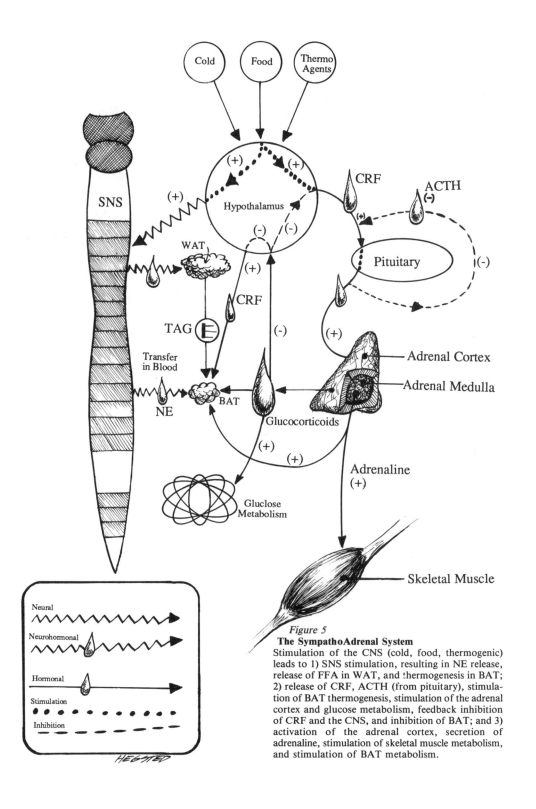

Figure 5
The SympathoAdrenal System
Stimulation of the CNS (cold, food, thermogenic) leads to 1) SNS stimulation, resulting in NE release, release of FFA in WAT, and thermogenesis in BAT; 2) release of CRF, ACTH (from pituitary), stimulation of BAT thermogenesis, stimulation of the adrenal cortex and glucose metabolism, feedback inhibition of CRF and the CNS, and inhibition of BAT; and 3) activation of the adrenal cortex, secretion of adrenaline, stimulation of skeletal muscle metabolism, and stimulation of BAT metabolism.

THE SYMPATHOADRENAL SYSTEM

The sympathetic division of the autonomic nervous system together with the adrenal medulla comprise the functional division of the nervous system sometimes called the sympathoadrenal system (SAS), or the adrenergic nervous system. While the major emphasis of this book is on the SNS and the major adrenergic neurotransmitter of the SNS, norepinephrine (NE) which is synthesized by, stored in, and secreted from the endings of sympathetic nerves, it is impossible to separate these actions from effects on and by the adrenal medulla. One reason is because both the SNS and the adrenal medulla arise from the same fetal material. Thus it could be reasoned that the adrenal gland is actually a highly specialized division of the nervous system.

Lewis Landsberg and James B. Young have synthesized much of the experimental work on the SAS and placed it within a comprehensive theoretical model.[740,742] Much of the material in this chapter is drawn from the Landsberg and Young model. The SAS (SNS plus adrenals) plays an intricate role in the regulation of both obligatory and facultative thermogenesis. It controls metabolism in adipose tissue, both white and brown. Defects in the SAS are thus implicated in the etiology of obesity.[55,53,710,790] The adrenal medulla synthesizes and secretes adrenaline (epinephrine), a hormone that circulates in the blood, producing certain reactions in cells throughout the body as it locates and interacts with cell membrane receptor sites. Adrenaline and NE are catecholamines, and are synthesized and released from the adrenal medulla or sympathetic nerve endings respectively in response to descending signals from command centers in central nervous system structures such as the hypothalamus, pons and medulla oblongata.[17,1169] The control of NE and adrenaline metabolism by higher brain centers is extremely complex; it may stimulate or inhibit both hormones simultaneously or independently.

Studies have shown that NE secreted by the SAS plays a critical role in the thermogenic response to stimulation from cold and/or ingested foods, thermogenic and medications.[53,63,386,438,583,739,744,745,1149,1394] Basically, the relative importance of NE is much greater than that of adrenaline.[737,767] Thermogenic stimuli cause an almost immediate reaction in the SNS which is sustained to some degree as long as the stimulus is applied. Besides increasing thermogenic activity in BAT, stimulation causes increased NE-mediated metabolic activity in the heart, pancreas, lung, spleen, skin and skeletal muscle, while activity in the liver, intestine and kidney is unchanged or reduced.[97,293,592,655,1245,1252,1328,1386,1396] SNS-mediated mobilization of fat in preparation for its role as fuel for BAT thermogenesis also occurs. Stimulation elicits a lesser, or no, response from the adrenal medulla, which response may decline substantially with prolonged exposure.[199,364,487,583,647,767,1254,1387,1389] Cold exposure and other thermogenic stimuli produce an immediate neural response followed by the gradual reduction in size of fat droplets in BAT cells; they also promote growth of brown adipose tissue, greater respiratory activity, increased concentration of protein in BAT mitochondria, conformational changes in intracellular components, and markedly increased activity among numerous enzymes.[195,297,737,765]

Cold exposure elicits an initial response in receptors in the skin which quickly recruit neural activity in the central nervous system.[140,187,513,991,999,1255] It appears that these signals are first integrated in the preoptic area of the anterior hypothalamus, and then relayed to the posterior hypothalamus before the efferent signals to the SNS are initiated.[80,406,412,1255] There follows an increased secretion of norepinephrine in brown adipose tissue. Blood flow through BAT is also substantially elevated. Increased flow accelerates the delivery of oxygen and free fatty acids to BAT and helps remove heat generated by that tissue. It is somewhat paradoxical that whereas thermogenic agents cause constriction of blood vessels in other parts of the body, they cause a dilation of blood vessels in brown adipose tissue.[1020]

With repeated exposure to cold, the body's capacity for heat production increases, and its need for CIST decreases. This increased thermogenic capacity is probably the result of hypertrophy of BAT.

A permissive level of thyroid hormone also appears to be necessary for cold acclimation to occur.[526]

In addition to integrating the complex neural and hormonal responses to cold, the SAS also regulates thermogenic responses to diet. Harvard scientists suggested in the later '70's that diet exerted a strong effect on sympathetic activity, an observation strengthened by the many features shared in common by CINST and DIT.

Finally, the SAS integrates the sympathetic and thermogenic responses to the consumption of thermogenic products. The specifics regarding these various attributes of the SAS will be explicated in the chapters that follow.

SNS Deficiency

Without an intact SNS, thermogenesis is impossible.[53] Chronic interference with the activity of the SNS by surgical lesions made to the SNS in animals, or by the administration of chemicals that block SNS activity completely, eliminates thermogenesis and leads to heightened metabolic/energy efficiency and an increase in body fat in lean animals.[13,54,323,324,901] Genetic lesions that inhibit or eliminate thermogenesis almost always create obesity. Parasympathetic involvement in thermogenesis is minimal.[1124] It would thus appear that when the SNS is underactive, fat stores increase, and fat accumulation may result. One would expect, therefore, that SNS activity would be found to be suppressed in obese individuals. It turns out that research findings are contradictory, with some finding diminished, others increased, others unchanged, SNS activity.[344,1292,1391] Landsberg has attempted to reconcile these differences on the basis of both procedural and physiological factors. Most significantly, Landsberg proposes that increased SNS activity in the obese may reflect an attempt by the body to compensate for obesity-inducing metabolic factors by increasing energy production, restoring energy balance and stabilizing body weight.[733] Thus, lowered SNS activity may produce obesity to which condition the body reacts by increasing SNS activity as a compensatory measure. Normally, however, the obese demonstrate lowered SNS activity. Formerly obese people show a lower metabolic rate.[722] Diminished SNS response to a meal has also been found in the obese.[89]

The activity of the adrenal medulla is also suppressed in the obese, a further indication that SAS activity suffers during obesity, though whether reduced adrenal medullary activity is a cause or effect of obesity has not been firmly established.[744,1292]

Alternatively, the administration of sympathomimetic drugs/nutrients restores thermogenic and fat-reducing capacity to animals lacking that ability.[322,325-327,329,330,562,795,1296,1390] The same results are observed when thermogenic agents are administered to DIT-deficient humans with a propensity toward obesity (see chapter on Pharmacology).

Both CINST and DIT depend on the SNS. The SNS responds rapidly to changes in the environment due to food intake and temperature changes, and quickly adjusts the level of thermogenesis to maintain the appropriate energy and thermal balance. Food in general, and carbohydrates and fats in particular, elicit immediate activation of the SNS.[946,1132,1325,1384,1385] This response is blunted in individuals with insulin resistance and heightened in people with insulin sensitivity.[89,118,617]

Dieting and the SAS

Caloric restriction, or fasting, suppresses SNS activity.[108,289,484,659,946,1193,1384,1385,1386,1393] Metabolic rate begins to fall 24-48 hours after a fast begins. Long term maintenance of a caloric restricted diet leads to further decreases in sympathetic activity, but with a mild adrenal medullary stimulation.[1387] Obese individuals who have repeatedly gone on restricted caloric diets may have experienced serious regression of BAT due to blunted NE activity.[86,1157] Consumption of a thermogenic agent tends to offset the suppression of the SNS and BMR that occur as a result of caloric restriction.[1156] Alternating short term fasts (24 hours or less) with restoration of high caloric intake seems to stimulate SNS and thermogenesis,[738] and may be viewed as tonic exercise for BAT.[739]

General Features of Sympathetic Nerves.

There appear to be two types of postganglionic SNS fibers.[1127] "Long" neurons are conventional SNS fibers. They originate in ganglia (groups of cell bodies) located outside of the spinal column but distant from target adipose tissue. They are situated between "preganglionic" nerve fibers and the brown adipose tissue cell or cells on which they terminate. They appear to directly innervate blood vessels in BAT. "Short" neurons, the second type of fiber, originate in tiny ganglia located within BAT itself. They are situated between "preganglionic" fibers (that bypass the normal ganglia) and BAT. They appear to innervate the brown adipocytes themselves.[292]

The combination of short and long neurons, originating in separate places and synapsing on both ends with different tissues (separate preganglionic fibers on one end and blood vessels or adipocytes on the other), imparts a great deal of flexibility to activity in BAT. Accelerated activation is thus possible, while such an arrangement makes it feasible to fine-tune BAT activity in response to numerous sources of activation and inhibition. Located in sympathetic nerve endings of long fibers and through the neurons of short fibers are the tiny sacs or vesicles that contain neurotransmitter substance, especially NE. When the neuron is excited, the vesicles burst, spilling NE into the synapse separating the neuron from target cells. Data is transmitted to the target cell via the neurotransmitter, whose action in the synapse is modulated by the presence of other chemicals.

Adrenergic Pharmacology of BAT Thermogenesis.

(To a great extent, the following paragraphs recapitulate and summarize material treated in the previous chapter.) Norepinephrine secreted by sympathetic nerve endings interacts with BAT at certain receptor sites on the BAT cell membrane.[171] BAT contains several kinds of receptor sites, the most important of which are called beta-adrenergic and alpha-adrenergic receptors.[581,663,686,717,765,1126] The beta receptors include beta-1, beta-2 and a novel subtype apparently specific to BAT, beta-3.[37,517,1034] NE released from sympathetic nerves interacts with all of these receptors. A synergism, as well as an antagonism, between alpha- and beta-adrenoceptors has been observed.[306,1178,1179,1223] Beta-blockers (drugs that block the activation of beta-receptors), used to treat hypertension, can produce significant weight gain, probably by reducing the metabolic events associated with thermogenesis.[5,107,360,660]

The immediate responses to NE binding at the receptor sites include increased blood flow and increased release of cyclic AMP (See Figure 1), increased adenylate cyclase activity, increased concentrations of FFA, increased synthesis of UCP, and increased thyroxine 5'-deiodinase activity.[361,375,945] In addition, NE promotes a sequential series of electro-chemical changes and the intercellular propagation of those changes via gap junctions.[1242] (Gap junctions are places where cells communicate directly with one another without the need of neurotransmitter substance; the impulse in one cell spontaneously generates an impulse in the neighboring cell.) A brief depolarization, mediated by alpha-adrenoceptors and associated with an influx of sodium ions, and an efflux of potassium ions, is followed by repolarization and a second, more gradual, depolarization mediated by beta-receptors.[438,523,577,578]

BAT membrane contains beta-adrenergic receptors coupled to an adenylate cyclase system which interacts with NE by activating adenylate cyclase and increasing cAMP production. Increased cAMP activates protein kinase that phosphorylates hormone-sensitive lipase (HSL), resulting in accelerated lipolysis. More than 75% of the resulting increase in thermogenesis is thought to be the result of the increase in lipolysis-generated free fatty acids.[529,930]

Increased cAMP and protein kinase production also lead to inactivation of acetyl CoA carboxylase, activation of pyruvate dehydrogenase and activation of phosphofructokinase-1. Another consequence of beta-adrenergic mediation of uncoupling of mitochondria is the production of adenosine which immediately increases blood flow in a positive manner but which acts negatively to inhibit the increase in adenylate cyclase activity generated by NE.

Jean Himms-Hagen has summarized the details of alpha-adrenergic stimulation as follows:[538] Alpha-receptors in BAT are stimulated by NE to increase the operation of the phosphotidylinositol bisphosphate (PIP_2) cycle. This, in turn, stimulates the production of two second messengers, inositol triphosphate and diacylglycerol.[880,882,912,1116] These substances initiate a chain of events that critically affects the cell's ability to handle ATP-linked ion modulation which serves as the background against which most of the other metabolic events occur and which restores ion gradients. It is felt that about 15-25% of the thermogenic response is mediated by the alpha-adrenergic receptors.[578,880,881,1120,1121]

Another product of NE stimulation of alpha-adrenergic receptors is an increased synthesis of thyroxine 5'-deiodinase that converts thyroxine to the thermogenically active hormone T_3.[178,202,582,654,665,718,766,776,-1065,1165,1235] (see the Chapter on Thyroid).

In summary, we have shown that NE from the SNS is the primary mediator of thermogenesis in BAT. It has both short term and long term consequences. The beta-adrenergic receptor in the plasma membrane of the BAT cell to which NE binds is possibly a novel subtype, beta-3.[517] The main effect of NE is to stimulate lipolysis and activate the thermogenic proton conductance pathway or thermogenic cascade. NE also increases the supply and delivery of oxygen and substrate (FFA) to brown adipose tissue by stimulating blood flow through BAT and by stimulating both glucose uptake and lipoprotein lipase activity which leads to an increased supply of FFA. Finally, NE stimulates the conversion of T_4 to T_3, a key enzyme in BAT under the control of the SNS.[780,1165]

The SNS exerts a long-term, chronic, tropic effect on BAT in many but not all species.[496,845,1287] Prolonged consumption of thermogenic leads to increases in cell number, mitochondrial count and concentration of UCP.[388,1009] These changes result in significant increases in BAT-mediated thermogenic capacity.

Other factors that affect BAT thermogenics are mediated through the action of insulin and corticosteroids. These influences will be discussed in subsequent chapters but are mentioned here for sake of completeness. Insulin apparently stimulates thermogenesis by stimulating the SNS and by activating specific metabolic processes in BAT such as glucose uptake and lipogenesis, while corticosteroids inhibit thermogenesis, acting in opposition to catecholamines.[369,415,566,742,841,1083,1165]

THE CENTRAL NERVOUS SYSTEM

Peripheral sympathetic activity is ultimately controlled by the central nervous system (CNS). Cold-, diet-, and thermogenic-induced thermogenesis can not occur until central neuronal mechanisms receive sensory input from pertinent areas of the body (e.g., thermal receptors at the body surface and in the cervical spinal cord) communicating changes in temperature, dietary intake or the presence of the thermogenic agent. The CNS then initiates the appropriate changes in SNS activity. Little is understood about what factors couple the sensory signals to changes in SNS activity, but insulin-mediated glucose metabolism within nerves situated in the *ventromedial hypothalamus* (VMH) is probably involved.[34,35]

The central nervous system is composed of the brain and the spinal column. The brain is anatomically a collection of dozens of nuclei, or centers, or groups of related neurons. Of these, one in particular plays a critical role in thermogenesis. It is the hypothalamus, and is itself a collection of distinct control centers. The hypothalamus is located deep in the center of brain, immediately above the pituitary gland, to which it is connected by a hypothalamic extension called the pituitary stalk. From here, the hypothalamus issues a continuous stream of command and control instructions and orders to several parts of the body, many of which are directed to the pituitary. Hypothalamic integration, control and regulation is autonomous in nature; it takes place almost completely on a subconscious level.

The neurosecretory cells of the hypothalamus can be stimulated by other nerve cells in higher regions of the brain to secrete specific peptide hormones into the blood vessels of the pituitary stalk: these hormones can then stimulate or inhibit the secretion of second level hormones from the pituitary. The pituitary hormones in turn stimulate other endocrine glands to secrete third level hormones into the blood. These hormones then exert specific actions on cells throughout the body.

149

The aggregates of neurons and fibers that make up the hypothalamus are identified as nuclei. Stimulation of certain nuclei is excitatory to the parasympathetic nervous system, causing sweating, vasodilation and a decrease in the rate and force of the heart contraction. Stimulation of other (*caudal*) nuclei is excitatory to the sympathetic nervous system, causing an increase in the rate and force of contraction of the heart, vasoconstriction, increased respiration due to dilation of bronchioles, dilation of the pupil and inhibition of peristalsis. Lesions in this area of the hypothalamus completely disrupt the ability to regulate body temperature, resulting in hypothermia. The SNS is the primary effector for carrying out the thermoregulatory commands of the hypothalamus.

Posterior nuclei of the hypothalamus act to increase the heat of the body and prevent its loss (while *anterior* and *rostral* nuclei regulate the dissipation of excess body heat). It is well known that the posterior hypothalamus (PH) is heavily involved in the activation of behavioral thermoregulatory responses that we would classify as obligatory. But it has also been found that direct stimulation of neurons in the PH can activate sympathetic mechanisms that arouse heat production in BAT, a facultative component.[18] The VMH is the major hypothalamic nucleus responsible for regulating descending signals from the brain that cause the sympathetic nervous system to secrete NE and stimulate thermogenesis in BAT.[399,563,564,609,693,874,1092] *Paraventricular* nuclei (PVN) have also been implicated.[379,563,564] There also appears to be an inhibitory pathway from the preoptic area of the hypothalamus through the anterior and *medial* nuclei to lower brain areas.[399] Thus, stimulation of the anteriour nucleus with PGE_2 produces the full blown thermogenic cascade in brown adipose tissue.[20] Studies involving the cooling of the preoptic area, or the skin, suggest that this pathway helps to regulate the control of CINST in BAT. Chemical stimulation of the supraoptic nucleus of the hypothalamus has also produced thermogenesis in BAT.[19] Thus, thermoregulation and thermogenesis are controlled by dozens of intricate interactions among hypothalamic nuclei which are in turn regulated and modified by both higher (CNS) and lower (brainstem) structures.

150

Lesions of the hypothalamus that cause obesity adversely affect the ability of the hypothalamus to control the SNS and BAT thermogenesis in response to feeding and other stimuli.[1313,1315] Lesions of the lateral hypothalamus that cause leanness do not alter SNS activity in BAT, or may even enhance it.[563,609]

As indicated above, insulin (and probably other hormones) help mediate events in the VMH. The role of insulin is discussed in the chapter on Insulin.

CRF

The hypothalamus also secretes a hormone called corticotrophin-releasing factor (CRF). (See figure 5.) The main function of CRF is to stimulate the pituitary gland to release adrenocorticotrophic hormone (ACTH) which in turn stimulates the adrenal cortex. Under stimulation by ACTH, the adrenal cortex secretes a number of hormones, among which are the glucocorticoids. In addition, CRF has an effect on the activation of thermogenesis in brown adipose tissue that may be hormonally mediated by endorphins and/or melatonin analogues.[290,1061,1063] In accordance with the concept of feedback control, CRF-induced thermogenesis is regulated by CRF-induced glucocorticoid inhibition of BAT metabolism.

Brainstem

The hypothalamus is not the only CNS structure that influences thermogenesis. Stimulation of certain areas of the brainstem have been shown to cause CINST and CIST in skeletal muscle, and may influence BAT thermogenesis as well.[17,110,1084,1158] The nucleus tractus solitarii (NTS) has been implicated in the regulation of BAT thermogenesis.[410] Lesions in the NTS attenuate the thermogenic response following infusion of prostaglandin E_1 into the brain. Alteration in the ability of beta-receptors to respond to pharmacological stimulation with the typical response is part of the problem.[409] Midbrain mechanisms are generally in a state of inhibition. It is not known if release of this inhibition occurs under normal conditions.

151

Sensory Integration: Cephalic Regulation

CNS involvement in thermogenesis also entails the integration of data about the sight, taste and smell of food. These sensory qualities have been shown to elicit thermogenic responses in the BAT of animals.[260,482] This cephalic phase of central regulation is mediated through the SNS and involves the stimulation of insulin secretion.[117,300,301,1201] Thus, it is clear that sensory qualities of the food we eat interact through CNS centers with the hypothalamic signals that modulate BAT activity. Tube-feeding in rats, which bypasses the gustatory senses of the diet, results in a marked attenuation of diet-induced activation of BAT.[1074]

Perhaps bland diets that restrict the number of calories consumed, but sacrifice the more pleasant aspects of food and eating, are actually counter-productive to the extent that they blunt the cephalic regulation of thermogenesis. A more effective alternative would be to assure that all meals are prepared attractively, with the goal in mind of exciting as many senses as possible.

In summary, the central nervous system regulates several aspects of BAT thermogenesis. First, it exerts an influence on several hypothalamic nuclei that affect feeding, thermoregulation, and related processes. The primary effect of the CNS is to coordinate sensory input that signals a changing external or internal environment with changes in the corresponding hypothalamic responses. The CNS integrates VMH signals with corresponding changes in levels of circulating hormones. CNS-directed combustion of free fatty acids requires coordination of fuel and oxygen recruitment with disposal of carbon dioxide and other wastes. Control of the rich vascular network serves these purposes. CNS-directed SNS activity supplies the hormonal catecholamines, that increase lipolysis and subsequent FFA oxidation; they also indirectly dilate blood vessels in BAT to allow the increased exchange of oxygen and carbon dioxide.

The second influence of the CNS is cephalic in nature, i.e., the CNS responds to sensory data from the sight, smell and taste of food (and even to the thermogenic capsule or tablet) either to increase or decrease the thermogenic response depending on how pleasant or unpleasant these stimuli are. The central nervous system probably regulates other aspects of sensory integration and efferent flow of controlling signals as well.

Feedback Model of CNS Mediation

George Bray of the Penington Biomedical Research Center in Baton Rouge, has provided a feedback model of CNS mediation of peripheral events and consequent sympathetic outflow that fits well with the systems, theories and data presented in this and other chapters of this book.[147,147a,148a,151] In this model, the CNS stands at the center of a string of signals arising from afferents (incoming) signals in the sensory system, including the gastrointestinal tract, and from the biochemicals derived from digestion and cellular metabolism, which are transduced by the CNS before giving rise to efferent (outgoing) signals that are mediated by the SNS and that consequently affect appetite, energy expenditure, and so forth.

In Bray's model, the specific central nervous system structure responsible for coordinating the afferent and efferent data is, as we might expect, the hypothalamus. Thus, feeding behavior may be enhanced or inhibited by stimulation of the appropriate hypothalamic nuclei by the appropriate chemical messengers, specifically a number of circulating peptides. For example, monoamines, such as norepinephrine, injected into the hypothalamus may either increase or decrease food intake behavior, depending on the nucleus affected. Another peptide, serotonin, generally tends to inhibit food intake.

Bray cites the work of Young and Landsberg on the effects of over- and under-feeding on the SNS.[1384,1385] Fasting initiates sensory signals that are interpreted by the CNS in such a way as to shut down SNS activity, with a consequent reduction in energy expenditure, as expressed by lower BMR and lower thermogenesis. Alternatively, drinking sucrose has the opposite effect on sympathetic activity, as discussed in an earlier chapter. Other studies confirm the inverse relationship between food intake and SNS activity.[1094a]

Bray's model superimposes a negative feedback loop control structure on these data. The model predicts that the intake of certain nutrients will stimulate the activity of NE and the peripheral SNS leading to satiety and subsequent inhibition of food intake. Nutrient intake also stimulates insulin release either directly by stimulating the pancreas, or indirectly through CNS activation of vagus nerve afferents/efferents to the pancreas. At this point, Bray's model seems to segue with Landsberg's theory of the importance of insulin-mediated glucose metabolism, discussed elsewhere.

Among the peptides cited by Bray that stimulate feeding are galanin, beta-endorphin, dynorphin, neuropeptide Y, and growth hormone-releasing hormone. Peptides that inhibit feeding are insulin, glucagon, cholecystokinin, anorectin, enterostatin, calcitonin, corticotrophin-releasing factor, neurotensin, cyclo-his-pro, somatostatin and thyrotropin-releasing hormone. From time-to-time, one or more of these peptides is touted by the popular press as the answer to obesity, as the panacea of anti-obesity agents. So far, nothing has come of these pockets of excitement.

Bray's feedback model nicely predicts the action of these peptides. Those that stimulate appetite ultimately reduce SNS activity. Conversely, those peptides that decrease appetite tend to increase SNS activity as the CNS attempts to restore homeostasis in the energy balance equation. This model thus helps resolve some of the paradoxical findings emerging from the thermogenic research. The degree of experimentally measured SNS activity level will depend on when it is measured relative to the status of CNS-directed and SNS-mediated feedback control of energy intake and expenditure.

Food intake definitely stimulates DIT.[1105] DIT is mediated by the SNS. However, consumption of agents that stimulate SNS activity tend to reduce appetite. Such substances, e.g., ephedrine, also tend to affect body composition, by increasing protein and lowering fat deposits. Applying Bray's model to these data, we find the afferent signals arising from food intake traversing the vagus nerve to the brain and back to the pancreas via the SNS, or going directly to the pancreas. In the CNS, the afferents inhibit excitatory feeding nuclei

154

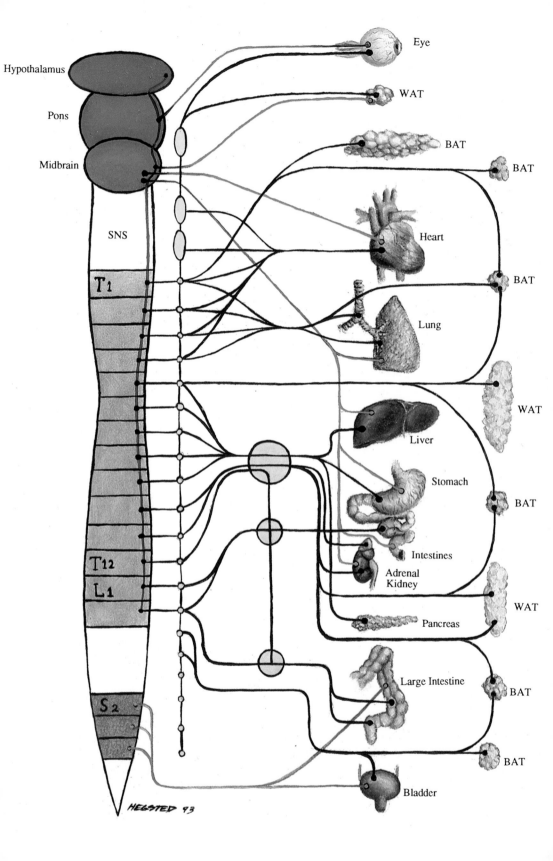

Hypothalamus

Pons

Midbrain

SNS

T₁

T₁₂

L₁

S₂

Eye

WAT

BAT

BAT

Heart

BAT

Lung

WAT

Liver

Stomach

BAT

Intestines

Adrenal
Kidney

WAT

Pancreas

Large Intestine

BAT

BAT

Bladder

HEGSTED 93

Plate #1   The Sympathetic Nervous System
Showing higher central control centers, spinal column, and innervation of key organs, including BAT and WAT, involved in the expression of thermogenesis and related sympathic processes in
humans.

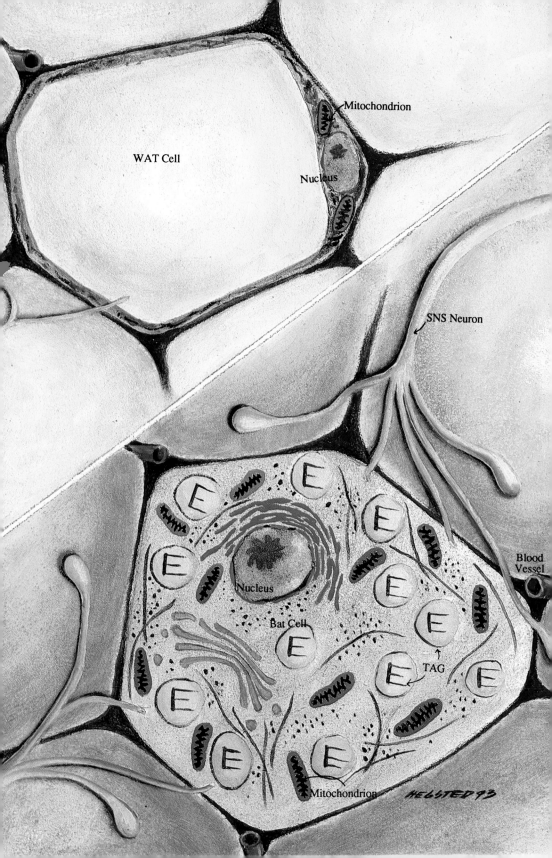

Plate #2   Brown & White Adipose Tissue
A comparison of a typical white fat cell and brown fat cell in cut-away view. Note large fat droplet in WAT cell that crowds other cell components to the side. Smaller triglyceride-containing fat droplets in BAT allow more even distribution of cell components. Note greater presence of blood vessels and nerves in BAT. The "E" shaped substances are triglycerides.

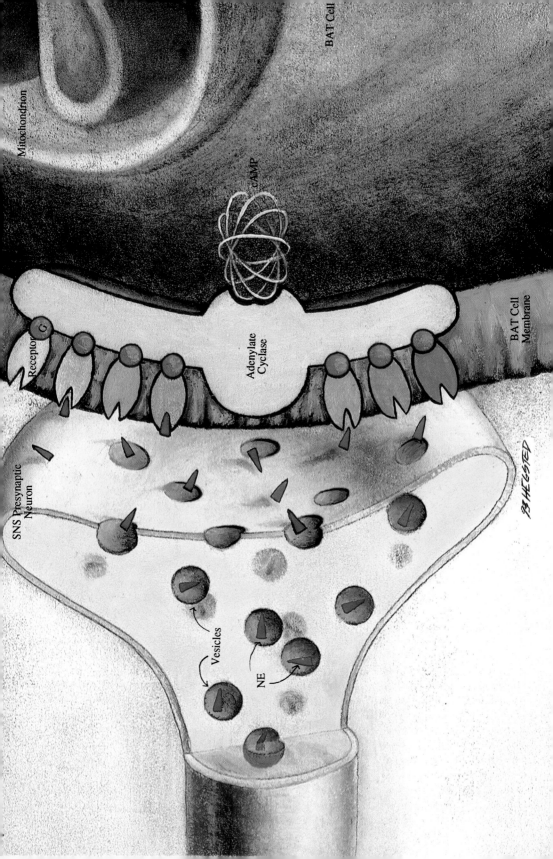

Mitochondrion

BAT Cell

cAMP

Receptor G

Adenylate Cyclase

BAT Cell Membrane

SNS Presynaptic Neuron

'83 HELSTED

Vesicles

NE

Plate #3   BAT Synaptic Junction
Cut-away view of SNS nerve fiber in proximity to a brown fat cell, showing release of norepinphrine and its key interactions with receptors in the BAT cell membrane, leading to activation of cAMP (cyclic AMP). These reactions constitute the early stages of the thermogenic cascade.

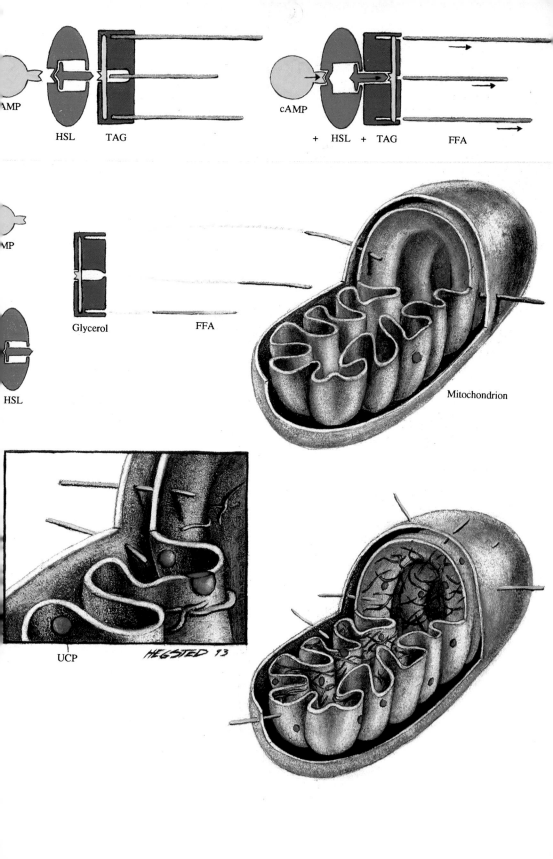

AMP

HSL   TAG

cAMP

+ HSL + TAG   FFA

MP

MP

HSL

Glycerol   FFA

Mitochondrion

UCP   *HEGSTED 93*

Plate #4   Lipolysis & UCP Activation

Middle and late stages of the thermogenic cascade. cAMP and hormone-sensitive lipase (HSL) combine to break down triglycerides (TAGs) hereby liberating free fatty acids (FFA) which easily enter mitrochondria. Accumulation of FFA eventually activate UCPs located on inner mitochondrial membrane. Thermogenesis is the final result of UCP activation.

in the hypothalamus. One consequence of the descending efferent SNS signals is an increase in skeletal muscle and BAT thermogenesis that relies on a concomitant increase in lipolysis. The increase in FFA may then initiate a reduction in appetite through reciprocal afferent signals to the CNS.

Bray also notes that some peptides tend to affect the intake of specific nutrients. Fat intake is inhibited by enterostatin and stimulated by galenin and opoids. Neuropeptide Y increases carbohydrate consumption, while cholecystokinin decreases it. Protein intake has been found to be inhibited by glucagon. Again, these data seem destined for exploitation by the pharmaceutical industry in coming years.

HYPERTENSION

Hypertension is a common side-effect of obesity. And whether they do or they don't have high blood pressure, almost all over fat people experience a fall in blood pressure when they lose a substantial amount of fat.[1032,1198] Since the consumption of thermogenic substances results in significant fat loss, we would expect to see a concomitant *decrease* in blood pressure. However, since thermogenic substances stimulate the sympathetic nervous system, they should also, theoretically at least, tend to raise blood pressure.[222,223,226] Since many over fat people have high blood pressure, perhaps they should not use the thermogenic for fear of aggravating their hypertension.

Practical experience tends to favor the first hypothesis. High blood pressure may be caused by the extra weight of obesity. Lowering the fat by activating thermogenesis might lower blood pressure. Alternatively, the proper composition of the thermogenic may impact on the product's ultimate influence on blood pressure. The presence of the appropriate amounts of aspirin, for example, is thought to decrease the tendency of NE to raise blood pressure.

The blood pressure paradox has been addressed theoretically, if not in any published clinically study. L. Landsberg of Harvard University, and M.G. Clark, et. al., of the University of Tasmania, have explored the experimental and theoretical relationships between several dietary and metabolic factors that may contribute to hypertension in obesity.[225,271,734,735] They discuss how sympathetic factors that stimulate thermogenesis should also produce high blood pressure. But, according to the investigators' theories, a closer inspection of the data indicates that just the opposite may, in fact, be true. Landsberg theorizes that hypertension in the obese is a by-product of the body's futile attempt to restore energy balance, limit weight gain and stabilize body mass. In instead of normalizing function, the battle creates a state of insulin resistance which simply makes matters worse and is the root cause for increased blood pressure. Restoration of thermogenesis, removes the stress, like opening the flood gates of a dam. Clark suggests that under normal, healthy conditions, several factors, including eating and insulin, tend to increase facultative thermogenesis both in skeletal muscles, through exercise-induced thermogenesis (EIT), and in BAT (DIT).

Should successful activation of facultative thermogenesis be achieved and energy balance restored, then most excess calories would be used up and energy and thermal balance maintained. More importantly, all of the effort that went into generating thermogenesis would also be used up, along with any secreted norepinephrine (NE). However, in the obese facultative thermogenesis fails, and the effort expended to make it happen creates insulin resistance and hypertension. Secreted NE is not being taken up in the thermogenic metabolic pathways; perhaps active competition for the receptor site is so intense that NE secreted into the synapse is not utilized appropriately in BAT or skeletal muscle. We have learned that NE non-selectively stimulates all types of beta-adrenoceptors. Only one of these receptor types is specific to BAT. The other two are found throughout the body and mediate all of the events representative of sympathetic stimulation. The beta-receptors mediating blood pressure would therefore be stimulated at least at normal levels, but perhaps at greater-than-normal levels. Assume that diet-induced secretion of NE does exert its normal hypertensive action. Now suppose that the entire mechanism that maintains a balance among the thermic and energy balances of the body is a homeostatic mechanism (i.e., operates as a negative feedback loop). If this is true, then what will happen when the hypothalamic control center fails to receive a signal confirming the utilization of excess calories in thermogenesis? It will, of course, continue to transmit the signals required to stimulate thermogenesis. These signals result in the secretion of yet more NE which tends to increase blood pressure still further, but still without impacting thermogenesis. The cycle could repeat itself for prolonged periods of time.

According this model, then, high blood pressure in obesity is the result of almost constant sympathetic stimulation due to frustrated thermogenic processes. Re-establishing thermogenesis might, therefore, destroy the forces maintaining hypertension. By extension, consumption of thermogenic, rather than increasing hypertension in obese persons with high blood pressure, might actually lower blood pressure by restoring the thermogenic cycle and reestablishing proper feedback control of thermogenesis by central nervous system command centers.

What if, however, defective thermogenesis involves defective NE secretion? In that case, the NE could not be maintaining hypertension. Do the models suggested by Landsberg and Clark handle this eventuality? The problem is resolved in one way by considering the fact that defective feedback control of NE metabolism would result in increased secretion of adrenaline from the adrenal medulla. That is, if *insufficient* NE is present in the system, feedback signals would be continually instigating circuits that should result in increased NE secretion (which, according to the model, would not occur). However, those same signals *would* result in increased secretion of adrenaline. Adrenaline would in turn activate sympathetic arousal in a nondiscriminatory manner, just like NE would.

Increasing NE secretion through the thermogenic would satisfy the demands of the feedback system, and reduce the signal to the adrenal gland calling for adrenaline output. Increasing BAT activity through the consumption of thermogenic, according to this hypothesis, would also effectively result in the creation of enough BAT and uncoupling protein that the tissue could effectively assimilate all NE being secreted, and would therefore reduce the tendency toward hypertension by yet another increment.

Another hypothesis that might account for an observed lack of hypertensive effect of thermogenic, is that thermogenic agents do not work by simply increasing the amount of NE in the tissue, but actually increase its utilization by BAT cells. This is, after all, the main reason for adding caffeine and aspirin: they tend to increase the effectiveness of ephedrine through actions complementary to, but not identical with, sympathetic hormones. Increased NE utilization by BAT cells would reduce hypertensive pressure otherwise exerted by circulating NE. Once BAT capacity grew to the point where it exceeded normally circulating NE levels, blood pressure would tend to actually decrease under the influence of the thermogenic, and a permanent reduction in blood pressure would also result. Some validation of this hypothesis has been achieved in ongoing research at APRL.

On a more practical level, hypertension in the obese is probably the result of many independent factors. Treatment should involve not only use of thermogenic, but restriction of alcohol intake, mild restriction of salt, eating more vegetables and fruits, increasing exercise, and cessation of smoking.[137,1170]

INSULIN AND GROWTH HORMONE

The role of insulin in the regulation of thermogenesis in BAT is not well understood. Yet insulin forms a critical bridge between dietary, CNS, and SNS considerations. (See Figure 6.) Insulin activates specific metabolic processes in BAT, including glucose uptake and utilization, and lipogenesis, which are important processes in the formation of substrates (free fatty acids) utilized as fuels for BAT metabolism.[340,369,471,472,635,984,1090,1398] Facultative thermogenesis is thought to be mediated by insulin-mediated activation of the SNS, and insulin also appears to directly stimulate the SNS,[6,287,1381] although it also tends to inhibit the lipolytic action of catecholamines secreted by the SNS. Insulin metabolism is dramatically upset by diets that severely restrict the number of calories ingested. Insulin metabolism and effective thermogenesis are both enhanced by sensible eating habits. Even slight caloric restriction adversely affects insulin metabolism, but it can be offset to some degree by consumption of thermogenic. It appears that disruption of insulin metabolism, especially insulin resistance, disrupts the thermogenic response to eating (DIT, TEF).[366]

Insulin controls the level of blood glucose. As glucose levels increase, insulin forces excess glucose into cells. *Glucagon*, another pancreatic hormone, has the opposite effect; as glucose levels drop, glucagon is recruited to stimulate the removal of glucose from cells, especially of the liver through a process called glyconeogenesis (which involves the now familiar cascading pattern of adenylate cyclase-cAMP-mediation of glycogen degradation to glucose which is shunted to the blood). Glucagon also stimulates GDP binding and the conversion of TAG to free fatty acids in BAT and WAT in a similar manner.[126] Administration of large amounts of glucagon to laboratory rats results in a substantial growth of BAT, independent of sympathetic innervation of BAT,[125,126,730] and to increase thyroxine 5'-deiodinase activity.[1167]

The action of insulin is to increase the activity of enzymes in adipose cells that synthesize TAGs. These lipid molecules may then be recruited under the proper thermogenic stimulus to yield up their FFA. Thus the actions of NE and insulin are opposite to a large extent, yet complement each other well, provided that insulin metabolism is proceeding at a normal pace.[209,291,812,1360] The primary site of insulin-mediated conversion of carbohydrate to fat is adipose tissue. The effect of insulin appears to be an enhancement of the glucose molecule's entrance into the lipid pool from extracellular spaces. Circulating insulin regulates the relationship between extracellular glucose and adipose tissue lipogenesis. Thus, a rise in serum glucose stimulates the release of insulin into the blood stream; as serum insulin levels rise, an eventual conversion of glucose to fat occurs in adipose tissue. This insulin-mediated action complements the normal conversion of circulating fat from the blood stream and chylomicrons from the lymphatic into TAG storage molecules in adipose tissue.

As obesity progresses, glucose-induced thermogenesis (GIT) decreases in non-diabetic persons.[448,450,637] In diabetic people GIT is already blunted and remains so as obesity increases.[448,450,452,1133] As GIT decreases, so does glucose tolerance and insulin responsiveness, although fasting insulin levels do not decrease. Thus, worsening GIT, decreased glucose tolerance, and the occurrence of diabetes, may all contribute to a reduction of energy expenditure during the evolution of obesity.[447]

Decreased GIT has been observed in all forms of obesity.[118,131,638,1021,1129] Obesity per se or insulin resistance could account for this action, but available research indicates that insulin resistance is at fault.[449,853,855,1022,1162] Impaired insulin action leads to decreased glucose substrate availability and a consequent reduction in thermogenic capacity.[118,131,490,1157]

As insulin loses its metabolic activity, as in diabetes, glucose metabolism diminishes, the ability of adipose tissue to store fat is compromised, and fatty acids are released. One might think that

162

the release of FFA is good, since that is what is required as fuel for thermogenesis. But the *premature* release of FFA is not good, since without TAG stores, the thermogenic response to thermogenic stimuli is blunted just when it is most needed. It is a matter of timing and intensity. A signal arising from NE requires an immediate and large response, a rapid release of FFA into circulation from WAT and an immediate increase in FFA from triglyceride stores in BAT. If these materials have been slowly dissipated due to insulin resistance, then the adipose tissue cannot respond with a rapid increase in FFA. The thermogenic response in the diabetic individual is therefore severely limited, leading to loss of BAT activity.[84,85]

Interestingly, it has been shown that the NE-stimulated effect on glucose and fat metabolism requires a functioning adrenal cortex. Glucocorticoids may directly stimulate free fatty acid release due to an inhibition of glucose uptake and a decrease in reesterification. Thyroid hormone also plays an important role in facilitating the events of lipolysis and may interact with the actions of insulin.

Hypoglycemia

Hypoglycemia, a common problem associated with dieting, slightly stimulates the adrenal medulla, but suppresses SNS activity.[736,1387] However, as Landsberg and Young have pointed out, it is not the low level of glucose that affects SNS activity, but the lack of glucose metabolism.[1018] Since glucose metabolism involves the presence and activity of insulin, it is this hormone that is intricately involved in the SNS events underlying thermogenesis. Research with animals has shown that insulin-mediated glucose metabolism is adversely affected by fasting, which in turn adversely affects SNS activity.[1382] On the other hand, the consumption of refined carbohydrates in experimental settings (as an experimental method, not as a chronic lifestyle) stimulates insulin-mediated glucose metabolism and also stimulates activity in the SNS.[1088]

163

CNS Integration

Insulin is thought to be the most important mediator between peripheral events and the CNS. An insulin-sensitive region of the CNS has been identified as the *ventromedial* hypothalamus (VMH).[1376,1388] Signals arising from the periphery, either through eating, or fasting, or cold exposure, are integrated in the VMH and outgoing signals are initiated that regulate SNS activity. Without insulin, these affects of dietary thermogenesis could not occur.[1308]

Disinhibition

Fasting, or dieting, appears to *stimulate* a hypothalamic pathway that *inhibits* SNS activity. Experimentally, it has been shown that acute feeding of sucrose and other refined carbohydrates depress the *inhibitory* pathway, thereby *disinhibiting* the lower brainstem centers that stand between the CNS and the SNS, thereby, in turn, promoting DIT.[742] Harvard's Landsberg and Young have offered an interesting model of insulin-glucose-mediated hypothalamic function that is based on these observations.[742,743] On the one hand, according to their model, during fasting or caloric restriction, the very low level of glucose and insulin circulating in the blood diminish glucose metabolism that is normally mediated by insulin. This action is regulated by neurons of the VMH. Reduced glucose metabolism in the VMH *stimulates* the *inhibition* of SNS activity. Carbohydrate consumption, on the other hand, stimulates glucose metabolism that is mediated by insulin; this effect is also mediated by the VMH. Stimulation of glucose metabolism *depresses* the *inhibition* of SNS, resulting in a net gain in SNS activity and thermogenesis.

The advantage of the Landsberg and Young model is that it accounts for both substrate (glucose) and hormone (insulin) influences on thermogenesis stimulated by the SNS. It is suggested that other substrates, including the critically important free fatty acids, may invoke a similar CNS-mediated mechanism.

Obesity and Insulin Resistance

One consequence of obesity is the gradual development of insulin resistance as fat accumulates and glucose tolerance diminishes. Insulin resistance interferes with the uptake of glucose into liver and muscle cells for oxidation or storage as glycogen (obligatory thermogenic processes),[1173] and interferes with thyroid function.[1368] As discussed above, insulin resistance interferes with facultative thermogenesis by depressing activity in the VMH.

In a recent study it was found that obesity and insulin resistance exerted independent inhibition of the thermic effect of food.[1138] However, if a person exercises before a meal, the blunted TEF is more related to the person's obesity than to insulin resistance. The implications of these findings are not clear, but they suggest that mild exercise before eating may improve the metabolic fate of ingested calories in the obese.

Overfeeding

Overfeeding produces two distinct responses in BAT. The first, a trophic response, does not depend on the presence of insulin. The second is a thermogenic response that does require insulin for the full expression of sympathetic activation.[946,1396] Like thyroid hormone, the role of insulin in facultative thermogenesis may be viewed as mainly permissive.[221,1070]

Insulin resistance can be induced in normal individuals simply by overfeeding.[1345] Glucose intolerance develops in response to frequent snacking on foods with a high caloric content which prevents insulin levels from returning to normal fasting levels.[512] It is possible, with such bad habits, to keep insulin circulating in the blood almost 24 hours a day.[377] Overweight individuals tend to exhibit the net effect of prolonged exposure to high levels of circulating insulin: glucose intolerance, insulin resistance, a tendency to toward diabetes mellitus, and blunted thermogenesis.[160,545,897,1025]

Consumption of thermogenic may tend to reverse the above effects. An interesting case was observed by physician using the APRL program. An insulin-dependent diabetic obese woman was placed on an adrenal support program involving vitamin C and pantothenic acid and certain herbs for two months. During the next five months, she ingested a thermogenic/adrenal support product five days per week. Her diabetes went into remission and has not returned. Her weight loss was minimal during this period (15 lbs).

The maintenance of insulin sensitivity can be improved by avoiding the chronic consumption of junk foods, and by incorporating foods in the diet that help prevent insulin resistance, including sources of essential fatty acids, chromium, niacin, manganese, and nicotinic acid.[24,359,721,750,1089,1301]

Growth Hormone

Body composition is affected by growth hormone (GH) secreted by the pituitary gland. Animals and humans with hypopituitary tend to exhibit a low lean-to-fat ratio. Treatment with GH tends to increase the lean body mass and decrease body fat. These changes are accompanied by an increase in metabolic rate.[1440]

Of apparently greater importance than the absolute level of GH is the ratio of GH-to-insulin. The obese are characterized by high levels of insulin and decreased levels of GH. Restoring the ratio to its optimum value decreases body fat. Relative to one another, GH is lipolytic while insulin is lipogenic.

The idea of ratios among body hormones, the brain-child of Steven Woods of the University of Washington, deserves special mention.[1441] Too often researchers ascribe experimental results to increases or decreases in individual hormones, when the results are actually due to experimentally disrupted ratios among two or more hormones. Maintaining equal ratios in experimental procedures may produce similar results over a wide range of individual concentrations of in-

166

dividual hormones. Ignoring the importance of balance (ratios) probably accounts for much of the confusing data emerging from research on the biochemical properties of physiology and behavior.

The proper way to restore the GH-to-insulin ratio is to normalize insulin secretion in the obese; this would restore the proportion of GH to insulin and reduce fat stores. Dietary supplementation with nutrients listed above would not only help normalize insulin, but would also increase GH release. Added to that list could be l-lysine and l-arginine.[596]

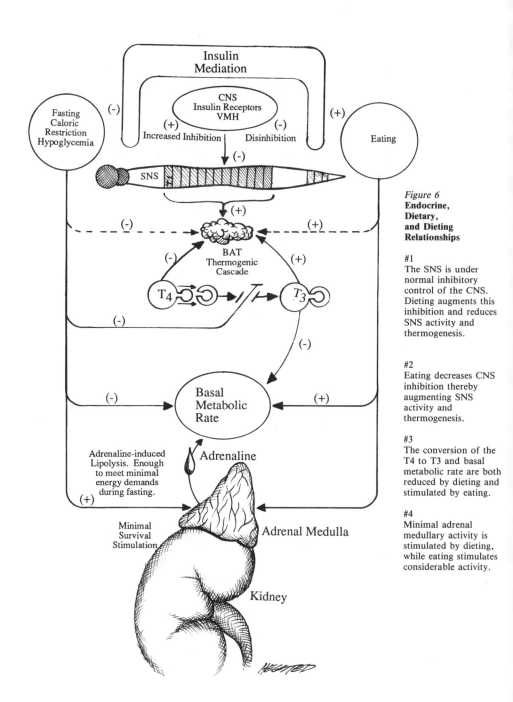

Insulin
Mediation

CNS
Insulin Receptors
VMH

(+) Increased Inhibition   (-) Disinhibition

Fasting
Caloric
Restriction
Hypoglycemia

(-)

(+)

Eating

(-)

SNS

(+)

(-)   BAT   (+)
Thermogenic
Cascade

(-)   (+)

T_4 T_3

(-)

(-)

Basal
Metabolic
Rate

(-)   (+)

Adrenaline-induced
Lipolysis. Enough
to meet minimal
energy demands
during fasting.

Adrenaline

(+)

Minimal
Survival
Stimulation

Adrenal Medulla

Kidney

Figure 6
**Endocrine,
Dietary,
and Dieting
Relationships**

#1
The SNS is under
normal inhibitory
control of the CNS.
Dieting augments this
inhibition and reduces
SNS activity and
thermogenesis.

#2
Eating decreases CNS
inhibition thereby
augmenting SNS
activity and
thermogenesis.

#3
The conversion of the
T4 to T3 and basal
metabolic rate are both
reduced by dieting and
stimulated by eating.

#4
Minimal adrenal
medullary activity is
stimulated by dieting,
while eating stimulates
considerable activity.

THYROID

Thyroid hormone is the principle hormonal regulator of obligatory thermogenesis. It also plays an important role in facultative thermogenesis.[526] Without thyroid hormone, obligatory thermogenesis is seriously impaired, and this deficit adversely affects the efficiency of facultative thermogenesis, both DIT and CINST. In ways not currently understood, lack of thyroid results in reduced responsiveness of BAT to the thermogenic effect of NE. The presence of thyroid, however, 'facilitates' thermogenesis.[1235,581,202,766,665,531] Stated another way, the presence of thyroid hormone 'permits' thermogenesis to occur. In the absence of thyroid hormone, animals die when exposed to the cold. Protection against cold exposure occurs primarily because of thyroid-augmented facultative thermogenesis.[206] In newborn humans, lipolysis is primarily under the control of the thyroid, shifting to catecholamines later in life.[830] Thyroid involvement begins in the anterior pituitary which secretes thyroid stimulating hormone (TSH). TSH acts on the thyroid in the typical manner, i.e., via a cAMP-mediated cascade that results in the production and release of thyroxine by the follicle cells of the thyroid gland.

T_4-T_3 Conversion

An interesting alpha-receptor mediated effect of NE on BAT involves the conversion of the major secretory product of the thyroid gland, thyroxine (T_4) to the thermogenically active form of this hormone, triiodothronine (T_3).[438,780,1165] BAT is rich in the enzyme that catalyzes that reaction, thyroxine 5'-deiodinase.[780] Both cold exposure and NE stimulate this reaction.[121-123,346] The concentration of the enzyme is directly under the control of NE, and can increase several hundred fold in a very short time when a person is exposed to cold or ingests a thermogenic agent.[674] Deiodinase activity is also influenced by insulin. Cold-adapted rats show an increased sensitivity of BAT to insulin that involves 5'-deiodinase activity.[870,1309,1310] In animals with insulin resistance, BAT shows a much lower reaction of 5'-deiodinase to cold exposure.[346,674,1368]

Within a few hours following cold exposure or consumption of the thermogenic, T_3 receptors in BAT become completely saturated. BAT is thus the target of T_3 circulating in the blood and of large quantities of the hormone produced in BAT itself.[1166]

Functions of T_3

Although the functions of T_3 have not been fully explicated, the hormone appears to play several important roles. It seems to regulate the hypertrophy or growth of BAT tissue under the influence of prolonged NE stimulation.[39,702,765,1148,1224,1366] Hypothyroid rats do not show the typical increase in GDP binding nor the normal increase in mitochondrial mass following cold exposure.[2] As previously noted, T_3 exerts a permissive influence on the thermogenic response to NE itself; it acts at several receptor and post-receptor sites in BAT to facilitate thermogenic biochemical events.[1148,1224,1366] T_3 accelerates the rate of oxygen consumption and heat production in most tissues of the body (exceptions being the spleen, brain and gonads). Thyroid hormones also stimulate catabolism of glucose, fats, and proteins by increasing the concentration of the many enzymes that catalyze these metabolic processes.[951] In BAT mitochondria, T_3 increases the amount of UCP.[121-124,1164] Thyroid hormone does not appreciably affect BAT metabolism when thermogenic requirements are small, but becomes increasingly involved as thermogenic requirements rise.[2]

Scientists are perplexed by some observations concerning T_3, however. For example, blood levels of this hormone are apparently sufficient for thermogenesis, and endogenous production is not necessary. So why does the proliferation occur? It is also apparent that T_3 does not exert a thermogenic effect on BAT by itself. Perhaps the action on UCP proliferation is the answer. Also, since T_4 prevents the activation of BAT under some circumstances, perhaps the conversion to T_3 is necessary to eliminate this suppression. Finally, although other aspects of BAT regulation exhibit circadian fluctuations, the activity of T_4 5'-deiodinase does not.[962]

Another question deals with the role supplemental thyroid hormone might play in activating thermogenesis. Hypothyroidism is a rare cause of obesity. Early weight gain in hypothyroidism is usually water retention, not fat accumulation. The use of supplemental thyroid hormone to promote weight loss is not a good idea, as this not only misses the point of the problem, but actually promotes the loss of lean body mass, instead of fat. Its inclusion in a thermogenic product would therefore be counterproductive for most people. Persons with diagnosed hypothyroidism, on the other hand, would probably respond well to restorative thyroid hormone therapy, as it would tend to restore obligatory thermogenesis against which the facultative thermogenesis could operate.

Furthermore, it has been found that hyperthyroidism, as might occur in normal individuals taking thyroid hormone, actually suppresses sympathetic nervous system activity in BAT;[711] this is a somewhat paradoxical finding since hyperthyroidism tends to increase basal metabolic rate and thermogenesis in skeletal muscle.[1244] Perhaps thermogenesis in skeletal muscle inhibits facultative thermogenesis in brown adipose tissue in the same way that a rise in core body temperature would. Another paradoxical phenomenon is that hypothyroidism is associated with the development of a strong synergy between alpha 1- and beta-adrenergic pathways that stimulate the activity of the enzyme that converts T_4 to T_3. How this is possible when hypothyroidism also attenuates the respiratory capability of brown adipocytes in response to NE is one of the intricacies of the thermogenic process remaining to be investigated by the research community.[206,940]

Unlike the influence of NE, which is short-lived, the influence of T_3 is long-lived. It may persist for hours or days. Also, unlike NE, which is water-soluble and can not penetrate the BAT cell membrane, T_3 is lipid soluble and hence can easily penetrate lipid-base cell membranes. T_3 does not interact with membrane-bound receptors. It is metabolically active only after diffusing into the cell and interacting with receptors on the inner side of the membrane.

Since T_3 does not interact with membrane receptors, its effects are not magnified like those of NE are. Thus, while just a few, short-lived NE molecules can exert an influence that is enormously out of proportion to their meager concentration, the effects of T_3 are limited to a one-to-one basis, depending directly on the amount of T_3 present. For this reason, perhaps, the concentration of T_3 in the BAT cell is typically quite large compared to other substances.

T_3 Enhances NE Influences on BAT

Ironically, NE actually exerts a positive effect on T_3 concentration and activity. Hence, this is yet another way that NE potentiates its own influence on brown adipose tissue. In fact it is the presence of thyroid hormone that may prolong the thermogenic action of a thermogenic agent. The initial action of the thermogenic would be to increase NE concentration. NE is short-lived but powerful in its action. It instigates a large, immediate thermogenic action, and then is inactivated via the numerous feedback mechanisms we have discussed in other chapters.

Meanwhile, under the influence of the thermogenic (or NE), conversion of T_4 to T_3 begins, and the thyroid gland begins to secrete additional hormone. Since this process is not under control of the SNS, but is a glandular function, it may take 1 -2 hours before enough T_3 accumulates in BAT to exert a significant effect. Then, the effect of T_3 may remain for several more hours, long after NE has been cleared from the system. It is known that thyroid hormone reduces the turnover rate of NE in cardiac and skeletal muscle, while an increased rate is found in hypothyroid rats. There exists, therefore, a reciprocal relationship between SNS activity and thyroid hormone activity. Since the production of heat in BAT is initiated by the release of NE from the sympathetic neurons, the general effects of T_3 on SNS activity should also influence brown fat thermogenesis. An important question is whether these T_3 effects are initiated quickly enough to impact the short-lived NE molecule.

172

The importance of possessing an active, healthy thyroid gland may be felt in continuing thermogenesis during the hours of the day in which no thermogenic is being consumed.

To enlarge upon a concept mentioned earlier, the increase in basal metabolic rate that occurs in hyperthyroid subjects decreases the need for BAT facultative thermogenesis. The demand for non-shivering thermogenesis is lessened and this is reflected in a reduced thermogenic capacity of BAT mitochondria. Thus, it is counterproductive for people with normal functioning, healthy thyroids to consume thyroid nutrients.

Persons with hypothyroidism, on the other hand, experience reduced thermogenic capacity mainly because thyroid-deficient BAT is characterized by diminished lipolytic responses to NE as a result of changes in membrane-bound receptor properties.[1026] Some research indicates that this is due to increased activity in the inhibitory alpha-adrenergic receptors, rather than decreased responsiveness of beta-adrenergic receptors. In other words, in the absence of thyroid hormone, T_4, NE has a tendency to combine more readily with the alpha-receptor that inhibits adenylate cyclase, rather than with the beta receptors that stimulate that enzyme. The end result is less conversion of TAGs to free fatty acids, and hence less thermogenesis. Again, the question of time-scale requirements for these actions remains unresolved.

In summary, it appears that both hypo- and hyper-thyroidism result in diminished thermogenic capacity.[2,590] In the absence of reasonable medicinal agents, the consumption of plant thyroid tonic (a substance that restores and maintains balance, or homeostasis) would be a logical dietary adjunct to the consumption of thermogenic in persons suspecting any type of thyroid problem. The best tonic candidate as of this writing is Fucus vesiculosis, or bladderwrack. This plant is a rich source of organic, or natural, iodine and other nutritive minerals, and has been used by Northern Europeans for centuries in the treatment of obesity and thyroid disease. Typically, there is not enough room in a capsule or tablet for all of the necessary thermogenic components and sufficient fucus to exert a noticeable effect on the thyroid. Hence, the use of separate fucus supplements is recommended. Kelp, another seaweed, is also an excellent thyroid tonic.

ADRENAL AND PITUITARY

The pituitary and adrenal glands are arguably the most important glands of the body. They are intricately involved in almost all life-sustaining processes. Their impact on BAT thermogenesis has not been thoroughly explored. Mainly, we know that if we remove them, the thermogenic process is disrupted. But the exact role each plays is still somewhat a matter of speculation.[146] In this section we will attempt to summarize the current body of knowledge concerning the impact of these glands on thermogenesis.

The adrenals are small kidney-shaped and kidney-sized glands situated on either side of the body just above the kidneys (they are sometimes called *suprarenal* glands because of their position atop the *renals* or kidneys). There are two major divisions to each adrenal gland: the cortex and the medulla. The cortex secretes corticosteroids, glucocorticoids, mineralcorticoids and sex hormones. The medulla mainly secretes adrenaline. When the secretions from the two parts impact on the same physiological process, they generally work in opposition to one another.

ADRENAL CORTEX

Along with the hormones from the thyroid and pancreas, adrenal cortical hormones also play an important role in BAT thermogenesis. Whereas insulin generally tends to stimulate thermogenesis, an excess of corticosteroids, especially the glucocorticoids (cortisol and corticosterone), released from the adrenal cortex tend to inhibit thermogenesis and cause atrophy of BAT, either directly or in a roundabout way through a CNS-mediated inhibition of the SNS.[34,35,415,566,574,575,576,701,773,897,1265,1375] The CNS component is undoubtedly mediated by suppression of corticotrophin releasing factor (CRF) in the hypothalamus, a known property of glucocorticoids.[35,770,797] (See Figure 5) The direct action of corticosteroids is currently the subject of some debate; glucocorticoid receptors have been found in BAT, but so far scientists have been unable to determine their role.[365,1373] The probable direct action of adrenal corticoids is to reduce adenylate cyclase activity in BAT. Clinically, the wasting of body fat and loss of body weight are associated with subnormal levels of adrenal corticosteroids, while high levels correlate with increased body fat storage, as in Cushing's Disease. Contrary to popular notions, true Cushing's Syndrome, caused by increased secretion of ACTH from the pituitary, is actually a rare cause of obesity. More common is iatrogenic Cushing's Syndrome that results from the prescription of glucocorticoids in the treatment of connective tissue problems, chronic lung and liver disease, and some blood disorders. Studies in genetically obese animals have uncovered deficiencies in the function and response to corticosteroids, and removal of the adrenal glands (adrenalectomy) is observed to help restore normal physiological parameters, such as lower weight gain, lower food intake, lower insulin resistance, and increased energy expenditure.[14,283,565,828-,829,1057,1081,1207,1267,1375]

By reducing activity in BAT, corticosteroids promote a positive energy balance that results in the accumulation body fat and to the accumulation of lipid in BAT cells;[415] adrenalectomy promotes negative energy balance by increasing activity in BAT. Adrenalectomy decreases glucocorticoids and increases CRF.[147,156] CRF in turn decreases food intake and increases SNS activity, perhaps through a serotonergic pathway.[769,1060] Thus adrenal corticosteroids appear to be very important energy-regulating hormones. Indeed, research results strongly support the idea that these hormones are influential in nearly all forms of obesity. Corticosterone, a principle glucosteroid, increases the supply of glucose to the blood. The brain is the primary beneficiary of this action. Blood glucose comes from the liver.

Cortisol, another glucocorticoid, inhibits glucose uptake by peripheral tissues, such as muscle and adipose tissue. It also causes an increase in the release of fatty acids from adipose cells. Under the influence of corticosteroids, dietary protein can be converted to glucose for transport, metabolism and storage as glycogen in the liver. The adrenal cortex is thus an important structure responsible for several metabolic processes that affect energy metabolism, adjusting hormonal and energy balances first one way and then the next. These

176

modifications are overlaid on thermogenic metabolism in brown adipose tissue. We are just beginning to tease apart how all of these processes interact to affect the onset of obesity. Interaction with sex hormones and other related factors are just beginning to be explored.[1192] Eventually, we will be able to manipulate adrenalcortical function through nutritional means in such a way to reduce fat stores and enhance BAT processes.

Sex hormones

Another important class of substances secreted by the adrenal cortex are sex hormones. These substances account for a great deal of the variance in the ability to gain and lose weight that exists between men and women. The differences between male and female hormones (androgens and estrogens, respectively) are profound. Because of the presence of androgens, male adipose tissue contains a greater amount of the enzymes that are involved in lipolysis than does female adipose tissue which, because of estrogenic influence, contains a disproportionate concentration of enzymes that promote lipogenesis (fat storage enzymes). The action of female hormone, therefore, is directly in opposition to that of most thermogenic substances. That's the bad news. The good news is that consumption of thermogenic, while it may not produce rapid fat reduction in women with high estrogen levels, will nevertheless counteract the tendency to fatten up in these women.

ADRENAL MEDULLA

During stress or exposure to cold, food or thermogenic, the adrenal medulla is stimulated to produce adrenaline, although this effect is not as strong as the effect on the SNS. It is generally thought that adrenaline levels increase dramatically after oral glucose administration in humans, yet some research shows that the ingestion of glucose (100g) actually suppresses adrenal medullary activity,[1294] perhaps through a negative feedback loop, as proposed by George Bray (see chapter on CNS). An increase would be elicited by a post-prandial decrease in plasma glucose. In contrast to NE secreted directly from SNS nerve terminals, the appearance of adrenaline is rather late in the thermogenic chain of events.[53] Yet it still makes an impact on facultative thermogenesis. However, it is not certain whether this component of facultative thermogenesis is present following administration of thermogenic agents that typically act independently of glucose metabolism.

Hypoglycemia and/or fasting have a stimulating effect on the adrenal medulla and a suppressive effect on the sympathetic nervous system.[225,689,736,1101,1387.1393] Apparently glucose utilization within the CNS is involved in the relationship between dietary intake and sympathetic activity.[741] However, since glucose levels are maintained within narrow limits in spite of widely varying dietary intake, it is more likely that insulin is the actual effector of CNS-mediation.

Stimulation of the sympathetic nervous system causes the release of NE from presynaptic sympathetic nerves and unavoidably also causes the secretion of adrenaline from the adrenal medulla. Circulating adrenaline plays perhaps a minor role in BAT thermogenesis, though its presence may help in the proliferation of lean body tissue in skeletal muscle. Also, an adrenaline-induced cascade regulates the synthesis and degradation of triglycerides, the storage form of fatty acids, throughout the body. The effects of adrenaline depend to some extent on the health of other endocrine functions. For example, if insulin levels are too high, adrenaline-induced thermogenesis, lipolysis and oxygen consumption are all inhibited. Thus under normal-insulin conditions, an infusion of adrenaline raised metabolic rate by 12.9%; this effect was reduced by one-third in the presence of elevated insulin levels.[1144] Adrenaline normally has only a minor direct influence on thermogenesis, though it may occasionally reach levels that do exert a direct thermogenic effect.[111]

In some genetic models of animal obesity (ob/ob mice), removal of the entire adrenal gland (medulla and cortex) increases thermogenesis in depressed BAT metabolism, with minimal effects in lean mice. This effect in the obese animals is completed reversed by administration of glucocorticoid, demonstrating that the adrenal cortex helps regulate BAT metabolism.[1326].

Adrenal Depletion

Prolonged adrenaline production may have adverse consequences on the human body. Adrenaline is essentially a stress hormone. Continuous adrenaline secretion could lead to depletion of adrenaline stores in the adrenal, thus blunting the body's ability to respond to the stresses of everyday living. Many Americans already have borderline adrenal exhaustion due to their hectic life style, management of life by crisis, consumption of less than wholesome foods, exposure to pollution of every conceivable variety, and so forth. The adrenal doesn't really need another form of stress. Yet that is exactly the risk posed by the consumption of thermogenic agents that lack an adrenal support component.

Furthermore, since one goal of thermogenic would be to prevent an excess of corticosteroids, continued consumption of such an agent could lead to atrophy of the adrenal cortex, a situation that could lead to hypoglycemia, and serious disturbances of electrolytes in the kidney and throughout other parts of the body. We feel that such a situations highly unlikely, but should be mentioned nonetheless.

Pharmacological/nutritional regulation of BAT thermogenesis should recognize the role of the adrenal cortex and medulla, and should provide support for the adrenal glands simultaneously with the stimulation of the SNS. Intervention should be aimed at normalizing adrenal functioning. Nutrients such as vitamin C and pantothenic acid, and herbs such as ginseng and licorice root, tone the adrenal cortex and also help maintain the health of the adrenal medulla.[980,1357] Failure to include adrenal support nutrients in thermogenic compounds is a common occurrence. One of our goals at APRL is to increase the awareness of the this problem in the minds of manufacturers as well as the public.

THE PITUITARY

Part brain and part gland, the pituitary is often called "the master gland" of the body. It is linked directly to the hypothalamus on the one hand, and to the entire endocrine system on the other hand. (See

Figure 5) It thus plays a pivotal role in interpreting outgoing (efferent) signals from the CNS, and initiating the appropriate hormonal changes. The hypothalamic/pituitary axis has the important job of maintaining homeostasis in both the psychological and biological sectors. Although no pituitary hormones have been shown to participate directly in BAT metabolism or hypertrophy in response to cold or diet, removal of the pituitary results in thermogenic activation and growth of BAT.[770] This suggests that under normal conditions, the pituitary exerts an inhibitory effect on BAT metabolism. However, it is likely that the positive effect that occurs following pituitary removal may be due to increased secretion of CRF in the hypothalamus that occurs when the stimulatory action of the pituitary on glucocorticoids is eliminated.

Under normal conditions the hypothalamus secretes CRF which stimulates the pituitary to secrete ACTH which in turn stimulates the adrenals to produce glucocorticoids. (See Figure 5) The glucocorticoids in turn inhibit CRF secretion from the hypothalamus. This mechanism is thus designed to be self regulating under normal conditions. Removing the pituitary from the loop eliminates the stimulation of the glucocorticoids and hence eliminates the CRF inhibition; CRF is therefore left free to stimulate BAT metabolism. Note that when the glucocorticoids are present, they also inhibit CRF effects on BAT metabolism.

By the same token, excess glucocorticoids can exert a greater than usual inhibition of CRF thus completely eliminating this one factor that normally produces some degree of thermogenic action. This is just one more reason why it is important to consume tonic nutritional substances that tend to stabilize or normalize adrenal activity, rather than substances that either just stimulate or just inhibit action.

Adrenal medullary activity is diminished in animal models of obesity and in obese humans. Likewise, the relationship of adrenaline to obesity is the converse of norepinephrine; the most obese have the lowest adrenaline levels. The significance of this observation has not been established.

Another important pituitary hormone, growth hormone or somatotrophin, is discussed in another chapter. Good pituitary health and functional balance can be helped through proper nutritional habits. Protein intake is important because several amino acids, especially arginine and lysine, are required for good pituitary health. Zinc is another important pituitary nutrient, whose lack in the diet can seriously upset the function of this important gland.

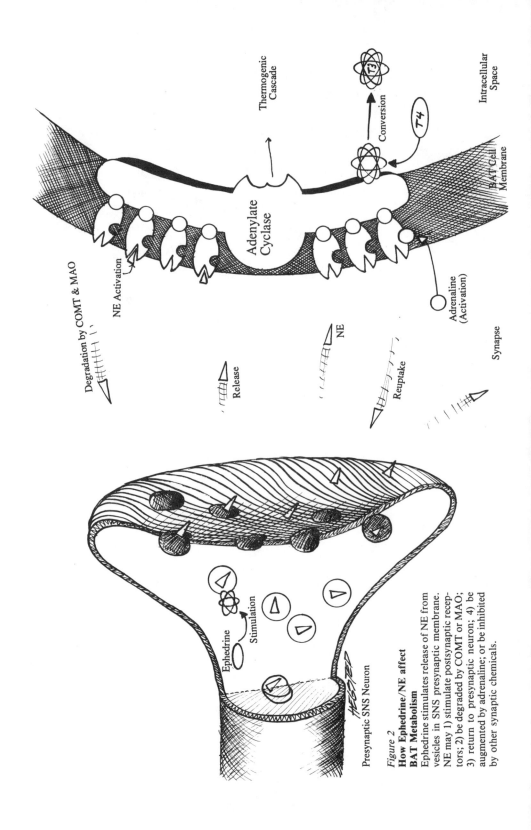

Figure 2

How Ephedrine/NE affect BAT Metabolism

Ephedrine stimulates release of NE from vesicles in SNS presynaptic membrane. NE may 1) stimulate postsynaptic receptors; 2) be degraded by COMT or MAO; 3) return to presynaptic neuron; 4) be augmented by adrenaline; or be inhibited by other synaptic chemicals.

Presynaptic SNS Neuron

Ephedrine

Stimulation

Thermogenic Cascade

Conversion

Intracellular Space

T3

T4

Adenylate Cyclase

NE Activation

Degradation by COMT & MAO

Adrenaline (Activation)

BAT Cell Membrane

NE

Release

Reuptake

Synapse

APPLIED
PHARMACOLOGICAL

The goal of the pharmacological manipulation of BAT thermogenesis is to reliably and safely improve body composition through the reduction in body fat. The search for substances that can achieve that goal is the subject of a large portion of the research in this area.[236,559,903] A wide assortment of natural and synthetic substances are contenders for the role envisioned by researchers. The current champions are ephedrine, caffeine and aspirin. In this chapter we will learn why these are the favorites, how they work, and what researchers have learned about their application in human clinical trials.

Under the direction of George Bray, the USDAFDA recently completed a review of the pharmacology of weight loss, with emphasis on appetite suppression. Bray's review of the pharmacological agents has become more-or-less the 'Bible' for the industry. It is noteworthy that he sees thermogenic drugs as among the most interesting agents currently being studied.[148b] He gives two reasons for his evaluation: 1) beta-2 and beta-3 adrenergic (read thermogenic) agonists tend to help repartition body composition, i.e., they increase the amount of lean tissue relative to fat deposits, or improve the lean/fat ratio; and 2) they increase energy expenditure and reduce food intake in much the same way that vigorous aerobic exercise does. Along the way, thermogenic agents enhance overall cardiovascular fitness, increase oxidative metabolism of free fatty acids by skeletal muscle cells, and slow bone loss. Since NE and adrenaline are the primary mediators of the effects of exercise, thermogenic agents that enhance NE and adrenaline metabolism would demonstrate similar properties.

Bray admits that ephedrine (ma huang) is the only available agent in this class, but bares his pharmaceutical soul in suggesting that the newer synthetic pharmacologic agents are more interesting than ephedrine. Truthfully, should we find a pure beta-3 agonist, be it syn-

183

thetic or natural, it would quickly replace ephedrine or ma huang as the agent of choice. While the pharmaceutical companies pursue the development of synthetics, APRL continues to search for natural alternatives.

On the whole, Bray is not overly gratified with how things have been. He sees the traditional pharmacological treatment of obesity has having not met with more than a meager amount of success. Aside from political and cultural factors, he admits that most drugs produce only minimal weight loss at the cost of a variety of unwanted side effects. Nor is he alone in this opinion Many other experts in the field of weight loss share Bray's views. The search for an ideal anti-obesity drug continues—a drug that has no potential for abuse or dependence, can be consumed safely for long periods of time, maintains effectiveness with long term use, and is truly effective at improving body composition through the reduction of significant amounts of fat with only minimal loss of lean body mass.

Current Drugs of Interest

Most current research focuses on a handful of drugs with appetite suppressant properties, while lesser amount of attention is devoted to non-prescription drugs that dilute calories. The targets of most anti-obesity drugs are the catecholamine neurotransmitters NE and serotonin. Amphetamines, methamphetamines, phemetrazine, phendimetrazine, benzphetamine, diethylpropion, mazindol, phentamine, phenylproponolamine, fenfluramine and dexfenfluramine account for the vast majority of research trials. Scattered research is done on the mixed agent sibutramine and the nutrient absorption inhibitors orlistat and acarbose. In addition to these drugs, an increasing amount of attention is being paid to new synthetic thermogenic drugs, such as BRL 26830A.[239]

Ironically, very little research is devoted to the one substance that most closely fulfills the criteria for an ideal anti-obesity drug: ephedrine itself. Why is this so? The reason is transparently simple: Ephedrine, as a proprietary drug, long ago entered the public domain. Hence there are few monetarily rewarding incentives for studying it.

Taking all of the above factors into consideration—namely, the difficulty of obtaining an ideal drug, the shortcomings of current drugs, the pertinence of thermogenics, the absence of profits to be had from ephedrine research—we can now understand why Bray is excited about the field of thermogenics and so unexcited about ephedrine as the agent of choice. Ma huang would be automatically discarded as an agent of choice for the same reasons. Being a plant material, it is not patentable and hence not a profitable subject for study. It is ironic that ma huang's chief active principle, ephedrine, is unattractive for similar reasons. Yet, these substances are currently the backbone of all current thermogenic clinical work, and have been proven to be both safe and effective.

Once the new thermogenic process that impacts so dramatically on fat management and weight regulation was discovered, it was only a matter of a very short time before scientists began investigating the subject in earnest. Actually, the finding that ephedrine produced losses in fat was observed *before* many fundamental aspects of thermogenesis were determined. At any rate, to date, literally hundreds of studies have been carried out that were dedicated to discovering pharmacological ways to stimulate thermogenesis in brown adipose tissue, and other tissues.

One important observation of these studies is the presence of a difference between the adrenergic events involved in thermogenesis and those involved in cardiac and pressor effects. Agonists, or stimulators of post-synaptic adrenoceptors have been shown to be much less effective in stimulating thermogenesis than agents which simply increase synaptic NE concentration (ephedrine, yohimbine, tranclcypromine).[89,331,687] The more specific agonists, however, exhibited greater cardiovascular effects at lower doses than required to produce significant thermogenesis. Elimination of classical adrenoceptor stimulants drastically reduced the number of viable agents for use in stimulating thermogenesis. So far, just one patent exists on a combination of agents that are both safe and effective: namely ephedrine, caffeine and aspirin and their equivalents. This state of affairs is, of course, very enticing to the pharmaceutical industry which attaches great significance to research aimed at devel-

185

oping novel synthetic drugs. Companies are no doubt seeking to find novel chemicals or natural plant materials that will match or exceed the effectiveness of that trio, agents that will promote thermogenesis without concomitant cardiovascular side-effects.

Sympathomimetics

For purpose of discussion, pharmaceutical thermogenic agents can be divided into two categories: sympathomimetics; and substances that improve the action of sympathomimetics. Fortunately, representative agents from these classes have been discovered that not only work, but which have a combined synergy that is nothing short of amazing.

The first class of thermogenic agents was an obvious choice: sympathomimetics, substances that enhance the activity of the sympathetic nervous system, or mimic its action themselves (true sympathomimetics only mimic actions of the SNS, but the class is often broadened to include substances whose primary action is stimulating the SNS). Since almost all aspects of thermogenesis are impacted by norepinephrine, it is logical to investigate substances known to increase the amount norepinephrine in brown fat. Ephedrine is the prototypical sympathetic nervous system activator, and after much research, remains the agent of choice. Synthetic analogues, novel beta-agonists and other agents have not proven as effective or as safe as ephedrine.[7,27,30,37,38,74,143,213,239,237,322, 521,694,1115,1327] Synthetic agents are continually being explored in pharmaceutical research, but history shows that man-made moleculed always have an uphill battle when competing with an effective natural substance.

Improving the Action of Ephedrine

The second class of pharmaceutical thermogenic agent is used to inhibit some of the natural biochemical events that reduce the effectiveness of NE. A problem with the early use of ephedrine in thermogenic research involved the fact that its effects on NE production were considerably attenuated by negative biofeedback mechanisms. In other words, natural cellular processes were in place that reacted to the presence of NE by shutting it down. These inhibiting processes were so effective that a good thermogenic effect could only be achieved by administering roughly five times more ephedrine than was safe. This fact hampered ephedrine research in humans for decades. Much research focused on ways of prolonging the activity of ephedrine on NE synthesis, release and half-life. Agents were sought which inhibited the substances that inhibited NE activity. This is frequently referred to as disinhibition of NE activity.

Since NE was under negative feedback control, those feedback mechanisms had to first be identified and elucidated. The next step was to find substances that disrupted the negative feedback loops, thereby prolonging the length of time a given NE molecule was able to effectively stimulate the events of BAT thermogenesis. Some of the first negative feedback mechanisms to be discovered involved the presence of adenosine and prostaglandins in the synaptic cleft, as well as cAMP phosphodiesterases inside the cell.

The following sections discuss the major research findings and mechanisms of action involving the major pharmaceutical-grade substances that affect thermogenesis: ephedrine, xanthine (caffeine) and aspirin. We will note indications of synergy, elucidate modes of action, and present summaries of pertinent research. These sections are not meant to exhaust theoretical thought on thermogenesis, but should give the reader some idea of the different approaches being taken by the major research teams.

EPHEDRINE

The first suggestion that ephedrine might be useful as an anti-obesity thermogenic agent was made in 1977 by Derek Miller of Harvard University.[358] This suggestion has had a dramatic impact on all subsequent thermogenic research. Miller's own observation confirmed that the administration of ephedrine decreased body fat accumulation in lean subjects and produced loss of body fat in obese subjects. Other researchers, especially Arne Astrup of Copenhagen, have reported similar results.[69,973]

As repeatedly stated throughout this book, the primary agent for reactivating and maintaining healthy thermogenesis is ephedrine (or its natural equivalent, ma huang, a Chinese herb that has been used for thousands of years by people throughout the world in the treatment of asthma and other respiratory conditions). No attempt to find an adequate substitute has yet succeeded. Ephedrine is so important because it directly stimulates activity in the sympathetic nervous system that leads to the secretion of NE in BAT tissue, and then initiates the major work of thermogenesis, including the stimulation of UCP and lipolysis.[62,64,65,72,358,896]

In the earliest clinical trials utilizing ephedrine to help promote weight loss, the positive effect was attributed to the chemical's ability to curb appetite.[825] However, it was apparent to other researchers that ephedrine administration elicited a definite thermogenic response.[59,358,865,896,970,974] The increased metabolic rate observed in early animal studies that consumed ephedrine was attributed to increased thermogenesis.[36,836,1371] This kind of work suggests that consumption of ephedrine, or perhaps other sympathomimetics, serves at least two useful functions: 1) stimulation of thermogenesis in BAT; and 2) stimulation of BMR which may offset the detrimental effects of dieting.[1156]

EPHEDRINE:
CELLULAR MECHANISM OF ACTION

The primary cellular actions of ephedrine include the following (note that all but the first of these is the result of action number one).

1. Stimulation of release of NE from SNS presynaptic terminals into the synapse, inhibition of reuptake of NE by the presynaptic membrane, and release of adrenaline from the adrenal medulla.[749,1289,1290] These two actions initiate the following series of reactions:

2. Activation by NE of beta-adrenergic receptors on BAT cell membranes,[174,1219] followed by:

3.   Activation of adenylate cyclase in BAT cell membranes, which results in:

4.   Increased production of cyclic AMP in BAT cell cytosol, which is significantly involved in:

5.   Lipolysis, the creation of free fatty acids (FFA) from triacylglycerol (TAG) stores, glucose, and chylomicrons, obtained from white fat stores, dietary fats and from TAGs in brown fat itself. FFA are, of course, the fuel for thermogenesis.[926,1202]

6.   Increased metabolism in the mitochondria of BAT cells (with the arrival of FFA)[64,65,825,835,898] which results in the automatic activation of the uncoupling protein.

7.   Accumulation of mitochondria, UCP and BAT tissue with repeated stimulation.

8.   Sensitization of thermogenic events, and tolerance to cardiovascular events, with repeated stimulation.

9.   Beta-2 adrenergic mediation (NE and adrenaline) of protein synthesis.[220]

10. Alpha-adrenergic mediated stimulation of thyroid metabolism.

11. Production of facultative and obligatory thermogenesis in skeletal muscle and the viscera which is, in fact, much larger than the BAT component.[233,251,485]

In addition to these cellular effects, mention should be made of other actions of ephedrine. One, the substance is a safe and reliable bronchodilator used successfully and commonly in the treatment of asthma.[1027] Two, it has a distinct vasoconstrictive action that helps dry up sinuses during hay fever season. Third, it helps curb appetite by redirecting the flow of blood away from the abdomen, and toward the muscles. The increased flow of blood to the muscles increases the supply of oxygen and blood-borne nutrients, resulting in enhanced energy capacity.

Ephedrine does not exert a direct effect on any of the events underlying thermogenesis. Rather the effects of ephedrine are mediated by norepinephrine released from terminals of SNS neurons, and by other catecholamines, especially adrenaline released in the blood stream by the adrenal medulla.[335]

Adrenergic Receptors

At the BAT cell membrane, NE interacts primarily with beta-adrenergic receptor sites. Much early work on thermogenesis was moderately confounded by belief that there were only two kinds of beta-adrenergic receptors, beta-1 and beta-2. Yet the effect of ephedrine did not conform to expectations, producing fewer cardiovascular side effects and greater thermogenesis than anticipated. It wasn't until the discovery of a novel beta-adrenergic receptor (beta-3) on BAT membranes that was selectively sensitive to NE, that scientists obtained a more complete profile of the actions of NE in BAT.[350]

The discovery of beta-3 receptors explained why several potential thermogenic activators were ineffective (they worked only on beta-1 and beta-2 adrenoceptors), and it helped explain why substances that worked directly on the postsynaptic membrane (the membrane of the BAT cell) were less effective than substances that simply increased the level of NE in the synaptic cleft. This discovery also required that pharmaceutical companies direct their research attention to the development of novel beta-3 agonists, or restrict attention to the development of compounds that behaved like ephedrine. So far, none of the novel beta-3 agonists has been able to match ephedrine for both efficacy and safety.[237,517,816]

Except for its initial beta-1 and beta-2 effects, ephedrine is an almost ideal substance for stimulation of BAT metabolism. Not only does it produce a rise in norepinephrine levels through stimulation of the sympathetic nervous system, but it also promotes lipolytic degradation of lipids into free fatty acids for use as the primary fuel of thermogenesis.

CAs Affect DIT, CINST, and TEF

Ephedrine apparently potentiates not only facultative thermogenesis (DIT and CINST), but also impacts on obligatory thermogenesis (TEF) and raises basal metabolic rate. The obligatory effect is probably the result of stimulation of the adrenal medulla to secrete adrenaline which exerts the obligatory thermogenic effect in skeletal muscle, liver and elsewhere. Adrenaline is also a specific agonist for all types of beta receptor types, though its effect on BAT cell thermogenesis appears to be quite small. The action of adrenaline is also exerted to a small degree on the beta-3 receptors. The thermogenic action does not fade with time, as do the cardiovascular properties.

Protein Synthesis

The combined effects of NE and adrenaline on beta-2 adrenoceptors probably enhance the thermogenic action and also result in another beta-2 mediated event—the stimulation of protein synthesis which helps counteract any tendency toward loss of lean body mass as might occur during a calorically restricted diet; the research indicates, in fact, that these effects help in the accumulation of lean body tissue to help replace the loss of white fat.[971]

Other Effects

Certain alpha-adrenoceptor actions are also stimulated by NE and adrenaline, including the activation of thyroxine deiodinases that lead to the conversion of the main thyroid hormone, T_4 to the metabolically active form T_3, whose influence would enhance adrenoceptor sensitivity to the thermogenic effects of NE.[64]

It should also be remembered that the majority of the thermogenic action of ephedrine is obligatory and comes from skeletal muscle and the viscera.[62] Only about 10% of the thermogenic action is facultative and occurs in BAT. But when BAT fails, it is the caloric energy from that same 10% that gets deposited as white fat.[318,327] Furthermore, as BAT tissue begins to grow and proliferate under the

191

influence of prolonged stimulation through the daily ingestion of thermogenic, that 10% may increase to 15 or 20%, or perhaps even more.

PHARMACOLOGY OF EPHEDRINE

Research has shown that ephedrine stimulates thermogenesis and lipolysis in laboratory animals[36,835,836] and in man.[59,64,65,62,174,358,331,835,896,969] It appears to reduce body fat content and thereby body weight.[36,322,325,1371] Ephedrine stimulates the uptake of oxygen in BAT,[391,174,1069] and thereby increases lipid oxidation.[65] Animals made obese through lesions, drugs or other manipulations and subsequently treated with ephedrine have been shown to lose body weight and body fat, accompanied by increased oxygen consumption, without increasing food intake.[36,322,835,836,1316,1371,1390] The weight loss promoting effect of ephedrine is restricted to body fat; loss of body protein is not observed.[322,36] Ephedrine administration produces secretion of norepinephrine from sympathetic neural tissue and performs as an agonist for adrenergic receptors.[174] These actions, of course, both lead to an increase in intracellular cAMP and the initiation of the thermogenic cascade.

It might be asked whether the obese are deficient in NE or simply insensitive to it. The studies cited above indicate ephedrine stimulates the accumulation of NE which acts as the trigger for the thermogenic cascade. Other studies show that obese humans and animals respond normally to the direct infusion of NE.[273,321,1064] It appears, therefore, that it is the absence of NE, not a lack of sensitivity that leads to deficient thermogenesis.

In man, ephedrine induces greater loss of body weight than placebo in diet-restricted obese subjects,[825,64,974,358,896] though not all studies have observed this effect.[968] In one study, the effects of ephedrine were studied in 10 overweight and obese women on low-energy adapted diet therapy (1000-14000 kcal/day).[972] A dose of 50mg ephedrine, 3 times a day, for two months produced significant weight loss in this group of patients. No clinically significant side effects were observed. In another study, the effects of ephedrine on the thermogenic effect of a single meal were studied. The substance elicited lower sympathetically-mediated thermogenesis than placebo. The authors concluded that energetically-efficient humans exhibit lower SNS-mediated thermogenesis in response to a meal than do energetically-inefficient ones. The idea that obesity is the result of greater efficiency and that staying thin is the result of greater inefficiency flies in the face of Puritan work ethic logic that demands that only good things can result from efficiency—but the logic fails when it comes to fat metabolism. Chronic administration of ephedrine enhances thermogenic respiration in BAT by stimulating beta adrenoceptors.[174] without adversely affecting blood pressure and heart rate.[825,64,974]

In one study, ephedrine was administered to human patients for three months. During this time, it elevated and maintained metabolic rate at about 10% above normal. Plasma adrenaline levels were elevated by 87% during treatment. Most importantly, substrate utilization switched from glucose to lipids. That is, body tissue began to select lipids as the primary source of energy rather than the more readily available glucose derived from glycogen stores.[65]

Chronic administration of ephedrine sensitizes rats and mice, and humans, to the effects of cold, diet and drug/plant induced thermogenesis. This increased responsiveness of BAT is due to the growth of BAT, and increased concentration of UCP.[83,169,898,1399] One of the most dramatic evidences of this in humans occurs in female Korean pearl divers acclimated to cold water stress; these persons exhibit a greater enhanced thermogenic response to NE.[670]

Obese humans exhibit a reduced sensitivity to ephedrine.[663,896] This suggests that the obesity may be due to impaired thermogenesis.[616,1066] However, until recently there was a question about whether impaired thermogenesis caused obesity, or obesity created a defect in thermogenesis, or whether both conditions were the result of a third factor. This question was addressed in research conducted in France, and described in the opening paragraphs of PART TWO. These scientists found that the defect in thermogenesis *preceded* obesity, and was present at the onset of obesity to about the same degree as later on.[729] The study didn't determine if some third factor created the defect in thermogenesis. Research on the genetic basis of thermogenesis has strongly suggested the presence of a predisposing hereditary trait as the initial instigator. Environmental factors arising from the lifestyle factors of food, dress, housing, exercise, etc. may also be predisposing factors.

Site of Ephedrine-induced Thermogenesis.

There is some confusion in the literature concerning the site of thermogenesis in humans induced by the diet and catecholamines such as ephedrine and adrenaline. Some of the confusion arises from differences in the subjects utilized. Some studies used lean people, while others used obese patients. If we assume that defects in BAT thermogenesis are responsible for obesity, then there will undoubtedly be differences in brown adipose metabolism between the two groups. Conclusions about the nature of thermogenesis based on the two samples will probably contain contradictions. Other theoretical and factual inconsistencies can arise from comparing data derived from animal studies with human-derived findings.[1044] Furthermore, the procedural problems involved in doing research on humans introduce systematic errors that can exceed any real differences that might exist between different experimental groups and between experimental groups and control groups.

These procedural and analytical problems have led the Danish group, under the direction of Arne Astrup, to conclude that the major site of facultative thermogenesis is skeletal muscle, liver, heart and white adipose tissue.[62,63] In their view, thermogenesis in BAT could account for a maximal 15% of ephedrine-induced increase in whole body oxygen consumption, or, in other words in BAT thermogenesis.[53] The Harvard/Swiss groups, including Landsberg, Young, Miller and Dulloo, meanwhile, have consistently followed the original theory of Rothwell and Stock who promoted BAT as the major site of clinically meaningful thermogenesis.[438,1066] Jean Himms-Hagen of the University of Ottawa has argued for a correlation between faulty brown adipose tissue thermogenesis and obesity.[527,537,534] Other re-

194

searchers typically line up on one side of the debate or the other,[814] or cite other anatomical areas for the occurrence of significant portions of the thermogenic response.[234]

One way to resolve the discrepancy has been continually promoted in this book; namely, that while it is true that most whole body thermogenesis occurs in non-BAT tissue, it is the increment occurring in BAT that is related to the onset of obesity. Non-BAT thermogenesis is a normal response to diet, cold and other forms of catecholamine stimulation. Any defect in thermogenesis in non-BAT tissue would probably result in serious hazards to health. Hormonal influences from the thyroid, pancreas, pituitary and adrenals would necessarily be corrupted. On the other hand, defects in BAT thermogenesis would only impact the energy balance or thermal balance of the body. Minor defects in BAT metabolism would contribute to slow fat accumulation such as is most commonly experienced by humans. The onset of defective BAT-mediated obesity would not produce concomitant symptoms of a serious physiological nature. The onset of defective non-BAT-mediated obesity, however, would probably be accompanied by additional signs and symptoms characteristic of serious health problems.

A common observation of the Astrup group is that body heat is elevated over the areas of the body thought to contain white fat, but not over the areas of the body thought to be the repositories of brown fat.[53,1066] They interpret this as a lack of evidence for BAT thermogenesis. However, it is known that BAT is much more heavily innervated by blood vessels than WAT. Furthermore, blood flow in BAT is greatly increased in BAT following ephedrine consumption. The observation that body surfaces over BAT deposits remain relatively cool following ephedrine stimulation may reflect a substantially greater capacity for dissipating heat in BAT as compared to WAT. As warmed blood is carried away from BAT, the temperature of immediately surrounding tissue would tend to remain normal. Astrup and other researchers have measured *increases* in temperature of BAT tissue following ephedrine administration. Astrup has interpreted this increase in temperature to be blood-flow dependent and not a function of thermogenesis.[61,62] However, as pointed out above, even Astrup admits to the possibility of a maximal 15% increase in BAT thermogenesis. This 15% increase, coupled with enhanced blood flow through BAT, could account for the heat measured in BAT.

In summary, while the debate over the site of thermogenesis continues, we believe that whole body thermogenesis is carried out in a variety tissues, including skeletal muscle, liver, heart, white adipose tissue and brown adipose tissue. While the majority of thermogenesis is generated by non-BAT organs, a physiologically meaningful amount occurs also in BAT. The relative contribution of BAT to the various forms of thermogenesis has not been determined. Thus, DIT may involve more BAT metabolism than CINST. Acute ephedrine-induced thermogenesis may encompass greater skeletal muscle activity than CINST, but chronic ephedrine-induced thermogenesis may depend more heavily upon BAT metabolism.

EPHEDRINE SAFETY: GENERAL CONSIDERATIONS

Although therapeutic doses of ephedrine are free of serious toxicity and carcinogenicity,[942] the administration of ephedrine typically produces an array of side effects that resemble sympathetic nervous system stimulation, though on a much smaller scale than observed with all other popular sympathomimetics.[26,123,216,272,833,858,985,998,1059,1122] The amazing thing in the thermogenic studies is that when administered to people who are over fat and whose thermogenic capacity appears to be insufficient, i.e., in people who are suspected of being obese because of deficient thermogenesis, ephedrine consistently fails to elicit the same set of side effects, or the same degree of intensity of side effects as would be expected of the normal population. For example, in one study of three months duration, involving daily consumption of 60mg, only transient hand tremor was reported in a couple of subjects during the first few days, but the treatment failed to increase heart rate or blood pressure.[64] In another study, the incidence of side effects was very rare, with only a few patients complaining of agitation, insomnia, palpitation, headache and giddiness.[974]

As impressive as those studies are, there remains a strong aversion among doctors to prescribe ephedrine for weight control, as it is clear that larger amounts of ephedrine would probably elicit stronger side effects.[520,595,806,1003] Therefore, until the synergy between ephedrine, caffeine and aspirin was discovered, most authorities rejected ephedrine as a safe thermogenic agent.

Differential Development of Tolerance

One positive feature of ephedrine that is difficult to match by synthetics is that tolerance develops to its already mild beta-1 and beta-2 mediated actions on heart-rate and blood pressure, but not to its beta-3 mediated thermogenic effects.[397,922,923] That is, with prolonged use, the body adapts to the cardiovascular effects by no longer exhibiting any tendency to high blood pressure or increased heart rate; but the body does not adapt to the positive effect on BAT thermogenesis. Instead, prolonged consumption of the thermogenic actually increases the body's thermogenic capacity.[83,1399] Prolonged exposure to the cold has the same effect.[169,670] Similarly, in endocrine disorders, such as pheochromocytoma, in which the concentration of serum catecholamines (e.g., NE) are high, the thermogenic response is enhanced.[755,1049] As stated earlier, the greater capacity is due to growth of BAT and increased concentrations of mitochondria and UCP.[898]

This selective adaptability was first observed by Arne Astrup's group at the University of Copenhagen; these researchers were conducting trials to determine the long-term effects of ephedrine consumption on thermogenesis in man.[64] The investigators administered ephedrine to five over fat women (14% over the average weight for women of comparable height and age) for three months. The amount consumed was 60mg per day (20m three times per day). The women complained of having a weight problem despite consuming low calorie diets. Before, during, and after, the treatment period, the subjects were required to ingest a test amount of ephedrine as a single dose (1mg/kg) and the thermogenic effects of the single dose were measured. Before treatment and 2 months after cessation of treatment, a similar thermogenic response to the single dose of ephedrine was observed. However, after 4 and 12 weeks of treatment, the single dose elicited a more sustained response, as measured by oxygen consumption. Thus, the thermogenic response to ephedrine actually increased during the treatment period. Meanwhile, all patients adapted to the cardiovascular and other metabolic effects of ephedrine in the expected manner. There was thus an up-regulation of the thermogenic response and a down-regulation of side effects during treatment.

This sensitization to ephedrine is of profound importance. It means that long term use of ephedrine will continue to produce thermogenesis. Furthermore, it means that most of the cardiovascular side effects of ephedrine will subside, making the product even safer to use for extended periods of time. The question of just how long humans can or should consume ephedrine on a daily or semi-daily basis has not been answered, but it appears probable at this time that people can use the product long enough to accomplish effective fat loss. Astrup's group observed that average weight loss among the five women was 5.5 kg after 12 weeks, and the body fat content was lowered by 5.2% of the body weight. However, 2 months after cessation of the treatment, there was an average regain in body weight of 0.5 kg.[64] This isn't a bad number, but the question of permanence of results has yet to be answered.

However, even though significant adaptation to the cardiovascular side effects occurs over time, there is still much concern over the effects on the patient *before* adaptation sets in. Finding the appropriate administration schedule for each patient could prove very difficult, and most people, even the very obese, hesitate to take something that would increase their anxiety, blood pressure and heart rate. This situation seems to call for an intensive search for a drug that would selectively stimulate the newly discovered beta-3 adrenergic receptors in BAT without stimulating the other beta receptors. Practically speaking, however, the urgency of that need has been seriously jeopardized by the discovery of the caffeine/aspirin/ephedrine synergy. Still, since the area is loaded with the potential for great financial gain, scientists continue the search for novel, and therefore patentable, beta-3 agonists.

EPHEDRINE + CAFFEINE

As mentioned above, one of the problems associated with the use of ephedrine in the activation of thermogenesis is the toxicity associated with a dose high enough to do the job. The high dose is required because of the presence in BAT tissue of negative feedback control mechanisms that dampen ephedrine-stimulated NE actions. These inhibitory feedback mechanisms are designed to enhance the efficiency of the body's energy storage mechanisms by keeping BAT activity under control. Prolonged activity in BAT would burn up more calories than the central nervous system control and command center typically feels is safe and healthy. Frustrating these control messages by decreasing the effectiveness of the feedback control mechanisms is the goal of pharmacological manipulation. In the short run, it is probably a much easier task than trying to understand and manipulate the CNS mechanisms involved.

One of the earliest substances with ephedrine-enhancing properties to emerge from the research was caffeine.[57,70,826] Caffeine is the most common member of the methylxanthine family of alkaloids.[177] These alkaloids occur in a wide variety of plants. Each differs from the next to some extent, and the overall action of the plant on human physiology is not only a function of the specific xanthines present in the plant, but also of the precise *ratios* among those xanthines. At the present time, caffeine appears to be the preferred xanthine for thermogenic enhancement. Preference for this substance is based on its inherent capacity for promoting thermogenic synergy while exerting a minimum of adverse side effects. Furthermore, it works at lower doses and exhibits a better synergy with ephedrine than do other naturally occurring xanthines, such as theophylline, theobromine, and various purely synthetic methylxanthine analogues of the natural substances.[64,326,883,896] Note: The xanthine that occurs in an herb called yerba mate (discussed in the next chapter) is very similar to, if not identical to, caffeine. It is sometimes called mateine. Although research at ARPL suggests that this xanthine may be better than caffeine from other sources, rigorous experimental trials have not yet been conducted on this material.

199

CAFFEINE:
CELLULAR MECHANISM OF ACTION

Following ingestion, caffeine is rapidly and completely absorbed from the gastrointestinal tract and remains active in the body for from several hours to several days, depending on age, sex, hormonal status, medication regimen, smoking status, prior use, and probably other factors.[3,12,713,1051,42,134,224,634,668] The fundamental cellular effects of caffeine (and other methylxanthines) include the following:[gen.ref.:41,42,92,185,479,1011,1229]

1.   Accumulation of cyclic AMP in the cytosol of BAT cells.

2.   Blockade of adrenergic receptors on BAT cell membranes, including alpha-1 and alpha-2 adrenoceptors.

3.   Translocation of intracellular calcium.[1012]

4.   Potentiation of substances that inhibit prostaglandin (PGE2) synthesis.[317]

5.   Reduction of uptake of NE by the presynaptic membrane of SNS neurons.

6.   Inhibition of substances that metabolize NE in the synapse, thereby prolonging NE half-life, and otherwise increasing plasma NE levels.[395,1053]

7.   Inhibition of adenosine and phosphodiesterase (PDE) in BAT cell membrane, which results in canceling the inhibitory effect of these substances on adenylate cyclase, which in turn results in the accumulation of cyclic AMP.[184,218,336,396,663,967]

8.   Stimulation of NE release from SNS presynaptic membranes.[8,105,115,433,664]

9.   Increasing fasting, or basal, metabolic rate by approximately 5 - 20%.[8,662,664]

10.   Stimulation of adrenaline release by the adrenal.[104,105,114,782]

11.   Does not interfere with fatty acid oxidation, but has conflicting actions on the net production of FFA. On the one hand, it stimu-

lates lipogenesis;[276,357,601,943,1019] on the other hand, it stimulates lipolysis.[103,129,1010] Supposing both actions occur simultaneously, the net effect would be heat production along the lines of futile cycling mechanisms.

Adenosine Inhibition

In several of the above actions, the effect of caffeine is to counteract the effects of adenosine, PDE and PGE2. Adenosine inhibits NE-induced lipolysis, reduces the release of NE from SNS terminals, inhibits the release of NE and other catecholamine, and even non-catecholamine in the CNS.[510,558,803] Adenosine also paradoxically increases certain alpha-adrenergic actions of NE, leading to the accumulation of cyclic AMP.[180] Two types of adenosine receptors have been identified on the basis of whether their activation produces stimulation or inhibition of cyclic AMP synthesis.[41,802,1208,1312,1332] Thus adenosine both stimulates and inhibits the accumulation of cyclic AMP.

A role for the importance of adenosine as a regulator of lipolysis in human obesity is supported by the following observations:

1. Adenosine exerts less antilipolytic action in visceral adipose tissue than in subcutaneous fatty tissue.[1319]

2. Visceral fat is more important than other fat in the etiology of abdominal obesity.[128]

3. The antilipolytic action of adenosine is reduced in obesity.[947]

4. Fasting and dietary composition fail to influence adenosine action on receptors.[681,682]

5. Adenosine definitely opposes lipolysis in human adipose tissue.[805]

NE also both inhibits and stimulates the production of cyclic AMP; stimulation comes via beta-adrenergic receptors on BAT cell membranes, while inhibition comes from an effect on certain alpha-receptors. Caffeine augments the actions of NE at both beta- and alpha-adrenoceptors, and competitively antagonizes the actions of adenosine.[25,269,395,398,431,478,664,683,1186,1187,1250,1362] (See Figure 3.) Thus, the net result of the influence of caffeine on BAT metabolism will be a factor of the sum of its opposing actions on adrenergic/adenosine receptors as well as a factor of other metabolic events that are either aroused or suppressed as a result of the adrenergic actions on receptor sites. For example, say that all adrenergic events cancelled out, would this not result in a greater weight or significance being imparted to caffeine-induced phosphodiesterase inhibition, or aspirin-mediated PGE2 synthesis inhibition? No data are currently available to answer this question, or any of the dozens of other questions arising from an examination of the interaction between adenosine and caffeine.[475,476,932]

While the major action of caffeine on thermogenesis is positive, the administration, or consumption, of large quantities of caffeine can definitely be counterproductive. Not only do large amounts tend to promote symptoms of caffeinism, but there are two other important considerations. One, large amounts will tend to disrupt the delicate balance between caffeine, aspirin and ephedrine and thereby disrupt the thermogenic action of the combination. Second, there is evidence that the consumption of large amounts of caffeine (more than 600mg/day) increases serum levels of corticosterone[42,1194] and decreases serum levels of growth hormone and thyrotropin (thyroxine and tri-iodothyronine and their precursors phenylalanine and tyrosine).[1194] Corticosteroids, it will be remembered, interfere with thermogenesis both directly by inhibiting adenylate cyclase, and indirectly through CNS nuclei. Growth hormone and thyroid both help potentiate thermogenesis; their inhibition would therefore lower thermogenic efficiency.

PHARMACOLOGY OF CAFFEINE

Because of its popular use as a stimulant, caffeine has received considerable attention from scientists interested in exercise physiology, or ergogenics. The study of caffeine for endurance training has yielded a mixed bag of results. Most findings support its ergogenic action,[188,250,284,339,356,604,605,807] while some find no support or negative support.[207,1356] Some experts have suggested that only well-conditioned athletes can use caffeine with benefit in exercise physiology,[186,356,1053] but others disagree,[207,372,480] and it is becoming increasingly evident that the ephedrine/caffeine mixture can aid the calorigenic and/or thermogenic response of almost all people.

Endurance training leads to exhaustion when glycogen stores in the muscles become depleted. Elevating serum FFA prior to exercise increases FFA oxidation and thereby has a sparing effect on muscle glycogen stores.[249,285,594,608,607,848,1013,1015,1033] This increases the amount of time it takes for glycogen stores to be depleted, and for exhaustion to set in. Overlaying these effects on thermogenesis, we could suppose that part of caffeine's effect on BAT metabolism is due to FFA mobilization.

Thermogenic Action

Caffeine augments the thermogenic response to a meal, exercise and to cold exposure.[211,327,847] Caffeine consumption causes an increase in urinary excretion of NE of about 20% and of adrenaline of about 80%. However, the adrenergic effects of caffeine may be somewhat blunted in people who are chronic consumers of caffeinated beverages. Studies have shown that habitual users of caffeine, in contrast to non-users, failed to exhibit a significant increase in catecholamine after consuming a single dose of caffeine.[64] Hence, some degree of adaptation to the caffeine has occurred with a resultant lowering of action on the fate of NE.

In addition to its stimulating effects on NE metabolism, caffeine exhibits certain other pro-thermogenic actions. It has a modest stimulating effect on glucose metabolism, slightly increases glucose and lactate concentrations in BAT, with concomitant increases in insulin and c-peptide concentrations.[69] Caffeine also stimulates lipolysis in WAT in a dose-dependent manner, that results in a increased concentration of serum FFA.[103,105,362,363,753,979]

Increased lipolysis may be due to either or both of a catecholamine release or cAMP phosphodiesterase inhibition. Increased BMR observed following caffeine consumption is probably the result of increased plasma FFA, but again this action may be mediated by caffeine-induced increases in catecholamine levels.[141] Furthermore, it has long been know that caffeine elevates oxygen consumption.[8,664]

By itself, caffeine has been shown to increase the proliferation of brown adipose tissue. Combined with ephedrine the impact of caffeine on the growth of BAT appears to synergistic or supra-additive in much the same manner as it is on increasing thermogenesis.[176]

It has been known since 1915 that caffeine provokes an increase in metabolic rate, an observation that has been confirmed in several subsequent studies..[8,483,555,573,605,606,664,668,850,869] The action of caffeine on BMR is thought to be partly due to the release of catecholamine (NE and adrenaline) from the SNS and adrenal glands, and partly the result of lipolysis and a calorigenic action.[112,250,493,608,653,1000,1216] Low doses (100mg) of caffeine increase BMR in lean subjects, while higher doses are required in obese persons.[8,555] Large amounts of decaffeinated coffee (equivalent to 20-40mg caffeine) raised basal metabolic rate in lean subjects but not in obese. Caffeinated coffee containing either 8mg/kg or 4mg/kg caffeine increased metabolic rate during the three hours following consumption, though a great deal of individual variation was observed.[8] Caffeine acts in part by disrupting the events that lead to accumulation of white adipose tissue.[175,176]

Caffeine exerts a greater effect in the RMR of inactive subjects than in exercise-trained subjects. Thus endurance training resulted in a reduced thermogenic response to caffeine.[994] However, caffeine consumed with a meal appears to increase DIT.[1400]

A 12% increase in overall cellular metabolism was observed in human volunteers one hour after ingesting 100-200mg of caffeine. Metabolism returned to normal 1 hour later. This effect was thought to be the result of circulating catecholamines, and was thus probably a thermogenic action.[21,894] Glucose involvement in caffeine-induced metabolism increase is thought to be minimal, though there are differing views on this point.[75,102,632,910,944,1019] When ingested prior to a light (aerobic) workout, caffeine improves the use of fats at the expense of glycogen.[250,1355]

Derek Miller of Harvard was the first authority to seriously pro- mote caffeine as a thermogenic agent.[869] Miller went on to initiate a series of investigations on the effects of caffeine and its interactions with ephedrine. Recently, Japanese researchers using rats found that 60mg/kg caffeine elevated GDP binding in brown adipose tissue, as well as oxygen consumption and RMR. These results strongly suggest that caffeine affects of the metabolic events of thermogenesis in that tissue.[1379] Astrup studied the effects of 100, 200, or 400 mg of caffeine in subjects who demonstrated a moderate habitual caffeine consumption. It is felt that habitual users respond less to a single dose of caffeine than do non-users or occasional users. Astrup found that caffeine increased energy expenditure in a dose-dependent manner. The thermogenic response was positively correlated with plasma caffeine level, plasma lactate and plasma triglycerides. In keeping with his bias, Astrup attributed the majority of these effects to actions occurring in non-BAT.

Caffeine Alone: Insufficient

In summary, while caffeine does not exert a direct effect on thermogenesis, other than a slight adrenergic effect, it does impact on several of the underlying processes involved in thermogenics. It stimulates the conversion of TAG to free fatty acids in white and brown adipose tissue, increases the rate of flow of blood, favorably impacts on the metabolism of insulin, adrenaline, and other hormones, and stimulates the proliferation of brown adipose tissue and the uncoupling protein. It must be emphasized that none of those factors alone qualifies caffeine for inclusion in a thermogenic compound. Even when all of those factors are taken together, the overall effect of caffeine is still too small to compensate for potential risks. The only justification for the inclusion of caffeine in a thermogenic compound is its synergistic effect with ephedrine and aspirin. Ephedrine and caffeine form two legs of a pharmaceutical trinity capable of eliciting safe and effective thermogenesis. The third leg, aspirin, makes the combination safe and effective.

PHARMACOLOGY OF EPHEDRINE + CAFFEINE IN BROWN ADIPOSE TISSUE

The combination of caffeine and ephedrine has been used by the lay public for some time as an unauthorized adjunct to dieting. Caffeine exhibits a supra-additive, or synergistic, thermogenic action when combined with ephedrine or any of its many analogues.[4,57,62,210,333,344,547,589,771,826,969,1010,1270,1304,1307,1347] The bronchodilator effect of caffeine probably also adds to that of ephedrine.[95] The combination also exerts a synergistic effect on glucose and lipid metabolism. One of the principle modes of action of caffeine is to inhibit intracellular phosphodiesterases (PDEs). This results, among other things, in the release of free fatty acids from adipocytes.

Increasing Cold Tolerance

In 1989, Canadian researcher Andre Vallerand and co-workers reported that a mixture of ephedrine and caffeine was effective in increasing cold tolerance in humans. The subjects were exposed to 10 degree centigrade conditions for one hour wearing nothing but bathing suits. Compared to subjects consuming a placebo control substance, subjects receiving 1mg/kg ephedrine and 2.5mg/kg caffeine were able to produce significantly more internal heat (19%) through thermogenesis.[1307] Vallerand's work is supported by the work of L.C.H. Wang using theophylline derivatives instead of caffeine.[1330-1334] Other studies have shown that these substances increase thermogenesis in subjects exposed to ambient temperatures as well.[8,62,64,322,763]

The increase in body temperature in cold-exposed humans probably involves a sympathomimetic action on peripheral thermogenic tissues. Incidentally, caffeine alone does not appear to exert a significant thermogenic action in cold-exposed organisms.[310,465,847] The action of ephedrine on beta- and alpha-adrenergic fibers and a release of NE from sympathetic nerve endings, combined with the antagonistic action of caffeine on cell surface adenosine receptors, would enhance adipose tissue lipolysis, liver and skeletal muscle glycogen breakdown, and liver gluconeogenesis.[771] BAT thermogenesis would undoubtedly be increased, and this, combined with enhanced carbohydrate mobilization and utilization, would have a significant effect on total body heat production.

Ephedrine + Caffeine Synergy

Arne Astrup's Copenhagen group has published extensive research concerning various combinations of caffeine and ephedrine,[22,56,70,71,73,1269,1270] and reported very recently about an observed synergy between caffeine and ephedrine on thermogenesis in humans using a double-blind, placebo-controlled protocol.[57] This is a landmark piece of research, for it clearly demonstrates the remarkable supra-additive synergy between these two components. The study compared the results of placebo, caffeine (C), and ephedrine (E) alone and in combination: E 10mg, E m, C 100mg, C 200mg; E + C 10mg/200mg; E + C m/100mg; and E + C m/200mg. All active doses exceed placebo. The combinations were ranked in effectiveness as follows: 20mgE/200mgC > 10mgE/200mgC > 20mgE/100mgC. The best thermogenic effect was achieved with the m/200mg regimen, and the results were 64% greater than would be expected by a simple additive action. Interestingly the thermogenic effect of the other two combinations did not exceed the expected additive effect; in other words, only the 20/200 combination exhibited synergy. It should also be noted that this study was done on <u>lean</u> subjects and did not address the issue of BAT metabolism. If we assume that lean subjects have normal BAT function, then the failure to include obese subjects in this study excluded the possibility of discovering differences in metabolism that may reflect a deficit in BAT function.

Over the years the Astrup group has demonstrated a plethora of effects in human subjects ingesting the ephedrine and caffeine mixture. A few of these findings are listed here:

1. An acute stimulation of energy expenditure.

2.   Greater loss of weight in more obese versus less obese people, when both groups are ingesting an equal number of calories.

3.   A supra-additive effect of ephedrine and caffeine combined versus either substance used alone.

4.   Improved weight loss in persons on a low calorie diet.

5.   Improved weight loss after major dietary weight loss achieved during a six month period.

6.   Inability to maintain weight loss after 6-12 months of treatment, though this result was confounded by time-of-introduction variables.

7.   Increase in loss of fat mass rather than lean body mass.

A Harvard Study

The Harvard group, lead by Dulloo and Miller, was the first to study the effects of a combination of caffeine and ephedrine in humans. In that initial study they measured the effects of a meal on the thermogenic response in lean subjects versus subjects with a predisposition to obesity.[327] The obese exhibited only one-third the response of the lean. Hence, the obese appeared to be more efficient metabolizers than the lean. Subsequently, the scientists administered an over-the-counter medication containing 22mg E, 30mg C and 50mg theophylline to both groups. They found that the ephedrine plus methylxanthine mixture was twice as effective as E alone in increasing metabolic rate. In a fasted state, the lean and obese exhibited identical responses to the mixture on an acute basis, but after 24 hours the groups tended to differ; the thermogenic response in the lean disappeared, but persisted in the obese. The compound tended to normalize the thermogenic response of obese subjects. In other words, it had no effect of 24-hour energy expenditure (24EE) in the lean, but increased 24EE in obese persons by 8%.

The failure of this combination to elicit a thermogenic response in lean persons stands in contrast to the findings of Astrup. The tendency of the compound to normalize thermogenesis in obese subjects agrees with the basic assumption of this book; namely, that thermogenic compounds tend to compensate for thermogenic defects that predispose a person to obesity.

The Harvard investigators reasoned that the initial thermogenic response to the ephedrine/xanthine mixture in lean persons is brought under negative feedback control, i.e., inhibited over time, through sympathetic nervous system mediation. Following a meal, an initial burst of thermogenesis is exhibited by both groups. This effect is intensified by the ephedrine/xanthine mixture. After a while, the SNS eliminates the thermogenic response in lean persons. In obese people, however, a defect in sympathetic tone is compensated for by the drugs which tends to normalize the thermogenic response. Because these persons lack a normally functioning SNS, however, the feedback mechanisms fail to respond in such a manner as to shut off the thermogenic process. Hence the response continues for a longer period of time. On the other hand, in the absence of the sympathetic stimulants, sympathetic mediation in the obese may fail to occur in part or in full following a meal, resulting in no net increase in thermogenesis, resulting in a tendency to store food calories rather than burn them.

The tendency for ephedrine and xanthine mixtures to restore thermogenic activity has been observed in animal models of obesity wherein the compound has been shown to reverse obesity in several cases.[823,326] Caffeine plus ephedrine and isomers of ephedrine increase BAT growth in lean as well as obese rats.[1348] This action is therefore a consistent property of the compound in all organisms.

The whole question of whether lean animals and humans respond to thermogenic in the same manner as obese, and if they do not, to what degree they may differ, or the basis upon which comparison is even possible, have not been satisfactorily answered.

Theophylline, as used in the previous study, is commonly employed in thermogenic research.[589,905] It is a xanthine, like caffeine, but exhibits greater bronchodilator action and is thought to impart greater side effects than caffeine. In some studies, however, a combination of ephedrine plus theophylline produced significant weight loss in humans in a controlled clinical setting.[822,825] In another study, the combination of ephedrine and caffeine and theophylline resulted in increased thermogenesis and weight loss compared to controls.[589]

Caffeine is by far the most frequent xanthine used to stimulate metabolism in clinical trials, by itself or in combination with ephedrine. This combination routinely raises BMR.[62,64,358,555,896,869] and augments the thermogenic response to a meal (DIT).[8,358,869]

Further Danish Studies

Astrup administered the ephedrine and caffeine compound to obese women on a fairly long-term, or chronic, basis (24 weeks).[74] These women were also place on an energy restricted diet. He observed a significant reduction in weight and fat mass. Based on his observations, Astrup reasoned that chronic administration of ephedrine plus caffeine stimulated the proliferation of brown adipose tissue, the reduction of white fat, and the growth of skeletal muscle. He attributes these effects to the ability of the compound to stimulate lipolysis in white fat, thereby reducing its mass, and the simultaneous increase in lean body mass and brown fat as these tissues utilize the energy liberated during lipolysis. Other studies have yielded similar results.[69,817,971]

In a subsequent study, Astrup observed that the same mixture of caffeine and ephedrine preserved the fat free mass (the lean tissue) while enhancing weight loss. He attributed the results 75% to anorexia, and 25% to thermogenesis.[58] The anoretic effect of ephedrine and caffeine are well known. The relative contribution of anorexia and thermogenesis to weight loss is the subject of much debate. Astrup's figures will undoubtedly help resolve this problem. Research from the same lab has shown that ephedrine plus caffeine, when administered to growing pigs, resulted in the same growth rate as control animals, but at a 20% lower energy intake. Also noted was the fact that the amount of muscle tissue was increased by 10%, and that intermuscular and subcutaneous fat deposits were lowered by 30%.[948] These extraordinary findings will hopefully stimulate further research along similar lines.

Deficient Thermogenesis in BAT?

Dulloo assessed the rate of oxygen consumption of interscapular BAT in rats following the consumption of low doses of ephedrine and caffeine.[335] He found that increased oxygen consumption (which indicated increased metabolism and presumably thermogenesis) due to ephedrine or caffeine depended entirely on the presence of intact sympathetic nerve endings in the tissue. Thus it depended on presynaptic mechanisms. Direct postsynaptic stimulation of thermogenesis only worked with high doses of ephedrine and caffeine. Combined, the substances demonstrated a synergistic action. Dulloo reasoned that ephedrine stimulated the presynaptic sympathetic nerves in BAT which in turn released norepinephrine. Postsynaptic depolarization of the BAT cells, and coupling with adrenergic receptors, resulted in the thermogenic cascade. Caffeine, meanwhile, acted by inhibiting adenosine and phosphodiesterase, in agreement with mechanisms of action we have discussed previously. Interestingly, fasting in these animals reduced the overall effect.

Contrasting results were found in a similar study, in which ephedrine, caffeine and the combination of ephedrine and caffeine created 32%, 48% and 50% increases in oxygen consumption as measured by VO_{2max} in lean and genetically obese rats, respectively.[1296] Duration of the VO_{2max} averaged 50%, 26% and 42% longer in the genetically obese than the lean animals. Average weight loss was greater for the obese than the lean animals and corresponded with the duration of the elevated oxygen consumption. The researchers concluded that these results were consistent with normally functioning BAT.

A partial explanation of these results may be found in the particular genotype used in this study. The LA/N-corpulent rat may be the result of a genetic obesity-inducing defect that does not involve brown adipose tissue metabolism. Most genetic models of animal obesity, on the other hand, have been shown to have a dysfunctional thermogenic component.

Research on alternative adrenergic agents occasionally uncovers the same kind of synergy with caffeine that exists between ephedrine and caffeine. In one study, for example, 10, 20, and 40mg/kg caffeine increased thermogenesis in BAT as measured by increased temperatures in surrounding tissue.[1348] Combined with 10mg/kg phenylpropolalanine (PPA), the mixture elicited a greater increase in thermogenesis than PPA alone; the effect was calculated to also exceed the additive properties of the chemicals in a synergistic manner. PPA is not usually considered a safe alternative for ephedrine.[936]

CAFFEINE SAFETY: GENERAL CONSIDERATIONS

Caffeine has been consumed by man for thousands of years, beginning with the intake of caffeine-containing plants and culminating in the manufacture of numerous synthetic caffeine-containing beverages, foods and drugs. Millions of people have come to depend on the daily consumption of caffeine for improved mood, concentration, impulse control, attention span, alertness and stamina.[218,219,229.241.-380,381,394,455,456,506,732,775,1029,1123,1341] Yet the risk to health of caffeine has only been addressed by science in the last few decades.[228,299,240,-370,388,423,727,815,1017,1206] It is generally known that it would require a considerable amount of caffeine to cause human toxicity. the fatal oral dose for caffeine is 10 grams, or 80-100 cups of strong coffee, or 200 cans of cola, consumed within a half-hour period. There has never been a reported case of caffeine-fatality due to consumption of a food or beverage. Only eight people have died from caffeine (from drug sources) in all of recorded history.[425,634,846]

Caffeine is an important substance in the generation of safe and effective thermogenesis. In and of itself, it impacts total body thermogenesis only slightly; BAT thermogenesis is impacted almost not at all. But it interacts with ephedrine/ma huang in an extremely important manner.

One of the most difficult questions modern science has had to deal with is whether caffeine is harmful or perfectly safe for human consumption. Like the questions of alcohol and tobacco before it, the caffeine question involves the reality of widespread popularity. Right or wrong, this factor has played a big role in research.

Though not all professionals agree on the subject of caffeine safety in humans, the general consensus is that uncontrolled caffeine consumption is not without some risk, especially in newborns.[12] The questions of how much risk and who is most at risk are not easily answered. While serious injury and/or disease from caffeine consumption are rare, the addictive syndrome known as caffeinism is experienced by thousands of Americans, including pregnant women and nursing mothers. Are their fetuses and infants at risk? Current data suggest an affirmative answer. Deleterious effects on the central nervous system, cardiovascular system, gastrointestinal system and reproductive system, arising from consumption of caffeine-containing substances are being established daily. The bottom line of these data is not caffeine per se, but the amount of caffeine.

Complicated socio-politico-economic factors still play a role in the amount of both negative and positive publicity caffeine receives. Few modern scientists (and their supporting universities) are willingly to risk their jobs and the possibility of considerable grant monies to examine questions that could upset a majority of Americans; therefore, when negative results are obtained, these are usually down played to a remarkable degree. Even if caffeine use had serious consequences, it is currently unlikely that the public would be adequately informed about them. It will take decades more of research and objective evaluation to overcome the subjective social and economic factors that presently contend for the safety of caffeine. Nevertheless, there is widespread and growing concern among doctors and scientists about the possible risks of chronic caffeine intake, especially as the possible sources of daily exposure to this substance increase, in the form of hot beverages, caffeinated soft drinks, chocolate bars, popular analgesics and prescribed medicines. Currently, two conclusions are inescapable. One, consumers should exercise control and common sense in their selection of caffeine-containing foods and drugs. Two, manufacturers must exercise extreme responsibility in the preparation, distribution and promotion of these substances.

It is felt that since caffeine stimulates the most common non-immunologically-mediated food reactions in humans, and since some people are especially sensitive, there is a risk that the incidence of severe tachycardia and gastric symptoms may increase over the next few years. Such considerations have led the Committee on Review of Medicines in England to recommend that the following warning be placed on labels of caffeine products where the dose provides 100mg or more of caffeine: "Excessive intake of coffee or tea together with tablets containing caffeine may make you tense and irritable." Warnings about tremors and palpitations are also recommended by the committee.[1408]

In light of public concern over the use caffeine, it is important that we understand, appreciate and accept the rationale that forms the basis for either consuming caffeine or not consuming it, to learn what the real and apparent health considerations are, and to understand when caffeine may be ingested for positive health reasons. To this end, the following review of medical findings is presented. Only the most important and seminal papers are reviewed here. In addition, possible positive effects of caffeine on thermogenesis and ergogenesis will be summarized.

Caffeine Abuse, Addiction, Caffeinism and Central Nervous System Disturbances. The primary effects of caffeine are on the central nervous system. These effects may be mild, moderate or severe.[1409-1410] The most noticeable effect following an acute administration is a behavioral stimulant action.[557,567] Caffeine does not accumulated in tissues or organs, including the brain,[120] even though it readily crosses the blood brain barrier.[821] Danger from cumulative effects is therefore low, yet the "caffeinism" phenomenon does occur. Certain effects are limited to first time or infrequent users, including elevated blood pressure, heart rate, plasma epinephrine, plasma norepinephrine, plasma renin activity and urinary catecholamines. These effects generally disappear with chronic consumption.[235,464,1054]

Headaches, restlessness and mental irritability are the most frequently reported side-effects of caffeine consumption,[1411] with increasingly higher doses producing more serious effects.[408,560,715,1031,1364,1412] The severity of caffeine-withdrawal headache is frequently noted by researchers.[1143] The population of Utah is frequently used as the data

214

base for these observations, since the frequency of week-end abstinence and week-day usage of caffeine is high in that state.[1413]

One-third of the adult population of Canada has been estimated to be physically dependent on caffeine, especially among the Canadian Northern Indigenous People, where the recorded daily intake is much higher than the levels known to create adverse physiological and behavioral effects.[1414]

In an evaluation of 124 hospital patients, comparing the effects of caffeine among low, moderate and high users, certain dose-response relationships were found. The most common symptoms were diuresis, insomnia, withdrawal headache, diarrhea, anxiety, tachycardia and tremulousness; this constellation of symptoms, together with anxiety, has become known as Caffeinism.[1415] These same symptoms have been noted by other researchers.[152,286,308,313,383,348,-435,454,458,473,474,504,584,591,676,703,732,1030,1106,1161, 1168,1336,1349,1363]

Caffeinism specifically refers to a state of acute or chronic toxicity resulting from the ingestion of high doses of caffeine. An intake of 500-600 mg of caffeine per day (approx. 7-9 cups of tea or 4-7 cups of coffee) is currently regarded as representing a significant risk. The complete caffeine syndrome can be described as follows: restlessness, anxiety, irritability, agitation, muscle tremor, insomnia, headache, sensory disturbances, diuresis, cardiovascular symptoms and gastrointestinal complaints. Other problems may be associated with chronic Caffeinism: peptic ulcer, myocardial infarction, fibrocystic breast disease, ulcer, diabetes, and psychosis.[299,401,956,1206,-1416,1417]

The renal effects of caffeine are well known. All xanthines exert a diuretic action by increasing glomerular filtration and by blocking tubular reabsorption of sodium ions.[308,342] Tolerance to this effect does eventually develop, so that chronic consumers of caffeine experience less diuresis than infrequent users.[906]

Caffeine consumption produces certain respiratory effects that have a neurological foundation. It increases respiratory rate.[33,779,840,921] However, as with diuresis, adaptation does take place.[495] Caffeine has been used as a treatment in chronic obstructive pulmonary disease,[1218,1365] apnea,[32,491] asthma,[93,987] allergic rhinitis,[1152] and atopic dermatitis.[675]

High daily intake of caffeine, then, can lead to the development of a negative constellation of symptoms that rather impressively offsets the benefits of the substance. Native users of gooroo nut and yerba mate seldom, if ever, experience these problems. This suggests that it is not the xanthine itself we need to watch, but the amount and type of xanthine or mixture of xanthines consumed that may be critical. As with many different foods, ingestion of excess quantities of caffeine is not wise. In an age where excesses are readily available, the watch words should still be moderation in all things.

Maternal Caffeine Consumption and Abuse. Caffeine readily passes from the serum into breast milk and readily crosses the placental barrier[457,965,1298]. Peak concentrations are reached in 60 minutes. One cup of coffee is not expected to yield a large dose to the infant, but repeated maternal caffeine ingestion would have a difficult to predict effect on the infant, since the accumulation of caffeine in the infant's tissues would depend upon the average concentration in maternal serum and breast milk over time, the volume of milk ingested, and the infant's particular clearance rate for caffeine. Hence it is important that a mother be able to calculate with some degree of accuracy the amount of caffeine she consumes from day to day.[1418] The safest course of action for pregnant and nursing mothers is to simply avoid the consumption of caffeine altogether.

Female Caffeinism and Fibrocystic Breast Disease. One area of consistently great concern for scientists has been the observed link between caffeine consumption and the development of fibrocystic disease in women. Fibrocystic disease is characterized by benign lumps in the breasts which may predispose the woman to eventual development of breast cancer.

One important study tested this hypothesis by eliminating caffeine from the diet of 47 patients with fibrocystic disease and observing the results. Mean prior caffeine consumption was 190 mg/day. Twenty of the 47 abstained completely from caffeine. All experienced symptoms of caffeine withdrawal, including headaches which lasted 1 to 7 days. Of these 20 women, 13 experienced complete disappearance of all palpable breast nodules and other symptoms within 1 to 6 months. Only one of the 27 women who continued to consume caffeine experienced positive results. It was concluded that caffeine was definitely an aggravating factor in women predisposed to fibrocystic disease.[1419] Other researchers have failed to replicate this work.[355]

More recently, it was established that women who consumed 31-250mg of caffeine per day had 1.5-fold increase in the odds of fibrocystic disease, while the daily consumption of 500mg (4-5 cups of coffee) increase the odds to 2.3-fold.[1420] Studies of women consuming lower amounts of caffeine have failed to link caffeine consumption to fibrocystic disease.[1114,1421]

The conclusion in this regard, then, is that normal consumption of coffee and tea probably has little impact on fibrocystic disease.[798] The consumption of high amounts of caffeine on a daily basis, however, may dispose some women to this condition.[875,876]

Caffeine and Fertility. Recent research suggests that women who consume the caffeine equivalent of more than 1 cup of coffee per day are less likely to become pregnant than women who consume less. Among the higher caffeine users, the failure to become pregnant was nearly 5 times greater.[1422]

Males are not completely immune to the detrimental effects of caffeine abuse either. Excess caffeine consumption may impact men with effects ranging from possible chromosomal damage to testicular toxicity.[1423,1424]

Teratogenesis/Birth Defects. The possible adverse effects of caffeine consumption on the developing fetus has been a matter of great concern, but opinions about the severity of the problem vary greatly among investigators. This relationship was the subject of an extensive study involving 800 households, 75% of which comprised members of the Church of Jesus Christ of Latter Day Saints. The major finding was that a daily caffeine intake of 600mg or more may predispose a woman to reproductive difficulty.[1425] This level of caffeine is quite high but it is estimated that 10% of the population may ingest that much caffeine on a daily basis.

In a retrospective survey study it was found that a high rate of spontaneous abortion, still-birth and prematurity occurred when either the father or the mother had ingested more than 600mg of caffeine per day. In spite of the clear relationship revealed, the investigators were reluctant to hypothesize a cause-effect relationship until prospective (instead of after-the-fact) studies had been done on this phenomenon.[1426] This same pattern has been observed by others,[1427] but not by all.[799,1058] In one case, it was found that a daily intake of more than 8 cups of coffee increased the frequency of congenital malformations.[1428]

High doses of caffeine have been shown to cause birth defects and mutagenesis in animals, including decrease in intrauterine fetal growth, a lower birth weight and skeletal abnormalities. FDA animal trials showed that while high doses of caffeine given in one dose to pregnant rats did cause birth defects in offspring, allowing the animals to sip the same dose over an extended period in a manner more typical of human behavior did not produce birth defects.[231,232] The implications of these kind of findings for humans are not clear at this time. Good long term studies have not been done, controls have been lax, and so forth. Doses that correspond with normal human consumption appear to be much less dangerous. Nevertheless, at least one study has shown that caffeine does have a depressive effect on cellular mitotic activity; this fact demands further investigation.[1429] According to some investigators, until more scientific data are available, moderate to excessive use of caffeine during pregnancy should be avoided, and mild use should either be limited, discontinued, or be left to the individuals discretion.[1430,1431]

Cardiovascular Effects. Much of the current literature about caffeine consumption centers around the possible adverse effects this substance may have on cardiovascular function, while reports of positive inotropic and chronotropic effects go relatively unnoticed.[288,1215] Caffeine can increase heartbeat and dilate some blood vessels and constrict others.[1051] These are the most basic cardiovascular properties. Beyond these, the situation is very complex; research is complicated by dozens of confounding variables, few of which are ever controlled for in experimental situations.[86,615] Periodic reports appear in medical journals of excessive cases of myocardial infarction in patients who report drinking large amounts of coffee each day, representing a 2-1/2-fold increase in risk among women drinking 6 or more cups daily.[649,1432] These kind of reports generally conclude that the possible causal role of coffee drinking needs to be further investigated. Indeed, other studies have failed to find any association between caffeine and cardiovascular disease, including coronary heart disease, angina pectoris, and myocardial infarction.[280,522]

A variety of tachy-arrhythmias and extrasystoles, as well as bradycardia, are believed to be caused by the medullary vagal, stimulant and toxic, cardiotonic effects of caffeine.[235,305,1293,1350,1433]

Mortality rates from ischemic heart disease in the province of North Korelia in Finland where caffeine consumption is highest (for Finland) are much higher than in south and west Finland, where caffeine intake is correspondingly lower.[1434] These kind of positive correlations are common, yet scientists are cautious about generalizing the results to the rest of the population of the world.

Consuming 250mg of caffeine produces a mean elevation in mean blood pressure of 14/10 mmHg one hour later, according to one study, leading the investigators to concluded that caffeine "might enlarge the population of hypertensive subjects by increasing the pressure of those with borderline hypertension."[1053] While the conclusion is certainly reasonable, based on the findings of that study, and others,[463,1053] the situation is confounded somewhat by studies that fail to find a link between hypertension, heart rate and caffeine or coffee consumption.[302,372,434,573,986,1000,1053,1191,1199,1206,1268]

Clinically significant effects, including an increase in mean blood pressure level, glucose and free fatty acid levels and urinary catecholamine excretion, were noted in human volunteers after ingesting coca cola with 150mg of caffeine.[101,1435] Another study found that 150mg of caffeine from coffee lowered the effective and functional refractory period of the a/v node; it was theorized that this effect was due to catecholamine release.[1436]

Ten volunteers were given caffeine equivalent to 2-3 cups of coffee which resulted in an immediate increase in the mean systolic blood pressures of 14.7 and 16 mmHg, and in an immediate increase in mean diastolic blood pressures of 7.4 and 7 mmHg.[1437] Elsewhere, a previous history of caffeine-induced symptoms correlated highly with the development of supraventricular tachy-arrhythmias.[1438]

Incidentally, on the positive side, it should be noted that caffeine constricts swollen vessels in the head that are the cause of common headaches and migraines.[751,752] The ephedrine-caffeine-aspirin combination should be especially helpful for some people.

In view of the variety of inconclusive and often contradictory findings about caffeine and the risk of cardiovascular illness, at least one pair of scientists have stated that we have failed to substantiate the claim that caffeine consumption increases the risk of heart attacks, or death from ischemic heart disease or cerebrovascular accidents.[264]

As a final remark, it should be noted that caffeine in high doses appears to raise serum cholesterol levels, while coffee and other whole plant materials do not,[10,579,610,684,911,1372] unless consumed in enormous amounts.[1249]

Carcinogenesis. Surprisingly little research has been devoted to investigating the carcinogenesis effects of caffeine in humans. Occasional reports appear however. For instance, it has been observed that the incidence of pancreatic cancer is higher among coffee drinkers.[1439]

It is clear that more studies on the possible carcinogenic effect of caffeine are urgently needed.

Conclusions. The evidence implicating caffeine intake in the etiology of numerous clinically demonstrable diseases and complaints is, when all factors are taken into consideration, sizeable, but not very impressive, and the number of proponents of caffeine safety appears to grow at a swift pace. Seldom has America witnessed such dichotomous rhetoric about a single substance. For several years the

Center for Science in the Public Interest, a United States consumer organization, has been urging the FDA to place warning labels on caffeine-containing beverages and drugs. The FDA has so far refused to do this.

We urge consumers to be careful of the products they consume, avoid consuming large amounts of caffeine every day, restrict your intake to the one or two compounds that provide the kind of health benefits desired. We likewise urge manufacturers to be responsible enough to provide the best possible product and to do the necessary research before bringing a caffeine-containing product to the marketplace.

EPHEDRINE + CAFFEINE: GENERAL CONSIDERATIONS

By inhibiting two of the major inhibitors of NE activity, caffeine potentiates the action of ephedrine on SNS activity, and hence on thermogenesis. Caffeine therefore significantly reduces the amount of ephedrine required to initiate the thermogenic response. However, the amount of aid offered by caffeine is still not sufficient to insure the safety of ephedrine consumption for thermogenic purposes. As we shall see, the combination of ephedrine and caffeine requires the addition of sufficient aspirin to be rendered safe.

The amount of caffeine required to form a synergistic relationship with ephedrine and aspirin is relatively small, ranging on the low end from 50mg to 200mg on the high end. Since a level of 600mg per day is considered the maximum safe dosage, the amount required for thermogenesis is relatively minor. Still, as mentioned above, persons using thermogenic are advised to eliminate other forms of caffeine from the diet, not necessarily for safety reasons, but in order to avoid disrupting the balance between ephedrine, aspirin and caffeine.

The combination of ephedrine and caffeine for promoting thermogenesis is safer than the use of either of the substances by itself. Furthermore, although caffeine has been found to potentiate ephedrine-induced teratogenicity when administered in doses high enough to disturb embryo development in chicks, combining ephedrine with caffeine in a manner consistent with good thermogenic action actually reduces the amount of ephedrine (and caffeine) to levels where chances of embryonic damage are virtually non-existent.[934]

EPHEDRINE + ASPIRIN

Much of what was said about the combination of ephedrine and caffeine can be repeated for aspirin. The major differences would simply be in regard to the specific cellular substances inhibited by aspirin. The purposes and result are the same; namely, to increase the effectiveness of ephedrine by inhibiting inhibitors of NE, and to detoxify the ephedrine. We noted the synergy between caffeine and ephedrine. A similar synergy exists between ephedrine and aspirin.

ASPIRIN: CELLULAR MECHANISM OF ACTION

Aspirin has been shown to demonstrate the following actions.

1.   Interference with prostaglandin E_2 biosynthesis, which produces

2.   Lowered inhibition by PGE_2 of NE release and of the adenylate cyclase system involved in lipolysis.[328,378,843]

Prostaglandins

Prostaglandins are so named because they were first isolated (in 1935) from seminal fluid produced by the prostate gland. Prostaglandins, derived from arachidonic acid, are acidic lipids possessing powerful pharmacological activity. They occur in most of the body's tissues, including nerves, uterus, lung, brain, iris, thymus, pancreas, kidney, menstrual blood. They exert action on the uterus, gastrointestinal tract, bronchi, platelets, nervous system, as well as inflammatory and immune mechanisms. Individual prostaglandins are usually designated by a letter that indicates a particular family group, and by a numerical subscript that denotes the degree of unsaturation in the side chains of the molecule. (More information on prostaglandins can be found in the chapter on BAT.)

The most basic effect of prostaglandins is in the functional control of cyclic AMP formation. The thermogenic program aims at increasing the activity of PGE_1 and inhibiting the actions of PGE_2.

PGE_1 and PGE_2 are both potent vasodilators that can lower blood pressure when injected into the blood stream.[77] For this reason, most doctors might hesitate to recommend the consumption of substances such as aspirin, that interfere with their metabolism, in conjunction with substances that tend to exert a pressor effect, such as ephedrine. Since there is a strong correlation between obesity and hypertension, the fear is that drugs that inhibit prostaglandin synthesis in WAT and BAT my cause or worsen already existing hypertension in obese individuals. Yet the body of research shows that ephedrine plus aspirin fails to create a hypertensive action.

PGE_1 and PGE_2 also dilate bronchioles, an action that parallels that of the action of ephedrine. Combined enhancement by ephedrine/PGE_1 and inhibition of PGE_2 results in a net dilatory effect in the thermogenic program.

PGE_2 plays a very important role in the negative feedback loops that control and modulate the action of NE on the lipolytic events of BAT cells.[1203] This effect is observed in research showing that prostaglandin synthesis in human fat cells can be increased by lipolytic substances and reduced by anti-inflammatory substances such as aspirin.[877,1039] In rats, the counteracting effects of NE and prostaglandin have been verified.[214] PGE_2 decreases adenylate cyclase activity in the BAT cell membrane by direct inhibition. Aspirin interferes with the synthesis of PGE_2 by inhibiting the enzyme prostaglandin H synthase (and perhaps also the enzyme cyclooxygenase) that transforms arachidonic acid to prostaglandins. Prostaglandins are continually being synthesize in body cells. They are extremely short-lived, exerting their biochemical effects almost immediately after being created. The effects of prostaglandins are typically so strong that finding agents that compete with or directly inhibit them without any other form of toxicity to the cell is almost impossible. Hence, it is necessary to find agents that block the synthesis of the prostaglandins. Aspirin is one of the very few known substances that possesses the ability to inhibit prostaglandin biosynthesis at safe dosage levels. PGE_2 is also inhibited in persons with obesity or hyperthyroidism.[1040,1042] Unlike adenosine, the antilipolytic action of PGE_2 is equally strong in visceral and subcutaneous adipose tissue.[1043]

Inhibition of synthesis results in less prostaglandin available to exert an effect on cellular metabolism, in this case on the activity of adenylate cyclase. Aspirin, therefore, does not exert a direct, activating effect on any of the events involved in thermogenesis.[304] Rather it helps eliminate one of the events that inhibit thermogenesis.

PHARMACOLOGY OF EPHEDRINE + ASPIRIN

The first published work on the effects of aspirin on ephedrine-induced thermogenesis was produced by Dulloo & Miller.[328] Long-term consumption of aspirin by itself had no effect on energy balance or body composition in obese mice. Ephedrine by itself produced an increase in energy expenditure and reduced body weight and body fat by 18% and 50% respectively. When aspirin was combined with ephedrine, the effect on energy expenditure was doubled, and the obese group lost more than 75% of body fat. This result was interpreted by the researchers as a sign of obesity reversal since the genetically obese animals reverted to a lean state. The aspirin/ephedrine ratio was calibrated to match those available in several over-the-counter preparations for asthma, cough and bronchospasm. But the amount administered was considerably more on a kg/kg basis than a human would consume of OTC product.

It is important to note that this treatment regimen produced no effect on body protein; the losses came almost entirely from body fat. The Dulloo & Miller study was the first successful attempt to completely reverse genetic obesity. Previous attempts, utilizing ephedrine alone, produced decreases in body fat, but the animals were still technically obese (possessing up to three times as much body fat as lean animals).

In 1991, Horton & Geissler[576] reported that a single dose of aspirin given to obese women augmented the effect of ephedrine on thermogenesis, following a liquid meal (250 kcal). In this study, 30mg ephedrine was administered alone or was combined with 300mg aspirin. A single dose of ephedrine produced a significant increase in thermogenesis. The combination performed better than ephedrine alone, but the researchers do not comment on a possible synergistic action. DIT, as measured by BMR, was lower in the obese than in the lean. Ephedrine reduced the difference in DIT and ephedrine plus aspirin normalized DIT.

In the above studies, aspirin performed as expected. These studies thus confirm the notion that aspirin augments the thermogenic action of ephedrine. It is likely that the mechanism of action also conforms to expectations based on the known physiology of aspirin as a prostaglandin inhibitor.

ASPIRIN SAFETY: GENERAL CONSIDERATIONS

Aspirin is one of the most misunderstood of all OTC substances. It seems that hardly a week passes without the announcement in the popular press of yet another earth-shaking attribute of aspirin being revealed, both for good and bad. After so many years of hearing this kind of rhetoric, many Americans are reduced to state of confusion and inaction, not knowing whether to take their daily aspirin or to avoid the substance completely.

Being a devotee of natural medicine, my own bias is generally to avoid aspirin if possible. There are certain plant preparations, relaxation techniques and nutritional practices that can be employed in place of aspirin most of the time. However, as discussed in earlier chapters of this book and most fully in the chapter on phytopharmaceutical regulation of BAT, salicin and willow bark are not effective substitutes for aspirin.

Since current research as revealed no effective substitute for aspirin, the consumer must decide on an individual basis if the benefits of aspirin on fat management do or do not outweigh possible side effects. On the negative side is evidence that some people are extremely sensitive to aspirin and shouldn't consume it even in very small quantities. There is also evidence that large amounts of aspirin can damage the intestinal mucosa. Furthermore, we must note the tendency of aspirin to thin the blood, making it an inappropriate substance for use by pregnant women and persons with a tendency to hemorrhage. In a more neutral area, aspirin may affect sleep, producing mild sedation as evidenced by changes in EEG patterns.[510,704,714]

On the plus side, most people can stand to have a little boost in the anti-coagulant area to help prevent heart attacks.[515] Furthermore, the amount of aspirin used in thermogenic agents need not be large, well below the amount needed to markedly irritate the gastric mucosa. Aspirin consumption may also impart antipyretic, analgesic, and immune-enhancing actions. Recent research suggests that the consumption of aspirin every other day helped ward off four types of cancer: stomach, esophageal, colon and rectal. The research, although based on the records of 635,000 Americans, is still being viewed as preliminary, requiring additional supporting trials before definite statements can be made.

All rhetoric on the relative safety of aspirin aside, the decision to include it in a thermogenic compound is based squarely on its ability to lower the effective dose of ephedrine. Eventually, other agents may be found that augment the action of ephedrine, but currently the safest are caffeine and aspirin. Elimination of aspirin from the product would most certainly increase the probability of serious cardiovascular side effects arising from the use of the thermogenic. Again, the amount of aspirin in the thermogenic will normally be small (daily dose less than 500mg, and typically less than 150mg).

In summary, although the inclusion of aspirin in the thermogenic renders it more effective and safe for human consumption, and therefore delivers this valuable fat management tool into the public arena, the consumer must decide for him/herself whether it is an appropriate or desirable product. Choosing to consume thermogenic which does not contain aspirin is not acceptable, prudent or wise, and will most certainly increase the risk of health problems more than will a little bit of aspirin.

EPHEDRINE + ASPIRIN SAFETY: GENERAL CONSIDERATIONS

The remarks made earlier about the safety of the ephedrine + caffeine combination apply generally here also. The combination of ephedrine and aspirin is safer than either substance used alone. The amount of aspirin required to exert the synergy with ephedrine is relatively small, starting at about half a standard aspirin tablet per day. Nevertheless, persons with extreme aspirin sensitivity are best advised to avoid the product unless directed otherwise by their doctor.

We prefer to use the smallest amount of aspirin possible in compounding a thermogenic product; other formulators have chosen to use what we consider excessive amounts. We feel that the consumption of an aspirin tablet in conjunction with the thermogenic is easily accomplished by persons with no sensitivity to aspirin; meanwhile, most aspirin sensitive individuals can consume a thermogenic produced with the minimal effective concentration of aspirin.

CAFFEINE + ASPIRIN: SOME OBSERVATIONS

Both caffeine and aspirin have been shown to possess minor pain relieving properties.[215,88,192,445,1200] However, there is some evidence that a synergy exists between these two substances.[1113] Thus, consumption of thermogenic may impart these additional benefits, independent of, or augmented by, the presence of ephedrine. More research on the possible interactions between caffeine, asprin and ephedrine is badly needed.

EPHEDRINE + CAFFEINE + ASPIRIN

As of this writing, the culmination of research on the pharmacological manipulation of thermogenesis in brown adipose tissue has been the creation of the combination of ephedrine, caffeine and aspirin. These agents are appropriate for the present time because of their ready availability, relative lack of side effects and long history of use by humans. These factors short circuit the necessity for spending decades of expensive research in the development of drugs that can be used to activate thermogenesis. Although this trio is certainly not the end of the story, their use can proceed with success and confidence while the search for better agents continues. The benefactors are the thousands, perhaps millions, of people who will meanwhile experience the benefits of thermogenic activation in BAT.

CELLULAR MECHANISM OF ACTION: EPHEDRINE + CAFFEINE + ASPIRIN

Combining the actions of ephedrine, caffeine and aspirin produces the full blown thermogenic effect of ephedrine at about 1/4 to 1/3 the dose required if ephedrine is used alone. These actions may summarized as follows.

1. Stimulation of release of NE from the terminals of SNS neurons in the vicinity of BAT.

2. Stimulation adrenaline secretion from the adrenal cortex.

3. Stimulation of the multiple events that culminate in the accumulation of in BAT of FFA through lipolysis in BAT, WAT, blood and liver.

4. Stimulation of blood flow to, through and away from BAT, ensuring the delivery of substrate and dissipation of heat.

5. Stimulation of BAT growth and UCP proliferation with the resulting increase in thermogenic capacity in BAT.

6.   Inhibition of adenosine inhibition of NE action on adrenergic receptors and adenylate cyclase.

7.   Inhibition of cAMP phosphodiesterase, the enzyme that degrades cAMP and thereby reduces the amount cAMP available for FFA production.

8.   Inhibition of the inhibiting action on adenylate cyclase by prostaglandin E_2, by interfering with PGE_2 synthesis.

Discussions of each of the above events have already been provided and will not be reviewed further here.

PHARMACOLOGY OF EPHEDRINE + CAFFEINE + ASPIRIN

The notion that the combined action of ephedrine, caffeine and aspirin (ECA) would make a greater impact on thermogenesis in humans than any of the substances alone or in binary combinations was first discussed in 1988 by Miller and Dulloo of Harvard.[332,333] The next year, they published the pilot study on this effect, using over-the-counter drugs.[726] Diane Krieger, speaking for the Harvard group which included Patricia Daly, Dulloo, B.J Ransil, J.B. Young and Lewis Landsberg (Miller having passed away), before the Association of American Physicians, presented a summary of work in progress, actually the results of a limited pilot study on the ECA combination that indicated that it produced weight loss in obese humans, and was worthy of further investigation.

In this 8 week randomized double-blind placebo-controlled trial, obese subjects were administered an ECA combination (75mg/150mg/300mg respectively) or placebo each day. After 4 weeks, the E was increased to 150mg per day. At the end of the two months the ECA group had lost an average of 2.2kg (4.84 lbs) while the placebo group lost an average of 0.67kg (1.474 lbs).

No attempt was apparently made to relate the loss in weight to body composition, but the results were encouraging. Plans for further study were announced. Perhaps the most important feature of this study was the lack of significant side effects. No meaningful change was observed in heart rate, blood pressure, serum glucose, insulin or cholesterol.

It can truthfully be stated that thermogenic drug research over the past few decades reached a watershed in 1992 with the presentation of the next two-year study done by the Harvard group, led by Patricia Daly, on the combined action of ECA in human volunteers. This paper was delivered to great applause at the first International Symposium on Ephedrine, Xanthine, Aspirin & Other Thermogenic Drugs to Assist the Dietary Management of Obesity, held in Geneva Switzerland (results were published in 1993).[270]

The study was an eight week randomized double-blind placebo-controlled trial, followed by an eight week crossover trial involving the placebo subjects now consuming the ECA compound. The progress of a few subjects was then assessed at regular intervals for the next two years.

The rationale for the study consisted of three complementary factors. One, the previous work of Dulloo & Miller strongly suggested that ECA was a sound strategy for managing body composition and weight.[328,333] Two, the Krieger study suggested that further research on the ECA combination was warranted. And three, ECA combinations had been used sporadically for decades in the treatment of respiratory ailments and asthma and were known to be generally safe and well-tolerated. What remained to be answered was whether ECA would be tolerated on the more continuous basis required for thermogenesis activation, and whether the ECA product would yield results in humans comparable to those obtained in laboratory animals.[328] The Harvard group was the first to use ECA in humans (or animals, for that matter), and the somewhat surprising decision to proceed with a human trial at this early stage in the overall history of pharmacological manipulation of thermogenesis in humans must have been based on the general notion of aspirin safety.

The results showed that during the first 8 weeks the ECA subjects lost significantly more weight than the placebo group (2.2kg vs 0.7kg). During the second 8 week period, mean weight loss for the ECA group was 3.5kg vs. 1.3kg for placebo, again a very significant difference. During the following period of seven to 26 months, ECA treatment produced a mean weight loss of 5.2kg vs. 0.03 *gained* for the placebo group. Tolerance for the ECA was good, and validated the researchers' hypothesis that long term continued administration of ECA would not create serious side effects.

Strangely, the authors failed to evaluate the other primary hypothesis—that ECA would be significantly more effective than EC alone. The failure to include an EC control group in the experiment suggests that the researchers were not really attempting to rigorously test that hypothesis. Yet one of the most important questions inherent in using ECA instead of just EC is the degree of improvement that is afforded by incorporation of aspirin. The scientists address this question in the study supposedly by using smaller amounts of ephedrine and caffeine than had been used in previous trials. Yet, the 75mg and 150mg doses of ephedrine used by Daly, *et.al.*, are equal to or greater than the amount used in several other studies.[73,327] Furthermore, it is not made clear how the 150mg dose of caffeine qualifies as a substantially lower dose than that used in other studies.

Research at APRL has addressed this question to a small degree. Initial pilot studies designed to discover the best ratio between standardized ephedra, caffeine and aspirin revealed that the addition of appropriate amounts of aspirin (instead of willow bark, for example) resulted in greater efficacy and fewer side effects. Interestingly, the amount of ECA utilized by APRL was substantially lower than the amounts used by the Harvard group. Perhaps the use of plant materials exerts a more profound, complex and well-rounded effect on thermogenesis than do the drugs. The inclusion of cayenne in the thermogenic compound used by APRL may have also increased the effectiveness of the product. Cayenne would be expected to increase lipolysis above that expected from the ECA compound alone. Further work is required to explicate these findings.

ECA SAFETY: GENERAL CONSIDERATIONS

One of the main findings of the Harvard studies was the relative lack of side effects even at higher doses. In the Daly study, during the first eight weeks—double blind phase—of the 11 patients in the ECA group, the following side effects were reported: transient jitteriness (3 subjects); dry mouth and constipation (2 each). In the placebo group (13 subjects), one complained of jitteriness, two of dry mouth, and one of constipation. In both groups, side effects tended not to persist. There were no significant differences between ECA and placebo groups relative to systolic blood pressure, diastolic blood pressure, mean arterial blood pressure, heart rate, insulin, glucose, or total or HDL cholesterol.

During the second—unblind—phase, the incidence of side effects was smaller, and no statistical differences were found between the two groups. During the final—long term—phase, all subjects complained of dry mouth and constipation from time to time, but all other indices remained normal.

At APRL, we have found that symptoms are generally worse during the first few days. An inspection of the protocol used by the Harvard group reveals that evaluation of subject progress in phase I was made at the end of weeks 1, 4, 6, & 8. Therefore, the first request for side effects came at the end of week one, seven days into the experiment. Furthermore, the subjects were only asked to report side effects occurring during the previous 24 hours. Thus, if these procedures were strictly adhered to, problems occurring during the first 5-6 days would not be reported. Since this is the period during which adaptation takes place, the incidence of reported side effects may have been higher had these data been included. This is, admittedly, a minor criticism, but most manufacturers of thermogenic aids will attest that the first week is when the majority of complaints are registered. We suggest that future reports of human thermogenic research include mention of side effects occurring during the first week.

In summary, the pharmaceutical manipulation of thermogenesis has been an exciting enterprise during the last decade. It holds the promise of great profits for the first company to develop a safe and effective substitute for ephedrine (probably a specific beta-3 receptor agonist). Meanwhile, effective results in humans can be achieved safely with ECA derived synthetically and naturally. Given the possibility of minimal government interference, the public can look forward to several years of free access to this combination of thermogenic agents.

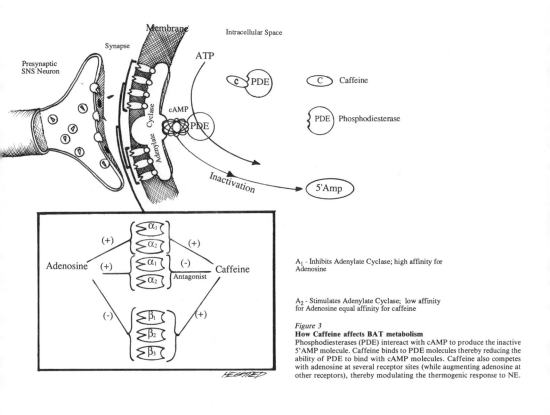

A_1 - Inhibits Adenylate Cyclase; high affinity for Adenosine

A_2 - Stimulates Adenylate Cyclase; low affinity for Adenosine equal affinity for caffeine

Figure 3
How Caffeine affects BAT metabolism
Phosphodiesterases (PDE) intereact with cAMP to produce the inactive 5'AMP molecule. Caffeine binds to PDE molecules thereby reducing the ability of PDE to bind with cAMP molecules. Caffeine also competes with adenosine at several receptor sites (while augmenting adenosine at other receptors), thereby modulating the thermogenic response to NE.

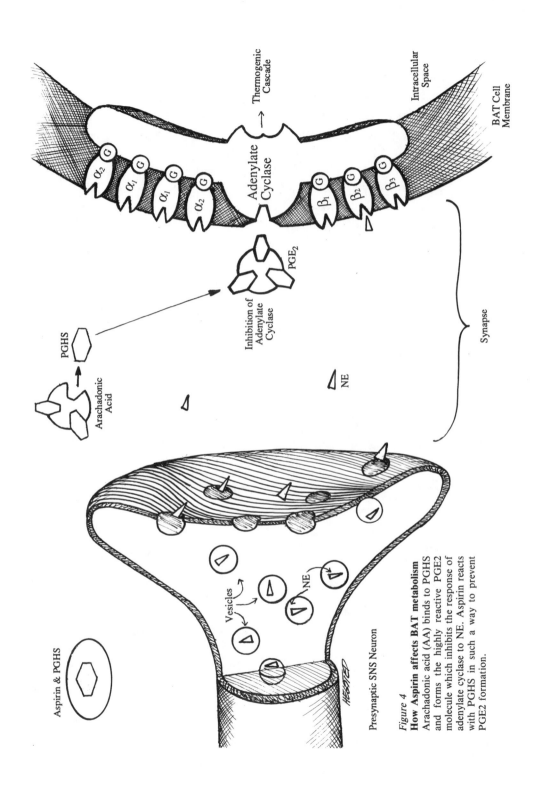

Aspirin & PGHS

Presynaptic SNS Neuron

Figure 4

How Aspirin affects BAT metabolism
Arachadonic acid (AA) binds to PGHS and forms the highly reactive PGE2 molecule which inhibits the response of adenylate cyclase to NE. Aspirin reacts with PGHS in such a way to prevent PGE2 formation.

PHYTOPHARMACOLOGY

The major part of this book has sounded druggy. So how is it that it should contain a chapter dealing with herbs and other plant materials? There is actually a very strong rationale for a discussion of plants in this book. It just so happens that at least two of the major thermogenic pharmaceutical agents are derived directly from the plant kingdom, and that most of the materials that nutritionally support it are plant materials. The study of herbs and spices is also a logical extension of past research in thermogenesis. The historical development of the thermogenic concept as it relates to diet extends back in time at least to Lavoisier who noted that the ingestion of food produces an increase in oxygen consumption and hence metabolic rate. This observation was the beginning of our understanding of TEF and DIT. Early experimental work was confined to macro-nutrients— protein, fat and carbohydrate. Later experimentation turned to certain micro-nutrients, such as caffeine and aspirin, derived from plants. More recently, research has studied the actions of other micro-nutrients in the diet, including active principles of herbs and spices. This chapter reviews available material on several major and minor herbs and spices that potentially affect the thermogenic process in human brown adipose tissue.

We feel this chapter is necessary for another reason. Perhaps never before has there been a greater need for a book directed at least in part to members of the alternative health community. On two previous occasions, we have sensed the need for a book targeted at this audience, as well as the general public. When we prepared the manuscript for The Scientific Validation of Herbal Medicine, we were trying to fill a void in American herbal lore regarding the average citizen's knowledge of scientific, experimental, clinical and medical research about our most popular herbs. Then, Guaranteed Potency Herbs: Next Generation Herbal Medicine was written to bring Americans up to date on the latest European research on standardization of herbal constituents to help overcome the inherent problems of variability of herbal materials. We coined the term "Guaranteed Potency" to differentiate these herbal materials from other standardized herbal extracts. The main difference was, and still is, a difference in the amount of basic and clinical research that has been done on the standardized product.

Anyone could standardize an herb; but would the new product be any better or safer than the original raw materials? Only several years of research could answer those questions. Once the research was done, and the results were favorable, then and only then, could the standardized herb be identified as "Guaranteed Potency". Interestingly, as soon as this phrase was coined, certain manufacturers, without knowledge of the concept or certain of its meaning, but convinced that it "sounded good," began identifying their products as 'Guaranteed Potency." This term has been used even on combination herbal products which could not possibly contain guaranteed potency herbs. Having been a strong supporter of wholistic herbal medicine for many years, we can feel good about herbal extracts and standardized extracts only to the extent that they are reliable and safe. Since the publication of the book on guaranteed potency herbs, many herbal manufacturers have taken steps to comply with the criteria for standardization, a few have taken advantage of the guaranteed potency herbs available, mainly from Europe, and increasingly from the East.

Unfortunately, some manufacturers have failed to sense the need or grasp the point. In addition to labeling regular herbal materials as guaranteed potency some manufacturers erroneously apply the term "guaranteed potency" to herbs that are really just standardized to some arbitrary constituent but lack good research on the preparation itself. For example, we are aware of an instance where a company is marketing a "gurmar" (*Gymnema sylvestre*) product that has been standardized to gymnemic acid. The label contains a circular stamp that designates this product as a guaranteed potency herb, yet the product as it appears in the bottle has never been subjected in any kind of research. Is gymnemic acid the most appropriate substance for standardization? There is no research available that addresses this question. The product may be standardized, but it is not a guaranteed potency herb. Ignorance and profit motive combine to dupe the public and create legitimacy where none exists. Hence, the consumer must once again rely on his or her own education and learning to discern the legitimate from the fake in the marketplace.

Like the book on Guaranteed Potency herbs, this book also addresses a new concept in the marketplace in need of definition and clarification. This occasion involves the need to understand how some very unique herbal materials and nutrients are being applied, since the possibility for misapplication is potentially large and the potential adverse impact devastating. Although most herbs, unlike their drug counterparts, are usually characterized by mildness of action, there are exceptions. While most herbs do not typically pack a large, immediate impact, some do have an immediate and fairly obvious impact on physiology. Most legitimate herbalists attempt to avoid the stronger-acting substances altogether, and it is indeed rare to find a legitimate use for them within the area of classical herbalism. So when the need for these substances does arise, it is not unusual to find some confusion regarding their proper application. Such a time is now at hand. We currently face a situation in which several so-called herbalists are not only promoting certain individual "strong-acting" substances, but are attempting to combine two or more such materials in products touted as dietary supplements.

Until recently, we have uniformly discouraged the use of herbs containing caffeine and ephedrine, and have always recommended the use of willow bark extracts instead of aspirin. This position conforms fairly closely to the feelings and practices of most American herbalists concerned with wellness and wholistic healing. Because of the growing need to address obesity-related problems[127] and for other reasons outlined in this book, we have been compelled to move beyond that position to a limited and very specific extent. Still, we routinely recommend abstinence from isolated ephedrine, caffeine and aspirin.

Ephedrine is the major constituent of chinese ephedra, or ma huang. The incorporation of a standardized ma huang extract in a thermogenic product can be easily engineered to provide a safe and effective quantity of ephedrine. In fact, it can be safely argued that ephedra is the plant equivalent of ephedrine. Caffeine is also found in plants. In this case, however, there are a variety of caffeine-containing plants from which to choose. Which one(s) is(are) selected is a function of personal preference, one's understanding of the differences among the plants, how best to combine them, and so forth. Caffeine-containing plants are not all equal. They each contain a variety of methylxanthine alkaloids of which caffeine is but one. Some plants contain a better constellation of xanthines than others.

Aspirin is not obtained directly from plants, but it is quickly derived synthetically from one of the main active constituents of willow bark: salicin. However, salicin and aspirin are not identical. Ignorance of the functional difference between aspirin and salicin led several herbal companies to produce thermogenic products that contain willow bark instead of aspirin. While this practice has beneficial consequences for most applications—antipyretic, anti-inflammatory—it is potentially a dangerous mistake in the field of thermogenics, as discussed later.

Cayenne is another plant that has the potential for strongly affecting several of the events of thermogenesis, especially the flow of blood to and from BAT, and lipolysis (the conversion of fats to free fatty acids, the main fuel of thermogenesis).

Because activation of thermogenesis by stimulation of the SNS adds to the stresses of everyday life, it is important that a thermogenic program include measures designed to minimize the impact of stress on the neuroendocrine system of the body. The plant kingdom contains dozens of substances that can perform these all-important functions. Plants that nourish and strengthen the nervous and endocrine systems and help restore and maintain balance in these systems are called tonics. While a discussion of herbal tonics is beyond the scope of this book, the reader is encouraged to read about them in my book Herbal Tonic Therapies, published by Keats Publishers, New Caanan CT.

American Phytotherapy Research Laboratory (APRL) is devoted to research on plant medicines ("phyto = plant). For the better part of the past two years, one of APRL's more important projects has been the intensive investigation of herbal thermogenics, nutritional thermogenics (see next chapter), pharmaceutical thermogenics, and combinations of the three disciplines. In so doing, we have discovered a handful of plants that appear to significantly affect thermogenesis in one manner or another, either by direct involvement in NE stimulation and lipolysis, or as modulators of adrenal, thyroid, cardiovascular and pancreatic function. While a good deal of experimental work remains to be done on these plant materials, it is not too early to recommend them as important primary and secondary substances for inclusion in sound plant-based thermogenic programs.

The following sections review the properties of several plants whose known properties probably impact on some aspect of thermogenesis and fat management. Since this book is not an herbal manual, a great deal of space has not been devoted to discussing the hundreds of plants that could possibly impact the thermogenic process. Instead, we have limited this discussion is the most likely candidates. Furthermore, with the exception of cayenne, references have not been included in this book. Those references are available elsewhere (see author's previous books, The Scientific Validation of Herbal Medicine; Guaranteed Potency Herbs: Next Generation Herbal Medicine; and Herbal Tonic Therapies, all available from Keats Publishing, New Canaan, CT).

ADRENERGIC/LIPOLYTIC PLANTS

The major plants that stimulate secretion of NE and adrenaline and affect lipolysis are ephedra, several caffeine-containing plants and cayenne. White willow bark and other salicylate-containing plants probably do not measurably affect thermogenesis. They are discussed in this section because of their popularity, and because they may augment, modulate, detoxify or otherwise affect the action of aspirin.

APRL has recently received a South American proprietary mixture of plants, called Mate Catalina, that purportedly exerts a thermogenic action without the aid of ma huang. This material is currently being prepared for clinical and experimental trials in our laboratory.

239

EPHEDRA SPECIES

Most species of ephedra are found in either China or India. The Chinese call it ma huang, and have used in medicinally for several thousand years. In India, ephedra is also a well reputed plant. Its juice is used to promote longevity and is even given to newborns, a practice belonging to Aryan custom that can be found in Rigveda. Even the Romans are said to have practiced this custom.[819] Ma huang, as well as its major active constituents, the alkaloids ephedrine and pseudoephedrine,[217,976,1370] have been used by the herbal and pharmaceutical industries for inclusion in products designed to ameliorate the symptoms of hay fever, allergies and asthma, as decongestants and bronchodilators.

The use of ephedra and ephedrine to stimulate thermogenesis is a relatively recent development in the history of ma huang use.[652] Yet this action depends upon the same pharmacological actions as do the bronchodilator and venoconstrictive actions. All of these effects are the result of stimulation of the sympathetic division of the nervous system, and all are mediated by the action of norepinephrine secreted from nerve endings of the SNS. Interestingly, however, different plasma cell membrane receptors for NE may be involved in the different actions.

Finding a substance that exclusively stimulates the receptors (beta-3) involved in BAT thermogenesis is one of the aims of current pharmaceutical and phytopharmaceutical research. Ephedrine nondiscriminately activates all types of beta-receptors.

Other effects of ma huang consumption include appetite suppression, increased basal metabolic rate, increased caloric burn rate during exercise, as well as certain mild side effects such as nervousness, insomnia, dizziness, hypertension, aggravation of an enlarged prostate gland, and nausea. Most of these side effects are dramatically reduced when ma huang is combined synergistically with aspirin, caffeine and other appropriate plant and nutritional materials.

As of this writing, it is becoming more difficult for herb companies to obtain ma huang. The importation of this plant is being severely curtailed by the FDA and U.S. Customs. Why? Because abuse of ma huang, especially ma huang standardized to higher than normal levels of alkaloids, is an ever-present danger,[94,1346] especially by individuals involved in creating, selling and using street drugs (the chemical similarity of ephedrine to methamphetamines (so-called "uppers," or "speed") makes this plant an ideal starting material for the synthesizing of the amphetamine)[667] and by health food manufacturers who fail to follow good manufacturing practices, especially of thermogenic agents.[959]

Herb Or Drug: Your Choice

The choice between using a good, high quality, standardized ma huang or synthetic materials such as ephedrine hydrochloride and pseudoephedrine hydrochloride can be reduced to a decision between drug and plant.[501] The implications of using drug are as follows: a uni-directional, simple, yet powerful action that operates within a single dimension to achieve a very predictable one-to-one effect on as narrow a range of metabolism as possible. The implications of using herb are as follows: a multi-directional, complex, yet milder action that will probably affect a broad spectrum of physiological systems in not very predictable ways. The herb effect will interact with the relative health of each body system more-or-less separately, yet the interaction between body systems will be highly individual and hence somewhat unpredictable.

The other prominent distinction between drug and herb, and one that arises naturally from the previous distinctions, is that the drug will have a more pronounced tendency to elicit the more serious of the possible side effects, while the herb will be much less likely to create these kind of side effects, yet will be more likely to precipitate so-called "cleansing crises" that represent an attempt by the body to detoxify itself. Part of the complex effect of plant materials is the stimulation of immune processes that result in a rapid influx into blood and lymph of toxins stored in cells that are then quickly expurgated. This crisis may be characterized by one or more of the follow-

241

ing symptoms: skin blemishes (as toxins are forced out through pores), nausea and headache, with or without diarrhea. The crisis may last from a couple of days to a couple of weeks.

The typical concentration of alkaloids in ma huang ranges from 0.5 to 2.0%.[500,652] But one can find 4, 5, 6, and 8% ma huang on the market. This means that the herb has been extracted and concentrated to contain increasingly higher levels of alkaloids. The extract is then spray dried back on to normal ma huang herb. A *standardized* extract is one that is guaranteed by the manufacturer to contain the percent of alkaloids found on the label. Unfortunately, in the case of ephedra, the standardizing processes used by several manufacturers are not reliable. Finding a high quality extract is sometimes frustrating for herb companies trying to maintain a superior product. Recent actions by the FDA are also making it more difficult to obtain high quality and effective material. Such short-sighted restrictive actions will ultimately impact plant-based thermogenics by making it extremely unlikely that products will contain the correct amount of ephedrine for significant impact on thermogenesis.

OTHER PLANTS WITH EPHEDRINE-LIKE ACTION

The use of yohimbe is sometimes recommended to stimulate thermogenesis. This herb contains the alkaloid yohimbine, which possesses sympathomimetic properties similar to ephedrine.[413,728,1401] Yohimbine is an alpha 2-adrenoceptor antagonist.[1401] It therefore increases sympathetic tone and stimulates lipolysis. However, there are at least three reasons why yohimbe is not suitable for widespread, general human consumption. First, it markedly increases heart rate and blood pressure. Second, it has psychogenic activity that could be troublesome for many people. Finally, yohimbe is the only known plant with true androgenic action. This would seem to make it unsuitable for widespread consumption by women.

Another herb that contains a thermogenic alkaloid is khat. Khat contains the alkaloid, cathine, which is essentially identical to norpseudoephedrine, an ephedrine-like alkaloid that also occurs in ephedra. Unfortunately, cathine is also a psychostimulant of some repute. Toxic psychosis is not an uncommon result of chronic khat consumption. Khat can not be recommended for general human use.[666,958]

Nicotine, from the tobacco plant, is yet another unsuitable thermogenic agent. Although it has been shown to activate thermogenesis in BAT at relatively high doses, the potential toxicity of this substance makes it of dubious value for addition to thermogenic compounds.[1377] On the other hand, persons who wish to quit smoking will be glad to know that consumption of a well-designed thermogenic will compensate for the thermogenic action of nicotine and will therefore help prevent the onset of weight gain following termination of the habit.[1335,1358]

Recently, a Chinese herbal combination called Mao-to was shown to inhibit cAMP, which would tend to retard thermogenesis. It was reasoned that the inhibitory action of this compound was derived from the presence of ephedra (ma huang).[933] This finding is directly contradictory to all that is known about the pharmacology of ephedra. Until more research is done, the results of this study must be viewed with some reservation. Perhaps other herbs in the mixture inhibited cAMP to a greater degree than ephedra could stimulate this enzyme, resulting in a net inhibitory action.

CAFFEINE/METHYLXANTHINE-CONTAINING PLANTS

One of the keys for getting ma huang to behave in an acceptable manner is to combine it with caffeine-containing plants. Caffeine is an alkaloid belonging the methylxanthine class. Perhaps no where else is the case for using whole herb instead of synthetic derivatives more clearly apparent than in the case of caffeine. The herb almost always produces a smoother, more beneficial effect than the synthetic. While the synthetic seems to work very well in combination with aspirin and ephedrine in the creation of thermogenics, restrict-

243

ing the formula to caffeine alone would eliminate the benefits of other nutrients in the plants, and the eliminate the possibility of healthy modulation and synergy afforded by whole plant material.

The differences in effects on the human body and behavior between different xanthine-containing plants are also easily discerned by anyone willing to consume sufficient quantities of each in succession. Not all xanthine-containing plants are equal. The field of thermogenics will benefit most significantly by restricting formulations to the use of the most beneficial plants, such as gooroo (kola) nut, yerba mate, and, to a lesser extent, guarana.

Gooroo Nut/Kola Nut/Bissey Nut.

One of the most important sources of caffeine is *Cola nitida*, the gooroo nut (kola nut, bissey nut, etc.), originating in Africa. The gooroo nut is considered one of the most important foods among many tribes in Western Africa. It is often given prominence reserved only for the most revered articles. Some writers have compared it to the 'Tree of Life' in Christianity, and its use is a part of many religious ceremonies and rituals.

According to legend, the Adam's apple is actually a gooroo nut. It seems that one day God, called away on important business, accidentally left his food, a gooroo nut, out on a tree stump where it was discovered by a passing native couple. Ignoring his wife's imploring to leave the food of the Gods alone, the man picked up the nut and began to chew it; whereupon the God returned, grabbed the man by the neck and began to choke him. The gooroo nut become lodged in the man's throat, and has remained so to this day as a sign of man's transgression against deity and fall from grace.

Like ginseng, gooroo has been regarded even as a panacea. Historical and anthropological writings by Greek, Egyptian, Arabian and Western observers suggest that natives used gooroo nut in the treatment of almost every debilitating condition. It was also used to ward off or prevent illness. Life extension was another common application.

When the natives were employed to work the early diamond mines, they exhibited an incredible degree of strength and endurance. Unaware of the importance of the gooroo nut in this regard, the mostly-European owners of the mines often forbid the 'heathen' and 'unhealthy' practice of chewing the gooroo nut and drinking the gooroo brews. Almost immediately, the performance of the crews dropped off significantly, depression set in, and near riotous conditions were everywhere observed. Realizing their mistake, the mine operators once again permitted use of kola nut and performance quickly rose to previous levels. The nature of this purported link between gooroo consumption and performance quickly became the subject of serious scientific investigation, and the use of kola nut extract became a world-wide practice.

The discovery of caffeine and other xanthines in gooroo helped to explain some of the plant's action on health and performance. The unique constellation of xanthines in the plant seemed to give the plant a superior position among similar plants, such as coffee and tea, for it imparted a large measure of health with minimal side effects. For example, it was used as an anti-stress, anti-fatigue substance, but continued use did not seem to impact the adrenal glands in a negative manner, or to create any of the typical signs of caffeinism. Additionally, kola was used to improve mental acuity, suppress the appetite, increase sexual performance, fight alcoholism, remove headaches, reduce the symptoms of asthma, improve convalescence, overcome depression, and improve cardiovascular performance.

Some of the beneficial properties of gooroo nut were explained with the discovery in the nut of a group of substances known as procyanidins. These substances impart to hawthorn berry its extraordinary cardiovascular tonic property. Finding them in kola nut was a significant event. Procyanidins improve the flow of blood to the brain and muscles, carrying with it an increased supply of oxygen and other nutrients that can be utilized to improve mental and physical performance.

Over a period of centuries the Africans have learned that gooroo nut is not just a source of caffeine, but is a food of supreme value if

245

used appropriately. The rest of the world knows benefits from the wide cultivation of gooroo nut. But it should be noted that so-called cola soft drinks do not contain the full spectrum of gooroo nutrients, and do not impart the same health benefits of whole gooroo nut. The entire constellation of nutrients and methylxanthines in whole gooroo nut make it an excellent candidate for inclusion in the thermogenic.

Guarana.

Another important xanthine-containing plant is *Paulina cupana*, guarana, a South American plant whose seeds are used by the natives in a manner similar to the African use of gooroo nut. The constellation of xanthines in guarana is not on an equal par with gooroo or yerba mate, but when combined with other xanthine-containing plants, the guarana xanthines, in small amounts, may contribute to the overall action of the compound in a positive way.

The tendency to put large amounts of guarana extract in a thermogenic product should be discouraged, since it has a marked tendency to overstimulate the nervous system and stress the adrenal glands.

Yerba maté.

One of the premier beverages of South America is *Ilex paraguariensis*, or yerba maté. There is an old Guarani Indian legend that relates the origins of the Guarani in the Forests of Paraguay. According to the legend, the ancestors of the Guarani at one time in the distant past crossed a great and spacious ocean from a far land to settle in the Americas. They found the land both wonderful yet full of dangers; through diligence and effort they subdued the land and inaugurated a new civilization. There were two brothers that vied for leadership of the people: Tupi and Guarani. Eventually they feuded and divided the people into two separate nations. Each nation, or tribe, adopted the name of the brother who was its leader. Guarani was the younger, fair-skinned brother, while Tupi, the older brother, had a darker skin.

The darker Tupi tribes adopted a more fierce, nomadic lifestyle, rejecting the agricultural traditions of their fathers. They engaged in the practice of drinking large quantities of a caffeine-containing drink prepared from the guarana tree.

The Guarani tribes became a stable, God-fearing people who worked the land and became excellent craftsmen. They looked forward to the coming of a tall, fair-skinned, blue eyed, bearded God (Pa'i Shume) who, according to legend, eventually did appear and was pleased with the Guarani. He imparted religious instruction and taught them concerning certain agricultural practices which would benefit them in times of drought and pestilence as well as on a day-to-day basis. Significantly, He unlocked the secrets of health and medicine and revealed the healing qualities of native plants. One of the most important of these secrets was how to harvest and prepare the leaves of the yerba maté tree. The maté beverage was meant to ensure health, vitality and longevity. The choice of favorite drink by the Tupi and Guarani came to symbolize opposition between the respective groups. The yerba maté of Guarani, reflecting the agricultural and domestic nature of these Indians, provided many more beneficial properties than the Tupi's guarana, which symbolized their preoccupation with running wild and free and their reliance on brute strength and the need to physically excel. Maté became the most common ingredient in household cures of the Guarani, and remains so to this day.

In current practice in industrialized, modern Argentina and Paraguay, maté tea is made from the leaves steeped in hot water. Actually, a large quantity of ground leaf is first soaked in cold water, then the hot water is added, over and over again, until all the good stuff has been extracted. In between each addition of hot water the tea is ingested through a special wood or metal straw, called a bombilla, that filters out the leafy material.

Among the native Guarani, the natural use of maté for healthful purposes has persisted. They use it to boost immunity, cleanse and detoxify the blood, tone the nervous system, restore youthful hair color, retard aging, combat fatigue, stimulate the mind, control the appetite, reduce the effects of debilitating disease, and so forth.

Several attempts to characterize part or all of the constituents of maté have been made during the last few decades. The one thing that unites the various assays is the consistent detection of numerous vitamins and minerals. There is the usual array of resins, fiber, volatile oil, tannins that characterize many plant substances. But then there is the growing list of vitamins and minerals, including carotene, vitamins A, C, E, B1, B2, B complex, riboflavin, nicotinic acid, pantothenic acid, biotin, vitamin C complex, magnesium, calcium, iron, sodium, potassium, manganese, silicon, phosphates, sulphur, chlorophyll, choline, and inositol. Different assays find different nutrients; there is probably no single assay that has found all of them.

One group of investigators from the Pasteur Institute and the Paris Scientific Society concluded that maté contains practically all of the vitamins necessary to sustain life. They focused especially on pantothenic acid, remarking that it is rare to find a plant with so much of this significant and vital nutrient. It is indeed difficult to find a plant in any area of the world equal to maté in nutritional value.

In addition, maté contains xanthine alkaloids. Though only small amounts of xanthine occur in maté, the presence of this substance has generated a huge amount of attention. The xanthine in maté is called matéine. Matéine is thought by most authorities to be identical to caffeine; but the effects of maté on the body are substantially different from those of simple caffeine. Doctors sometimes find it useful to give yerba maté to their patients who need to stop using caffeine products for health reasons. As mention in the chapter on Pharmacology, it seems to be the *pattern* of xanthine constituents in a plant that lends to it a distinctive profile.

There is only one xanthine property that seems to be shared by all xanthines: smooth muscle relaxation. It is this action that makes them, with the exception of caffeine, whose smooth muscle relaxant effects are diminished by other properties, good clinical dilators of the bronchi and hence useful in the treatment of asthma.

The effect of yerba maté may not even be attributable to any degree to the caffeine, yet its stimulant nature is well known. Researchers at the Free Hygienic Institute of Hamburg, Germany, concluded that the amount of caffeine in maté is so tiny that it would take 100 tea bags of maté in a six ounce cup of water to equal the caffeine in a six ounce serving of regular coffee (an opinion not shared by all experts). They make the rather astute observation that it is obvious that the active principle in yerba maté is not caffeine!

At any rate, maté appears to possess the best combination of xanthine properties possible. For example, like other xanthines, it stimulates the central nervous system, but unlike most, it is not habituating or addicting. Likewise, unlike caffeine, it induces better, not worse, attributes of sleep. It is a mild, not a strong, diuretic, as are many xanthines. It relaxes peripheral blood vessels, thereby reducing blood pressure, without the strong pressor effects on the medulla and heart exhibited by some xanthines. We also know that it improves psychomotor performance without the typical xanthine-induced depressant after effects.

Dr. Jose Martin, Director of the National Institute of Technology in Paraguay, writes, "New research and better technology have shown that while matéine has a chemical constituency similar to caffeine, the molecular binding is different. matéine has none of the ill effects of caffeine." And Horacio Conesa, professor at the University of Buenos Aires Medical School, states, "There is not a single medical contraindication" for ingesting maté. Clinical studies show, in fact, that individuals with caffeine sensitivities can ingest maté without adverse reactions.

Summarizing the clinical studies of France, Germany, Argentina and other countries, it appears that we may be dealing here with the most powerful rejuvenator known to man. Unlike the guarana of the Tupi, the coca of the Incas, the coffee of India, or the tea of China, maté rejuvenates not by the false hopes of caffeine, but simply through the wealth of its nutrients.

Dieters use maté to suppress the appetite, while providing necessary nutrition, energy and improved elimination.

Better than any other xanthine alkaloid, maté has the ability to quicken the mind, increase mental alertness and acuity, and do it without any side effects such as nervousness and jitters. These observations have been made time and again by qualified medical experts as well as by the lay user. The effects of maté on the nervous system are varied and not very well understood. The best guess is that it acts like a tonic, stimulating a weakened and depressed nervous system and sedating an overexcited one. Certainly the nutritional value of the plant cannot be overlooked as a possible substrate for improved health and function.

Our knowledge of maté's effects are currently limited almost exclusively to observations on gross changes in behavior: more energy and vitality, better ability to concentrate, less nervousness, agitation and anxiety, increased resistance to both physical and mental fatigue. One consistent observation is the improvement in mood, especially in depression, that follows the ingestion of the tea. This may be a direct result, or it may be an indirect result of increased energy.

Anecdotal reports of improved memory have not been substantiated experimentally or clinically, but are logical, and may again be attributed the nutrients, especially choline, and important central nervous system neurotransmitter. One of the remarkable aspects of maté is that it does not interfere with sleep cycles; in fact, it has a tendency to balance the cycles, inducing more REM sleep when necessary, or increasing the amount of time spent in delta states. Many people report that they require less sleep when using maté; usually such an experience is accompanied by a deeper more relaxing sleep.

Heart ailments of all kinds have been treated and/or prevented through yerba maté use. Yerba maté supplies many of nutrients required by the heart for growth and repair. In addition, it increases the

supply of oxygen to the heart, especially during periods of stress or exercise. The metabolic effects of maté appear to include the prevention of anaerobic glycolysis and the resulting build up of lactic acid during exercise. Reports of maté reducing blood pressure are not uncommon.

A consistent observation in most South American literature on maté is that it increases the immune response capability of the body, stimulating the natural resistance to disease. This also involves a nourishing and strengthening effect on the ill person, both during the course of the illness and during convalescence, sometimes dramatically accelerating recovery times.

Maté has long been known to prevent and reduce fatigue. The most logical mechanism of action at this time seems to be a direct stimulating effect on metabolism in muscle cells. Additionally, there is growing clinical evidence that maté stimulates the adrenal glands to produce corticosteroids. This mechanism of action may account for the commonly observed action of maté to decrease the severity and incidence of allergy and hay fever. The adrenocortical action may help explain reported cases of hypoglycemic patients responding to maté. It is possible that maté, by stimulating the adrenal cortex to secrete glucocorticoids, helps balance blood glucose levels. Similarly, it may also stimulate the production of mineralcorticoids, thereby helping to regulate electrolyte metabolism. These hypotheses are attractive, given certain clinical observations, but need to be scrutinized more closely in experimental settings. maté reduces the effects of stress on the body; this property probably involves a combination of effects on the endocrine system, the nervous system and the immune system, but is one of the most important of the herb's actions.

The combined effect of all of the above properties of maté on thermogenesis is to augment the action of ephedra, tone the adrenals, nurture the entire body, increase energy, mildly suppress the appetite, and improve cardiovascular function. None of these actions is extremely powerful, but added together, they may contribute a large measure of beneficial synergy with other thermogenic substances.

WILLOW BARK AND RELATED PLANTS

A survey of purported thermogenic products in health food stores reflects a bias toward including willow bark (Salix alba) and/or other plants containing salicin-like substances. This is a recommended practice. Excluding aspirin is not. <u>Willow bark is a fine addition to aspirin but an insufficient substitute for aspirin.</u> Since willow bark contains salicin—the immediate precursor to aspirin—the practice of us-

251

ing willow bark instead of aspirin, on the surface, appears to have merit.

However, according to the best current theory, advanced by the Harvard group, salicin and aspirin would probably not behave similarly. Aspirin is thought to enhance the action of ephedrine by inhibiting prostaglandin biosynthesis. Salicin probably does not inhibit prostaglandin biosynthesis, since it does not produce several of the side effects of aspirin, such as gastro-intestinal irritation and interference with blood-clotting. In this regard it behaves much like sodium salicylate. That is, the normal therapeutic effects of willow bark extract and/or salicin do not require the inhibition of prostaglandin biosynthesis, and arise in spite of its absence; but it is precisely this action that is responsible for the enhancement of ephedrine-induced thermogenesis.

We must note that the possible lack of activity of salicin on thermogenesis is currently just a theory. Even though there is good theoretical reasons to believe that salicin and aspirin possess widely disparate properties regarding thermogenic processes, a definitive experimental proof has not been forthcoming. Until research on this point has been done, it would wise to combine aspirin, whose positive action has been verified, with willow bark or willow bark extract, whose action remains unknown.

Aspirin itself is about as close to a natural ingredient as you will find among synthetic substances. It is created essentially by combining salicin with acetic acid (the main component of vinegar). More expensive aspirins are created in this manner. Less expensive brands are made through a multi-stage process that begins with aniline dyes, and is hence less natural.

History of Aspirin

"Natural" aspirin, that created from salicin has an interesting history. In 1763, a certain Rev. Mr. Edmund Stone of Chipping-Norton in Oxfordshire, England, wrote to a friend, "There is a bark of an English tree, which I have found by experience to be a powerful astringent, and very efficacious in curing aguish and intermitting disorders." Mr. Stone had discovered salicylates in an extract of the willow tree, though he didn't know it at the time. Over the next several years the willow bark extract would be shown to reduce fever and relieve the aches caused by several acute, shiver-inducing illnesses, or agues. The astringency of willow bark is due to the presence of salicin, the glycoside of salicylic acid. Other salicylates are found in the bark of white willow as well as several other willows.

Today, the most frequently consumed salicylate is aspirin, or acetylsalicylic acid. It has been estimated that Americans consume 16,000 tons of aspirin tablets a year; that's 80 million pills. The variety of applications for aspirin is truly astounding: to treat and prevent heart attacks and prevent cerebral thrombosis, reduce pain and fever, reduce the redness and swelling of joints in rheumatic fever, gout and rheumatoid arthritis, dissolve corns on toes, provoke loss of uric acid from kidneys, kill bacteria, inhibit clotting of the blood.

The earliest modern observers of willow bark activity noted that it tasted bitter, much like Peruvian bark (Cinchona spp.), the major source of natural quinine. Since quinine at the time was the major treatment for the ague, it was natural to suspect that willow bark would exert the same action. Actually, willow bark was used as long ago as the dark ages to treat feverish conditions. In those days, narrow branches of the tree were cut and stacked in the room of a patient, the idea being that the branches gave off an "aura" or fume that healed the afflicted. In prior centuries, Dioscorides (about 60 A.D.), Galen (about 200 A.D.), Hippocrates, and Gerard (1597), among others, used the willow tree bark in the treatment of corns, gout, aches and pains.

Following Stone's discovery, not much practical work was done with willow until the early 1800's when both French and German pharmacologists competed to find the active molecule. In Germany, J. A. Buchner isolated salicin. One year later, H. Leroux in Paris improved the extraction process and obtained considerable quantities of pure salicin. In 1833, a Darmstadt pharmacologist obtained a pure, cheaper extract. In 1838, Raffaele Piria of Pisa, working in Paris, named the substance *l'acide salicylique*, or salicylic acid. Meanwhile, a Swiss chemist Karl Jakob obtained a salicylate from mead-owsweet (*Spiraea ulmaria*; *Filipendula* u.) that he named *"Spirsaeure"*. This was later determined to be identical to salicylic acid. In 1844, A. Cahours demonstrated that wintergreen (*Gaultheria procumbens*) also contained considerable amounts of salicylates.

The Germans were the first to produce a cheap, synthetic sali-cylic acid. The Germans were the world leaders in the dye industry (I.G. Farben, e.g.); when it was discovered that salicylic acid could be made from cheap aniline dyes, this became the preferred method of preparation. The less expensive variety was 1/10th the cost of salicylic acid made from salicin. Aspirin itself was first discovered in 1898. Felix Hoffman worked at the Bayer division of I. G. Farben. Hoffman's father was suffering from severe arthritis, but reacted badly to sodium salicylate because of chronic and acute stomach irritation. Hoffman discovered that acetylsalicylic acid was tolerated better by his father, and was purportedly more effective. Bayer acquired the rights to the new compound and called it aspirin, from *acetyl* and *spirin*, a derivative of the German *Spirsaeure*.

Continued experimentation with aniline yielded substances such as acetanilde, created by reacting aniline with vinegar—and phenactin, the "P" in "APC" that since 1939 has been the product of choice by U.S. Army doctors (the "A" and the "C" are aspirin and caffeine respectively). Both acetanilde and phenacetin are bro-ken down in the body to form N-acetyl-p-aminophenol, or acetami-nophen. Neither acentanilde nor phenacetin have proved as useful and effective as aspirin in treating rheumatoid arthritis or rheumatic fever.

In spite of its long history of use, until recently aspirin remained an enigma as to its mode of action. Even now, there are competing theories about its mode of action. Early work suggested that it's anti-fever and analgesic actions were the result of prostaglandin synthe-sis inhibition. This theory needed revision when it was discovered that sodium salicylate, which possesses most of the analgesic prop-erties of aspirin, failed to inhibit prostaglandin biosynthesis. Cur-rently, scientists are predisposed to consider membrane transport dis-ruption as the basis for aspirin's action.

The failure of sodium salicylate to inhibit prostaglandin biosynthesis has immediate implications for thermogenesis. Since salicin has more in common with sodium salicylate than it does with aspirin (see above discussion), it is not reasonable at this time to assume that salicin inhibits prostaglandin synthesis. Since the thermogenic effect depends on inhibition of PGE_2, it is not prudent at this time to use salicin (or white willow bark) in place of aspirin. Further research may indeed vindicate the use of salicin, but until such time, aspirin is the only safe choice. Remember that without effective PGE_2 inhibition, it takes roughly twice as much ephedrine to evoke an effective thermogenic response in humans; that level of ephedrine approaches the dangerous realm and definitely possesses a risk to the adrenal glands.

CAYENNE

Though cayenne is not currently recognized as a primary thermogenic agent by the scientific community, we believe that it is only a matter of time until the importance of this plant is validated. Research currently underway at APRL is investigating the impact of appropriate amounts of cayenne on the thermogenic process and its impact on fat management. Thus far, we have noticed that the herb dramatically improves the thermogenic action of ephedrine-based compounds.

Few herbs have been used, abused, maligned and misunderstood as frequently as cayenne (*Capsicum annuum*). Cayenne has enjoyed a rich heritage. Hot red and green peppers have been the most commonly used spices in the world for hundreds of years.[200] In earlier times, cayenne was used to mask the flavor of spoiling food. It was also used to help preserve food. Over the centuries, it came to be identified with certain ethnic cuisines, and was incorporated in numerous curry blends.

History of Cayenne

Cayenne had its origins in Central and South America. It was then introduced to Europe, Asia and Africa by Columbus, who discovered it while searching for a new spice route to Asia. He took samples of cayenne, which he called "red pepper," back to Spain. On his second voyage, Columbus took Chanca, the physician, who brought the uses of cayenne as both a food and a medicine to the attention of the world. As the expansion into Central and South America continued, hundreds of observations and dozens of species of cayenne were recorded. Within a short time, because of the Spanish desire for world dominance, species of the plant were being cultivated in some parts of Europe as well as in Asia and Africa and other tropical and subtropical regions. The soil, climate, hybridization techniques and methods of cultivation in those areas combined to produce dozens of varieties.

It didn't take long after cayenne's introduction to civilized Europe for its medicinal and culinary properties to be experienced and researched. The major medicinal systems of the day soon incorporated the herb into their respective systems. Gerard mentions cayenne in his writings as early as 1597, but the herb was likely used even earlier than that.

One important historical development was the break with "regular" medicine that occurred in the early decades of the 19th century in frontier America under the leadership of Samuel Thomson, a self-taught, self-proclaimed doctor and herbalist. Rebelling against the tortuous "heroic" medicine of the day, Thomson sought a better, more gentle, mode. He eventually fell into the use of two primary herbs: cayenne and lobelia. He considered cayenne to be the best stimulant substance, and lobelia to be the best sedative. The proper combination and application of these two herbs, sometimes buffered by additional herbs, formed the foundation of Thomsonian medicine. Although the system eventually fell out of favor, one can still find practicing Thomsonians to this day. The late John Christopher was an outspoken proponent of Thomsonian ideas, especially as they involved the use of cayenne and lobelia.

Thomson did much to popularize cayenne in America. But the biggest impact on the modern use of cayenne for therapeutic reasons was the development of the gelatin capsule. The invention of the gelatin (digestible) capsule revolutionized the application of medici-

256

nal plants. Powdered whole herb in a capsule was basically a new way to take herbs that had far ranging consequences. One of the major obstacles to cayenne use was it's natural pungency. Natives of countries where the plant originates or where it is cultivated develop a tolerance for the acrid pungency. Newcomers to the herb seemed to have trouble taking it in any quantity. With advent of tablets and especially capsules, it became easy to ingest large, therapeutically effective, amounts. The method avoided the often unpleasant contact of herb to tongue, mouth and throat. People were now free to ingest a greater variety of plant materials and in larger quantities without being hampered by the plant kingdom's propensity for providing its nutrients in disgusting, sometimes painful, taste and touch sensations. Unfortunately, people may still experience adverse reactions in the stomach and bowel if they ingest too much caustic herb in this manner. Nevertheless, it is very easy in today's world to take any number of pungent herbs in large amounts.

The above discussion is pertinent to our discussion of thermogenesis. If cayenne should be included in the thermogenic compound, how much should be included? What guidelines can be follow? Can it be added in amounts that are both non-irritating and effective?

The Pharmacology of Capsicum/Capsaicin.

Before we answer those questions, we present a brief digression into the various physiological benefits of cayenne consumption. In 1876 the pungent principle of cayenne was isolated and called capsaicin. Capsaicin bears certain structurally similarities to eugenol, the principle in the oil of cloves, to nicotine and to lobeline, the active constituent of lobelia.[889] For one reason or another, capsaicin has been the subject of a large amount of research.[161,374,569,818,837,893,-907,11230,1321] Most of that research is immaterial from our point of view since the doses involved were huge compared to what can be obtained from a reasonable and typical human dietary dose. In many ways, the pharmacology of capsaicin is opposite to that of whole cayenne.[1263] Large amounts of capsaicin, for example, create hypotension, bradycardia and apnea;[119,133,153,230,886,935,996,997,1262,1336,1337] yet

257

the consumption of cayenne itself produces none of these effects. Whole cayenne can be used to promote thermogenesis, whereas capsaicin in large amounts leads to the loss of thermoregulation.[247,492,554,621-631,656,824,955,995,1213,1239] Preliminary studies by APRL in Mexico, where the use of extremely hot chilies is very prevalent, suggests that it is possible to "burn out" certain regulatory centers involved in thermogenesis. Other differences could be listed. The point is that the action of capsaicin cannot be equated to that of whole plant material even though capsaicin has received much more research attention than the plant. Only whole plant should be used in a thermogenic product.

General Pharmacology

Whole cayenne has been shown to have the following effects in animals and humans:

1. Lowers cholesterol and prevents rises that typically follow the ingestion of cholesterol-rich foods.[489,699,690,890,977,1099, 1097,1100,1195,-1196,1221]

2. Lowers the risk of blood clotting by inhibiting platelet aggregation and enhancing fibrinolytic activity.

3. Exerts a cardiotonic effect expressed as a normalizing of blood pressure, blood flow characteristics, and heart rate; strengthens and tones cardiac output.[442]

4. Increases activity rates and mental performance.[908]

5. Suppresses and prevents and infections.[392,859,1300]

6. Possesses secretagogue, diaphoretic and diuretic activity.[691,1190,1263]

7. Exerts a topical pain-killing action, especially on mucosal surfaces of throat and on gums.[448,593,619,1297]

8. Increases flow of digestive juices; increases tone of stomach and intestines.[1190]

Thermogenic Effects of Cayenne

In addition to the above properties, cayenne promotes thermogenesis through a number of means, which are briefly outlined below. Research has shown that small amounts of cayenne exert the following actions on human physiology.

1. Stimulates lipid mobilization in adipose tissue.[1197,690,692,938,1098]

2. Lowers peripheral adipose tissue weight and serum triglyceride concentration. Also lowers plasma phospholipids and total cholesterol values. Possesses a hypocholesterolemic action similar to that of curcumin (from turmeric), whole turmeric, gingerol (from ginger) and whole ginger root.[920,690,938,1098]

3. Stimulates release of catecholamine (adrenaline, NE) from adrenal medulla and SNS.[690,884,885,975,1220,1260,1261,1263]

4. Induces both obligatory and facultative thermogenesis.[891,938]

5. Increases glucose metabolism.[880]

6. Mimics and the depletes substance P, a putative neurotransmitter found in sympathetic ganglia and in the adrenal medulla. May play a role in postganglionic sympathetic neurons by modulating excitatory stimulation.[163,307,417,418,420,421,568,580,620,645,777,778, 904,909,997,1102,1103, 1180,1240,1241,1251,1322,1323,1324,1369] Nerves containing substance P are found in brown adipose tissue.[937]

Thermogenic Effects of Capsaicin

To review, cayenne contains a substance known as capsaicin, which is generally considered the principle active principle in the plant. Whether this is true or not, capsaicin has definitely been shown to help correct factors associated with obesity in animal studies. The effects should be observed at the concentrations found in therapeutic doses of whole cayenne.

1. Helps correct loss of brown adipose tissue.[692]

2. Causes a proliferation of new beta-adrenergic receptor sites on BAT cells.[690,692]

3. Raises resting metabolic rate.[519,692]

4. Increases the ability of mitochondria to use oxygen more effectively in oxidizing FFA, the fuel of thermogenesis.[690,692,785,1378]

5. Increases the concentration of adenylate cyclase (the intermediary between NE receptor and cAMP, the intracellular second messenger), cAMP itself, and GDP binding.[692,941,1378]

6. Elevates PGE_1 concentration; this is the prostaglandin involved in the metabolism of unsaturated fatty acids.[658]

7. Elevates hormone-sensitive lipase (HSL) and inhibits cAMP phosphodiesterase.[1095]

All-in-all, the addition of capsicum to the thermogenic program can have very positive results. Manufacturers should resist the temptation to use large amounts of pure *capsaicin*, since overdoses can desensitize temperature sensors in the hypothalamus to the point where they cease to function and may impair other organ functions.[79,444,294,887,892,1100,1183,1190] When this happens, atrophy of brown adipose tissue and depletion of uncoupling protein is certain to follow, and the person may lose all control over BAT thermogenesis.[257,258,259] To be safe, in preparing thermogenic compounds, only whole, powdered capsicum should be used, and the capsaicin concentration should never exceed 0.1%.

ADRENAL TONICS

The possible stress to the adrenal glands exerted by the thermogenic has been repeatedly emphasized throughout this book. Both the adrenal cortex and the adrenal medulla are affected by stimulation of the sympathetic nervous system. It is both necessary and unavoidable. If we lived in a stress-free world, the mild stress induced by ma huang and caffeine would probably be insufficient to worry about. But most of us live on the edge of our stress limits already. Add to that yet another stress and it may be just enough to create serious problems.

A good thermogenic should therefore embody at least enough adrenal support to compensate for the average estimated stress arising from the consumption of the thermogenic principles. Fortunately, there are a wide variety of adrenal support herbs and nutrients available from which to choose. The following pages certainly do not exhaust the candidates, but do discuss some that we consider the very best. An adrenal <u>tonic</u> is an herb that helps restore normal function to the adrenal gland. Tonics are bidirectional. They can either increase or decrease certain physiological events. Thus, their tendency is to maintain homeostasis in a given biochemical system. Tonic plants contain constituents that act in opposition to each other; the body selects which group to use.

SIBERIAN GINSENG, AMERICAN GINSENG AND ASIAN GINSENG

Ginsengs (Siberian, *Eleutherococcus senticosis*; Asian, *Panax ginseng*; and American, *Panax quinquefolium*) are among the most intensively research plants in the world. Species of ginseng grow on most continents, and though they may not be botanically related to one another, the properties of plants that go by the name 'ginseng' are very similar, differing mainly in degree of overt action, susceptibility to seasonal variation in effectiveness due to climatic variables, and the amount of political hype that surrounds them. Thus, proponents of Siberian ginseng are quick to point out that the action of Asian Panax is much shorter than that of Siberian, and that Asian is more prone to creating excitability in users. Asian ginseng users are just as quick to point to the centuries old reputation of Panax as the greatest tonic herb in the Chinese materia medica, and to its amazing ability to mediate stress reactions in the human body. Proponents of American ginseng emphasize the superior ability of this particular species to calm and soothe an excited central nervous system and its ability to restore tone to an imbalanced endocrine system.

By grouping all species of ginseng together under one heading, we are not trying to suggest that they are equivalent in all aspects.

However, we do contend that the use of any one, or two, or all three species is an excellent approach to maintaining the health of the sympathoadrenal and endocrine systems. These herbs have been shown to help maintain homeostasis in these systems during times of stress. Since the SAS must respond quickly to stress, but must not be allowed to deplete the adrenals, and must furthermore be able to recuperate quickly during non-stress periods, the herbal tonic should help in each of these goals.

Ginseng species generally exert offsetting catabolic and ana-bolic tonic actions on carbohydrate, lipid and protein metabolism. They exert both stimulating and sedative effects on the nervous system, increasing and decreasing effects of blood sugar levels and blood pressure, and act to stabilize immune system factors.

The wide range of tonic actions of the ginsengs has led to the practice of labelling these plants 'adaptogens,' or plants that help the body adapt to changing environmental and physiological con-ditions. The primary adaptogenic action is to increase the body's non-specific resistance to disease. This effect fits easily within Hans Seyle's General Adaptation Syndrome (GAS) theory, which pro-vides a much more comprehensive model than the 'adaptogen' theory for the manner in which the body deals with stress. The GAS describes in some details how the nervous system interacts with the glands of the body during times of stress. According to the theory, any agent that affects those interactions will affect the body's ability to successfully deal with stress. Some agents tend to disrupt the interactions, while others such as ginseng tend to stabilize and support the interactions.

While not all experts agree on how and where ginseng exerts its effects, most investigators have observed direct effects on the hy-pothalamic-pituitary-adrenal-sympathetic axis. Studies have shown that ginseng significantly reduces the typical signs of SAS-medi-ated stress: stomach ulcers, spleen enlargement, thymus enlarge-ment, anxiety, sleep disorders, etc. It also exerts a tonic effect on blood lipid levels and lipid metabolism that might help regulate the lipolytic requirements of BAT metabolism.

The often-observed antifatigue property of ginseng is thought to be mediated by a tonic effect on the adrenal glands. Stress is controlled by the adrenal glands through the secretion of adrenaline and glucocorticoid hormones. Ginseng produces an immediate increase in adrenal cortical capacity and output in response to stress. Increased adrenal weight and reserve capacity in ginseng-treated mice, as measured by both vitamin C and cholesterol content has been demonstrated, as well as the ability of ginseng to prevent depletion of adrenal capacity by repeated stress. Further, ginseng has been shown to increase the level of corticosteroids in the serum and to increase the level of pituitary adrenocorticotropic hormone (ACTH).

In one study, rats were subjected to cold or heat exposure and the ascorbic acid content of the adrenals was measured over the time course of the experiment. Ginseng groups were compared to non-ginseng control groups. The results showed that ginseng stimulated the initial depletion of ascorbic acid from the adrenals as the adrenals began to respond to the stress. Thereafter, ginseng increased the rate at which the ascorbic acid level of adrenals returned to normal. Thus, ginseng aided the response to stress and aided in the recovery from stress, exactly in accordance with what one would expect from an SAS tonic. The degree of influence of ginseng is always modest, suggesting a homeostatic mechanism rather than a purely stimulating action.

Now, relating these data to thermogenesis, it will be recalled that the initial stimulus of a thermogenic agent is probably to the hypothalamus, which secretes CRF which in turn stimulates thermogenesis and stimulates the pituitary to secret ACTH. ACTH, acting on the adrenal cortex, results in the secretion of glucocorticoids that inhibit both the hypothalamic secretion of CRF and the action of CRF on thermogenesis. The action of ginseng would therefore be to stabilize, normalize and maintain the homeostasis of this system.

LICORICE ROOT

Licorice root (*Glycyrrhiza glabra*) is one of the most important plants in the world. Among its properties is a capacity for dramatically improving the health of the adrenal glands. Though it has received an unjustifiably bad press at times, the use of *whole* plant material as an adrenal tonic is well supported by research.

The clinical relationship between licorice root and adrenal function actually comprises a little known chapter in the history of medicine. It is worth reviewing the research generated during this period (@ 1950-1965) not only because the information is both interesting and valuable, but also because there is some mystery about why that particular line of investigation ceased.

Though licorice root has been used in Europe and Asia for hundreds, if not thousands, of years, its scientific history actually began in this century, in Europe, where Dutch physicians noticed that a particular therapy used for peptic ulcers by a doctor from a small Netherlands town, though very effective, produced side effects closely resembling actions produced by deoxycortisone, the prescribed treatment for Addison's disease. This observation provided the impetus for a great deal of intense investigation on the properties of the substance used by the small town physician—licorice root (actually, a purified, highly concentrated licorice root extract). The research took many directions. Two important avenues that we do not have room to discuss involved its application in the treatment of ulcers wherein various licorice root extracts became arguably the best ulcer treatments in the world, and its application in the treatment of various liver and immune system disorders wherein its effectiveness rivals, and in many cases exceeds, that of the currently popular milk thistle extract (silymarin), another area wherein orthodox medicine has little to offer.

One of earliest and most interesting avenues of licorice root research involved the use of various extracts of this root in the successful treatment of many cases of Addison's disease. Since the progression of Addison's disease is directly related to deteriorating adrenal gland function, it was felt that somehow licorice was affecting the health of those vital glands. In early trials a crude mixture of all the various constituents of whole licorice root was used. Gradually, researchers began to refine the extract, eventually arriving at a concentrated solution of one or more of the following constituents:

264

glycyrrhetinic acid, glycyrrhizinic acid and/or glycyrrhizin—all closely related chemicals which, for the sake of simplicity, we will group together and call GLA. (3-4 gms of GLA = 20 grams of crude licorice root extract).

The success of this line of research led one investigator to write the following:

> "The discovery of the activity of licorice in the treatment of Addison's disease fills the physician with feelings of both amusement and modesty. Pernicious anemia...was for many years an incurable disease, while liver was available at every butcher shop. It now appears that one could have bought in every Dutch candy shop the substance that may save the life of a patient with Addison's disease. It makes one wonder how many more medical treasures of this kind are on exhibition in the windows of our grocery stores. Medical research is surely wonderful, but it sometimes achieves simple aims in complicated ways."

How many cases of Addisons's disease were treated is hard to tell, but cases reported in the literature numbered only about two dozen. These were usually women who had been exhibiting all the standard signs of Addison's disease; most were hospitalized, undergoing regular cortisone therapy.

A pattern emerged from the clinical research. Generally, it was found that in order for licorice root to be effective in reversing the symptoms of Addison's disease and maintaining those improvements, a bit of cortisone was also required. The exception to this rule was during the early period of the illness, when the adrenals were still somewhat active. Apparently, certain people are "weak" reactors to licorice root which probably reflects a complete lack of adrenal steroids in the body. When the adrenal glands stop functioning, so does licorice. As long as there is some activity in the adrenals, licorice can bring about remarkable improvements.

The research implies that licorice may be all that is required in cases where adrenal failure is not the result of some other long standing illness. In cases where licorice by itself does not seem to be working, a little bit of cortisone, administered along with the licorice, may be all that it is required; this way, the full potential of the licorice may be realized. Medically, the critical point is that such a treatment regimen avoids the many severe, eventually lethal, side-effects of traditional cortisone therapy. Even in the short run, patients using the combined licorice-cortisone therapy experience greatly superior metabolic, physiological and subjective health compared to the orthodox treatment involving just cortisone.

And so what has become of this revolutionary idea? Why do we not find licorice root extracts combined with cortisone in hospitals around the world for application in Addison's disease? What has subsequent research revealed? Following the initial investigations, research along these lines just stopped. And the potential benefits of licorice root extracts in the treatment of Addison's disease have never been realized. This answer to the licorice root/Addison's disease puzzle is so simple and exasperating as to almost defy belief, yet it is the truth: The men engaged in the original work simply went on to other fields of endeavor and never again wrote on the subject. Interest in herbal medicine in the Western world was ebbing and nobody bothered to follow up on these early leads.
On the other hand, lay people have continued to use licorice root to self-treat their Addison's disease. We don't hear much about the successful trials. Occasionally a case is reported in a medical journal—almost always because the person's adrenals were so exhausted that licorice root alone failed to completely reverse the course of the disease. Yet we are aware of numerous successful cases of self-treatment.

The implication of this research for thermogenesis is that licorice root may be used to offset stress to the adrenals arising from the stimulation of this gland by the thermogenic. The one qualification is that adrenals stressed to the point of exhaustion may not respond to licorice root alone, but may require the presence of a bit of cortisone (or some other agent—perhaps even ginseng would fulfill this role—the research literature is silent on this critical point).

The relationship between licorice and corticosteroids suggested other areas of application. Investigators asked, for instance, what effect licorice root would have on the course of rheumatoid arthritis? When all else fails in the treatment of arthritis, many doctors unwillingly resort to the use of cortisone. This brings relief to the patient and with it a psychological dependence that strains credibility, and all the unpleasant physiological side effects. In doses large enough to exert anti-inflammatory effects, cortisone also inhibits ACTH secretion by the pituitary gland, thereby leading to the cessation of adrenal gland functioning. The metabolism of carbohydrates, fats and proteins is disrupted, leading to symptoms of Cushing's syndrome. There is excessive retention of sodium and water and loss of potassium.

Research shows that licorice root may possess at least three different kinds of anti-arthritic effects. One, it may increase the levels of naturally occurring corticosteroids (produced by the adrenal glands). It does this by decreasing the body's ability to break down and inactivate the adrenocorticoids. The hormones thus stay active for a longer period of time, gradually increasing in concentration and effectiveness. The anti-inflammatory activity of the corticosteroids is thus enhanced. Two, GLA may block some of the side effects of cortisone without inhibiting the anti-inflammatory effects of this drug. Thus a cortisone-licorice combination may be a winning team in the fight against arthritis. It has been demonstrated that GLA prevents cortisone-induced atrophy of the adrenal glands. Three, licorice root has an anti-inflammatory action of its own. This action involves several metabolic pathways and generally targets the activity of a few important enzymes. However, there appears to be little current research interest in this potentially revolutionary treatment.

Again, there are implications of this line of research for thermogenesis. By prolonging the effects of circulating corticosteroids, licorice removes some of the stress being placed on the gland, giving it a chance to recover. We believe this action is tonic in nature, that is, licorice will not unnecessarily prolong the action of corticosteroids; once a normal level of activity is reached, the effect should subside.

A third avenue of research on licorice root also has implications for thermogenesis. In the folklore literature, licorice root is often recommended for hypoglycemia. However, one must realize that hypoglycemia is a symptom, not a disease, and can result from a number of different physiological problems. Of all the possible causes, licorice root may be helpful in just one: adrenocortical insufficiency. Without sufficient adrenocortical hormone (mainly the glucocorticoids), the protein building blocks for glucose (blood sugar) are not available, the blood glucose level drops and the glycogen supplies of the liver are depleted (glycogen is the main source of glucose). Licorice root contains principles that may reverse the process leading to hypoglycemia due to adrenal insufficiency. This form of low blood sugar often appears in Addison's disease.

Since the effects of stress are primarily on the adrenal glands, it is possible that hypoglycemia due to stress is one of the most common varieties of low blood-sugar related diseases. It that case licorice root could play an important role in the lives of many persons. It would act primarily by increasing the effectiveness of glucocorticoids already circulating in the liver and by mimicking the action of these hormones itself. These actions should improve adrenal health in a tonic fashion, and should impact favorably on thermogenic mechanisms, mainly by helping to increase the conversion of substrate fats to free fatty acids for use in the processes of UCP-dependent uncoupled respiration in brown adipose tissue.

Potential Licorice Root Toxicity

Since licorice root and its various extracts can have a powerful effect on adrenal functioning, one may legitimately pose questions about potential toxicity. After all, wasn't the whole line of research on licorice opened up because of observations of side effects? Indeed, there is such a thing as licorice abuse. But to the question, Does whole licorice root pose a threat of toxicity in the normal individual?, the answer is an unqualified "NO."

Perhaps no other herb has been as thoroughly and unjustly maligned as licorice root. The reason for this can be found in the literature. Medical researchers are often careless in their use of terms. When a child reports to a medical center suffering from severe potassium depletion, hypertension, edema, and salt retention (the usual indications of licorice abuse), and it is found that he has ingested several pounds of licorice candy over the past several days or weeks, the subsequent journal article will be certain to warn the public against potential licorice abuse. But it wasn't licorice root that was being abused at all—it was concentrated licorice root extract, even pure GLA.

It has been found that about 20% of the population is sensitive to GLA. The reason for this sensitivity is simple to understand. In most tonic herbs there are usually constituents that push the body's chemistry in two opposing directions; the tendency of the whole herb is toward stasis, balance, or normalcy. If you isolate one of the constituents there is a chance, without the presence of countervailing constituents, that it could push the chemistry one way only. That is what GLA does. Most licorice root extracts are some form of concentrated GLA. The stuff left over when the extract of root is made is generally considered to be inactive, even though research has proved beyond doubt that the dregs are almost as effective as the GLA in healing ulcers.

Licorice root extract is sometimes incorporated into habituating laxatives. There are reported cases of severe toxicity from using these kind of laxatives two or three times daily. The potassium-depleting action of the extract is greatly exacerbated by the practice of habitual laxative abuse. As in all cases of such licorice root extract-induced potassium depletion, the symptoms are completely reversible through potassium supplementation: there is no long term damage.

The Chinese have used great quantities of licorice root down through the centuries; it is included in the majority of their herbal preparations; it is called "The Great Detoxifier." They do not complain of toxicity, but experience just the opposite effect.

In the West, the isolation of licorice root constituents has presented a mild problem. But the use of whole licorice root has never been a verified or verifiable cause of toxicity. The 20% of the population that appears to be sensitive to the aldosterone-like activity of GLA should avoid overuse of true licorice-extract flavored candy (American licorice contains no licorice by the way), laxatives containing large amounts of licorice extract, and anti-ulcer drugs that are nothing but concentrated licorice derivatives.

As discussed in a previous book, <u>The Scientific Validation of Herbal Medicine</u>, the bad press on licorice root is exactly that. Inadequate familiarity with the research coupled with a characteristic susceptibility to scare tactics have led some writers, lecturers and other so-called health experts to be overly cautious about a situation that really needs very little caution at all.

THYROID TONICS

As discussed at length in the chapter on the thyroid, a normal functioning thyroid gland is critical to the thermogenic process. Although thyroid hormone does not directly participate in the thermogenic reaction, the presence of this hormone is necessary to permit certain of the underlying processes to occur in a reasonable manner. On the other hand, too much thyroid hormone will tend to inhibit thermogenesis. Therefore, the choice of thyroid-sustaining herbs should be limited to thyroid tonics, that is, to herbs that normalize thyroid function. Only in the case of an individual with diagnosed thyroid insufficiency would a thyroid stimulant be appropriate. The best thyroid tonics are seaweeds. The most well-known and best researched seaweeds are bladderwrack and kelp.

BLADDERWRACK

Due to its content of natural, organic iodine, bladderwrack (*Fucus vesiculosis*) is recognized worldwide as one of the best weight-reduction plants available. Iodine maintains a healthy thyroid, especially in people with myxedema (adult-onset hypothyroidism), thereby significantly reducing one of the major causes of obesity. The production of the important hormone, thyroxine, is bolstered by iodine, as found in fucus. The presence of thyroid hormone in brown adipose tissue helps to insure that effective lipolysis will take place. In addition to decreasing the chances that the thyroid gland will develop goiter, the iodine in fucus helps regulate the texture of the skin, and helps prevent dull hair.

Often overlooked in discussions of fucus is the ability of this plant to directly affect energy metabolism. By significantly affecting the rate of oxygen consumption in the heart muscle and in other muscle tissue, fucus increases the body's ability to burn off fat during work or exercise. Stamina is boosted, allowing cells to consume energy more efficiently. Iodine is also essential for the proper regulation of energy metabolism.

Fucus also contributes to weight regulation by preventing the absorption of dietary cholesterol. The mechanism of action of this effect is related to the presence of fiber, and its ability to bind bile acids and bile salts. In studies, fucus increases the concentration of, and daily excretion of, cholesterol, deoxycholic acid, lithocholic acid, and total bile acids in the stool.

In addition to direct effects on weight loss, fucus possesses properties and nutrients that augment the health of organs and systems that are often adversely affected by excessive weight. For example, it provides several nutrients, including vitamins and mineral salts, that contribute to the health of the heart and the blood vessels. There are also antibiotic principles that help prevent infectious and inflammatory disease. It has also been postulated that fucus activates the body's immune system.

Fucus further helps to regulate the thermogenic action of ephedrine by providing a certain degree of hypotensive action that helps offset any tendency that NE might exert on beta-2 receptors. Hypertensive people experience significant improvement in blood pressure readings, subjective well-being and cardiac efficiency from the use of fucus. No side effects are experienced.

Fucus aids in protein synthesis, carbohydrate absorption and the conversion of carotene to vitamin A. Finally, the fucus mucilage has a distinct appetite suppressing action.

KELP

The nutritive value of kelp (*Laminaria spp.*) contributes to the health of all the major organs and glands of the body, but the thyroid is singled out for special support. Like fucus, kelp is a major source of good, natural, organic iodine, along with numerous other trace elements and minerals. Kelp not only absorbs iodine from seawater but sponges up an enormous supply of essential nutrients, and delivers them to the thyroid gland and the rest of the body. These nutrients include protein, essential fatty acid, carbohydrates, fiber, trace elements, sodium and potassium salts, and a variety of other chemicals, such as alginic acid.

Kelp and fucus are two of the best weight-reduction plants available. The iodine in both plants helps to maintain a healthy thyroid. All of the material pertaining to thermogenesis, thyroid and weight management provided in the section on fucus apply almost equally to kelp. The only difference is historical. Fucus has the edge on kelp in terms of the reputation among reputable European herbalists in terms of weight management. However, in the Orient, kelp is used for much the same purposes.

Kelp also contributes to the health of the cardiovascular system in three primary ways. First, it provides several essential nutrients, including vitamins and mineral salts. Second, kelp's antibiotic principles help prevent infectious and inflammatory disease. Third, and most importantly, kelp is hypotensive. The Japanese preparation called "kombu," made from the leaves of several species of kelp, is routinely used in the treatment of hypertension in Japan. Research on kombu has yielded confirming results. For example, in one study hypertensive patients experienced significant improvement in blood pressure readings, subjective well-being and cardiac efficiency after drinking a kombu preparation. No side effects were observed. Several subsequent studies have reported similar observations in experimental animal models, as well as in humans.

The active principle in kelp, responsible for the hypotensive effect, is believed to be laminine, but histamine is probably also involved. Additionally, some investigators believe that the primary active constituents have not yet been isolated.

Kelp possesses other cardiotonic principles. For example, it increases the contractile force of muscles in the atria. In animal models, constituents of kelp stimulate the overall activity of the heart. So far, these findings have not been experimentally observed in humans, but there are clinical reports of improved cardiac health following kelp treatment.

Kelp possesses several other important properties. Although these do not pertain directly to thermogenesis, they are briefly discussed to provide the reader with an introduction to the full range of values obtainable from seaweed. Kelp is able to bind with several kinds of pollutants and prevent their absorption from the gastrointestinal tract. In experimental situations, kelp binds with radioactive strontium, barium, cadmium and other heavy metals. The salts that are formed in the binding process are inert or insoluble, that is, they cannot be digested or absorbed. The prevention of strontium-90 toxicity is particularly important in today's world, since this metal has a great affinity for calcium and accumulates in food substances that are high in calcium (such as milk and green leafy vegetables). Contaminated calcium carries the strontium directly to the bones where it damages the marrow. Kelp actually strips the metal ions from the calcium molecule and forms the insoluble salt which is harmlessly excreted in the urine and feces. Incidentally, brown kelp is the only species capable of supplying the sodium alginate principles responsible for this kind of anti-toxic activity.

Many kinds of kelp have been found to possess antioxidant, anticarcinogenic and general antitoxic properties. By way of summary, kelp prevents cancer and reduces the risks of poisoning from many sources of environmental pollution in the following ways: (1) by providing a source of non-digestible fiber; (2) by reducing cholesterol levels; (3) by altering the nature of fecal flora; and (4) by a direct cytotoxic effect.

Until recent years, kelp was eaten almost exclusively by the Japanese. Cultural studies have revealed interesting attributes of health that appear to be directly related to kelp consumption. Heading the list is a dramatically lower rate of breast cancer. Less obesity, less heart disease, less respiratory disease, less rheumatism and arthritis, less hypertension, less thyroid deficiency, less constipation and gastrointestinal ailments, and less infectious disease—all of these positive attributes of Japanese health seem to be related to using large quantities of kelp.

Among the interesting properties of kelp is its ability to ward off infections and fevers. In one study, the diets of one of each of seven pairs of monozygotic (identical) twin cows was supplemented with kelp for seven years. During that period the cows getting kelp yielded more milk and had a much lower incidence of mastitis than did the controls. Since mastitis is usually caused by unsanitary conditions, the antibiotic property of kelp is probably responsible for the control of this condition.

PANCREATIC/INSULIN TONICS

The important role of pancreatic hormones was reviewed in the chapter on Insulin. The proper balance between insulin and glucagon is extremely important in maintaining the efficiency of the metabolic events underlying the fat of dietary glucose, fat, and protein, as well as the conversion of foodstuffs from active molecules to storage molecules and vice versa. Once more, the object of proper health is to achieve homeostasis among these events so that the body can respond rapidly to changes in cellular needs. Certain herbal materials encourage this homeostasis,; their inclusion in a thermogenic program will benefit the end user by maintaining his or her ability to supply the substrates for BAT metabolism in ways conducive to a reduction in body fat stores. Additionally, these herbs will encourage the proper metabolism of fats and carbohydrates by all other cells of the body, not just BAT cells. Skeletal muscle metabolism, including thermogenesis, will be specifically impacted.

The following herbs are among the best pancreatic tonics in the world. In addition to their action on the pancreas, they often affect a wide variety of tissues throughout the body in a tonic and healthful manner.

SIBERIAN GINSENG/PANAX GINSENG
Ginseng exerts several effects that support the thermogenic process. Several of these were summarized earlier. In keeping with the tonic action of ginseng, it has been observed that ginseng consumption tends to stabilize or normalize blood sugar levels. The exact manner in which it does this is not known, but probably involves some homeostatic effect on pancreatic hormones. The tonic effect of ginseng on the adrenal glands, already discussed, would be expected to impact glucose metabolism in a favorable manner also.

GURMAR
Gurmar (*Gymnema sylvestre*) leaf is employed as a treatment for obesity and diabetes in Indian Ayurvedic medicine. After two thousand years of folklore use, gurmar has finally been subjected to the scrutiny of scientific examination. The results of these studies have

been heralded as miraculous in some popular circles. Scientists, while not willing to use terms like "miraculous," have nevertheless described gurmar's actions as unique and "startling."

The common name "gurmar" means sugar destroyer, in reference to the fact that the leaf has the peculiar property of neutralizing the sweet flavor of sugar placed on the tongue following a few seconds of chewing the gurmar leaf. It is believed that gymnema causes a temporary anesthesia of the tongue, which somehow selectively blocks the taste mechanism for sweetness.

Gurmar markedly affects the fate of dietary glucose or sugar. It insures that the maximum amount is metabolized and used properly in the energy producing cycles of the body. Through a complex mechanism gurmar maintains blood sugar levels at optimum levels even when dietary sugar intake is higher than it should be. Therefore, gurmar is a potentially excellent adjunct to the diets of diabetics and dieters.

The main activity of gurmar is due to the presence of a brittle, black, somewhat complex acidic resin known as gymnemic acid and a glucoside known·as hentriacontane. The leaves also contain chlorophyll, xanthopylls, carotene, phytol, pentatriacontane, and lime salts.

Among the basic pharmacological properties of gymnema are the following: antiviral; increased oxygen uptake and blood pressure; increased secretions of liver and pancreas. The leaves have also been found to stimulate the heart, uterus, circulatory system, and to raise urine output.

Most studies confirm the blood-sugar lowering, or anti-diabetic, property of gymnema. The word "control" best describes the action of gymnema on blood sugar and diabetes. It is not a cure, and does not totally substitute for proper dietary habits, but its use will significantly help to keep blood sugar levels within acceptable limits, even when the dieter "falls off the dietary wagon."

In one of the first studies on gurmar in diabetic patients, the herb significantly lowered blood glucose levels and total urinary glucose excretion in many, though not all, of the patients. Similar results have been found in almost all subsequent clinical trials.

Animals studies have also yielded very positive results. For example, in a very thorough study, gurmar greatly reduced the blood glucose levels in diabetic rabbits. Another study showed that gurmar was significantly more effective in correcting high blood glucose levels in moderately diabetic rats than in animals with severe and toxic diabetes. The anti-diabetic effect in the moderate group lasted for more than two weeks after treatment stopped. But while gurmar did not work in the severely sick group, it did significantly prolong survival time. The lack of hypoglycemic effect in the severe and toxic groups suggests that alloxan destroyed the pancreas "B" cells in these groups so completely that these cells could not be rejuvenated by treatment with gurmar (or anything else).

In another study, involving both animals and humans, gymnema reduced fasting blood glucose levels, together with serum cholesterol and triglyceride levels, while improving serum protein levels. However, the blood sugar lowering action took several weeks to develop and normalize, suggesting that, in some patients at least, it may take several weeks for gurmar to effectively stabilize the conditions producing the typically extreme fluctuations in blood sugar as seen in diabetes.

One important finding to emerge from the research is that gurmar apparently has no significant blood sugar lowering effect in normal individuals; the action is only apparent in individuals displaying diabetic symptoms. Furthermore, while gymnema does not lower blood sugar levels in normal subjects, it does appear to prevent a rise in blood sugar levels in normal subjects. Because of this tonic action, non-diabetic individuals can enjoy the other benefits of gurmar without worrying about experiencing detrimental effects to their blood sugar from either dietary sugar or from the gurmar itself.

Of course, most of the above material has important implications in the control of weight. With better glucose utilization, there are less "empty" calories to be disposed of by the body. In numerous clinical trials, gymnema has been used as an adjunct to normal diets, with highly favorable results.

The ingestion of gurmar does not mean that the dieter can ignore other important dieting habits, but it does mean that not quite as much attention need be paid to counting calories. And it does mean that ingested sugar stands a better chance of being properly metabolized in body cells.

Much work has gone into the attempt to elucidate the method of action of gurmar, for it appears to hold the keys to revolutionary advances in the management of diabetes and obesity. So far, scientists have only been able to speculate about possible modes of action. It may work by mediating the direct or indirect stimulation of insulin release, or by inhibiting intestinal absorption of glucose. Gurmar appears to reverse the normal depletion of glycogen and protein seen in diabetes, and to inhibit unwanted lipid accumulation.

The idea that gurmar works by increasing the levels of circulating insulin is supported by evidence that it increases the activity of insulin-dependent enzymes in the liver, kidney and muscles. It also increases the activity of enzymes that affect the utilization of glucose by insulin-dependent pathways. Furthermore, gurmar reverses signs of liver and kidney damage, and corrects metabolic derangements in liver, kidney and muscle tissues caused by diabetes. Since gurmar is just now being directly tested in thermogenic experiments, it is not known to what degree these properties affect the events of thermogenesis. Preliminary results at APRL suggest that gurmar does definitely enhance the thermogenic activity of a correctly formulated thermogenic agent.

The ability of gurmar to prevent rises in blood sugar levels has been tentatively attributed to a pancreotrophic effect due to sensitization of beta-cells of islets of Langerhans for secreting larger amounts of insulin in response to glucose. In addition, gurmar markedly inhibits somatotropin- and corticotropin-induced elevations in blood sugar levels.

While there is some indication that gurmar directly neutralizes some ingested sugar, the idea that it works by somehow binding with the sugar and thereby preventing it from even being metabolized, let alone being available for involvement in pancreatic and blood sugar functions, is not supported by the experimental literature. Such simplistic notions rarely turn out to be correct. Right now the mode of action of gurmar appears to be very complex, and at this time remains a wonderful mystery for some future nobel-prize aspiring scientist to unravel.

It is of interest to note that some authors, while not directly addressing the issue, are essentially describing the "adaptogenic," or tonic, nature of gurmar, since it increases the body's ability to adapt to the presence of sugar. The increase in longevity noted above was ascribed to "cardiotonic and adaptogenic characteristics produced by increasing resistance and immunity in diabetic animals." It would have to be a characteristic of a true adaptogen to "normalize" function, not just prolong life. Recently, researchers have reported that gymnema prevents death due to hypoglycemia in rats injected with beryllium nitrate. That the herb prevents both rises and falls in blood sugar and causes no significant change in normal blood sugar levels is in full harmony with the concept of a tonic, as exemplified by such famous plants as ginseng.

Gurmar possesses no known toxicity, and is perfectly safe to use in even large quantities. Diabetics currently using anti-diabetic drugs or diuretics, should consult with their physicians before, during, and after treatment with gurmar, as some adjustment in insulin dosage may be required. Daily monitoring of blood glucose levels is advised.

CARDIOVASCULAR TONICS

Cardiovascular tonics make good adjuncts to a thermogenic program as they tend to stabilize any tendency a person may have to hypertension or heart arrhythmias. Small amounts of these tonics may be combined directly with the thermogenic, but a more dramatic and health-promoting impact can be obtained by consuming them separately, i.e., in another capsule, in a tea or extract, or in a variety of other ways.

279

GINSENGS

The tonic cardiovascular properties of the ginsengs are well known, and are discussed in an earlier section. Please refer to that section for additional information. By way of review, ginseng has a tonic effect of blood pressure and cardiac output. It promotes oxidative metabolism in the heart, improves enzyme action, substrate utilization and generally strengthens cardiovascular health. Soviet athletes have relied on Siberian and Asian ginseng for years for helping to condition themselves for endurance and performance events. While Soviet research in these kinds of areas is not known for its objectivity, the shear volume of data so far produced argues in favor of the cardiotonic action of ginseng.

VALERIAN ROOT

Valerian root (*Valeriana officinalis*) has been used by man for centuries to calm upset nerves, treat mood problems, pain, headache and cardiovascular problems such as hypertension. It reduces the severity of both physical and the psychological symptoms. That is, it acts as both sedative and tranquilizer, for nerves and muscles. It is a good day time and night time sedative. Unlike sedative drugs, valerian actually improves coordination, improves concentration, and antagonizes the hypnotic effects of alcohol. Large doses are not any more effective than small or moderate doses, except to extend the duration of the effect. Thus it is almost impossible to overdose (though in some manner as yet unexplained, it appears that large doses may act in a stimulating, rather than relaxing, manner in some people). Valerian quiets the hypertensive patient and helps children overcome behavior control disorders and some kinds of learning disabilities.

Valerian root preparations have direct neurotropic effects on higher centers of the central nervous system. Its use in the treatment of sleep disorders such as insomnia is supported by an extensive library of research. It has also been effective in treating rhythm disturbances of

280

the heart that just precede going to sleep. Valerian improves measures of sleep quality and sleep latency but leaves no 'hangover' the next morning as is often observed with traditional sleeping aids.

Valerian increases coronary flow and produces distinct hypotension. It is included in some German heart tonics to provide inhibition of reflex hypersensitivity and to help maintain neuro-coronary equilibrium. Hypotensive, anticonvulsant and antiarrhythmic properties have been observed in several studies. Also, valerian root has been shown to prevent the appearance of acute coronary insufficiency. Moderately positive inotropic (ability to improve the muscular contraction of the heart) effects have also been observed.

On the basis of the above observations, we would expect valerian to improve thermogenesis by reducing any preexisting tendencies toward hypertension. In addition, it probably affects the level of circulating catecholamines in a positive manner.

HAWTHORN BERRY

Hawthorn (*Crataegus oxyacantha*) is, without doubt, one of the top three cardiotonic plants in the world. It is also one of the most researched. German scientists, in particular, have been studying hawthorn preparations in human patients for dozens of years. According to one noted expert, hawthorn has three basic healing properties, which combine to complement each other.

1.   It improves the flow of blood to and through the heart itself; clinically this means, among other things, that it will decrease the incidence and severity of anginal attacks.

2.   It improves metabolic processes in the myocardium; this gentle action means that general heart activity and health are improved.

3.   It abolishes some types of rhythm disturbances. This is in opposition to certain ill-founded gossip in this country that hawthorn will cause arrhythmias. Nothing could be further from the truth. Haw-

thorn is a tonic, and tonics, by definition, do not push systems, organs or cellular mechanisms of the body into abnormal activity; instead, they stabilize or normalize physiological activity. The totality of research on hawthorn has confirmed its tonic nature.

My own conclusions concerning the main actions of hawthorn, based on a review of all available research, are also threefold and overlap to some degree with the three listed above:

1.   Hawthorn dilates peripheral blood vessels, i.e., those outside of the immediate area of the heart. This action reduces hypertension and acts synergistically with various herbal sedatives, such as valerian root, to help one relax and successfully adjust to stress.

2.   Hawthorn stimulates enzyme metabolism in the heart muscle. This is the primary cause of its cardiovascular tonic quality when used for any length of time.

3.   Hawthorn improves the efficiency of oxygen utilization by the heart. This results in dramatic improvement in all kinds of physical exercise and work.

It is generally acknowledged that hawthorn is well-suited for use in the treatment of disorders related to the following conditions:

1.   Old age-related vascular problems.

2.   Hypertension; Nervous conditions.

3.   Coronary artery and perfusion disorders.

4.   Rhythmic disturbances of the heart.

Fifty-two men and women, patients in a sanatorium, who were experiencing coronary perfusion disorders, mainly due to coronary sclerosis, were given hawthorn extract every day for several weeks. These patients were also maintained on a "standard treatment regimen." Each day the subjects were required to undergo an exercise regimen that would typically result in some form of anginal attack. It took an average of less than two weeks for the majority of patients (77%) to exhibit significant improvement, both on subjective and objective measures.

In a thermogenic program, we would expect hawthorn berry to increase both obligatory and facultative thermogenesis, help stabilize blood pressure, raise BMR, and increase energy metabolism in general. It is difficult to estimate the amount of help that hawthorn would provide. It is still much to early in the history of thermogenic research to expect science to devote much attention to how herbs like hawthorn influence the metabolic events associated with the dissipation of excess calories.

CAYENNE

Cayenne is often recommended as a cardiovascular tonic. It has a very beneficial effect on general circulation and would therefore be expected to interact favorably with the stimulatory effect of thermogenic on blood flow. By increasing lipolysis in adipose tissue, cayenne would also improve this aspect of thermogenesis. In general, most of the properties discussed in the previous section on cayenne, contribute to this cardiovascular toning properties. Please see that section for details.

Summary

The selection and use of the proper plant materials can significantly affect the thermogenic process in humans. Herbs and spices abound in a great variety, each possessing its own special ability to maintain the health and vitality of the various body systems involved in, and affected by, thermogenesis. In addition, and most importantly, primary thermogenic agents are derived directly from the plant kingdom. Phytopharmacy is therefore a critical discipline in the overall science of thermogenesis. It has made very important contributions thus far, and promises to yield further secrets related to energy and fat metabolism.

Nevertheless, because of the novelty of the thermogenic concept and the relative lack of sophistication in the herbal industry, manufacturers should be careful to produce acceptable products. The following chart is provided to help both the consumer and the producer identify certain critical elements of a good prouct.

THERMOGENIC PRODUCT
COMPARISON GUIDE

Use this guide to help evaluate thermogenic compounds
encountered in the marketplace

REQUIREMENTS FOR SAFETY AND EFFICACY	GOOD PRODUCTS	POOR PRODUCTS
Aspirin Required.	Contain enough aspirin to meet criteria set out in research (90mg or more, with appropriate label warnings)	1. Contain no aspirin. May contain white willow bark which is not a source of aspirin. 2. Contain too much aspirin; inappropriate for aspirin-sensitive people.
Adrenal Support Required	Contain siginficant adrenal support.	Contain little or no adrenal support.
Use ma huang as source of ephedrine alkaloids, in amounts consonant with thermogenic research.	Use standardized ma huang; emphasis on fat reduction through BAT thermogenesis; view energy production as a side-benefit only.	May contain added synthetic ephedrine HCL; emphasis on so-called energy "buzz" rather than on fat loss through BAT thermogenics.
Adaptation should be minimized. (Refers to a tendency for a product to lose effectiveness over time.)	Minimize adaptation in two ways: 1) by following good formulation practices; and 2) by recommending two days of rest each week.	Adaptation is very likely. Do not recommend intermittant use. Promote adaptable "buzz" rather than non-adpating thermogenesis.
Must increase BMR in BAT; must promote fat loss.	Promote BAT metabolism and fat loss.	Minimal BAT activity; and minimal fat loss capability.

NUTRITION

Unlike the pharmacological manipulation of thermogenesis, the nutritional approach seeks to optimize the effectiveness of several individual biochemical events involved in thermogenesis as well as to improve the background metabolic picture against which the pharmacologically stimulated events unfold. It is fairly certain that nutritional manipulation by itself does not affect BAT thermogenesis to a large degree; but it is probably just as true that pharmacological activation can be seriously hampered if the various glands, organs and cells involved in thermogenesis do not receive proper nutritional support. Nutritional support can best be viewed as a natural facilitation of a pharmacologically- or phytopharmacologically-mediated thermogenic process.

Diet-induced thermogenesis (DIT) is the primary type of facultative BAT-mediated thermogenesis. The consumption of the right foods can markedly impact this kind of thermogenesis.[639] Alternately, the ingestion of the wrong foods may inhibit or frustrate thermogenesis. Attention must be given not only to the broad classes of nutrients, namely carbohydrates, proteins and fats, but also to the specific nutrients contained therein.

Many of the specific nutrients may be obtained from whole foods; others must be supplemented. Some nutrients are effective in very small amounts; others need to be consumed in rather large amounts. The effectiveness of almost all nutrients is dose-dependent—to a point. We want to avoid the fallacy of thinking that if a little is good, a lot must be better. The metabolism and utilization of most nutritional substances is under the control of tightly regulated negative feedback loops. Consuming too much of a good thing can upset these control mechanisms and become counterproductive. Dietary factors not only affect thermogenesis, but they also make a major impact on other physiological conditions that may be related to the obese state, such as hypertension.[137]

Natural Thermogenic Substances
(These substances either promote thermogenesis directly or they act synergistically with primary thermogenic materials materials. P = primary; S = secondary.)

Constituent	Source	Effect	
Ephedrine	Ma Huang	Sympathomimetic	P
Caffeine	Kola Nut (Gooroo nut, Bissey Nut) Guarana, Yera Mate	Blocks phosphodiesterases; Inhibit adenosine	P S
Aspirin (not a purely natural substance)	Synthesized from salicin (from white willow bark or aniline dyes)	Inhibits prostaglandin PGE1 biosynthesis	S
Capsaicin	Cayenne	Promotes lipolysis	P
Flavonoids	Ginseng	Promotes adrenal, pancreatic, cardiac health	S
Glycosides	Licorice root	Promotes adrenal health	S
Gymnemic acid	Gurmar	Promotes pancreatic health	S
Glycosides	Hawthorn	Promotes cardiac health	S
Misc.	Valerian root	Promotes cardiac, neural health	S
Iodine Minerals	Kelp Bladderwrack	Promotes thyroid health	S
Essential Fatty acids	Oils from Evening primrose, Purslane, black currant seed, borage seed, flaxseed, linseed, etec.	Promote lipolysis Promote thermogenesis Promote pancreatic health	P S P
Vitamins C, B, E	Dietary supplements	Promote adrenal health	P
Zinc, magnesium, manganese, chromium, selenium, phophate, potassium	Dietary supplements	Promote adrenal health Promote thermogenesis Promote pancreatic health	P S P

In the broadest sense, dietary manipulation of thermogenesis involves the number of calories consumed. For purposes of this chapter we shall assume that caloric restriction is not a factor. That is, we shall assume the diet contains enough calories to meet the basic metabolic needs of the body plus excess calories requiring effective BAT processing.

EFFECT OF A SINGLE MEAL

Considerable research has been devoted to examining the effects of a single meal on BAT metabolism.[643,644] The following results have been observed

1. Increased blood flow through BAT; this enhances the delivery of nutrients, especially free fatty acids (FFA) and oxygen to the tissue and helps dissipate heat created in BAT.

2. Increased binding of GDP in BAT mitochondria. As explained in an earlier chapter, the binding of GDP molecules on the membranes of mitochondria is a good indication of receptor activation and activation of subsequent intracellular events.

3. Increased thyroxine 5'-deiodinase activity, indicating the conversion of T_4 to T_3.

4. Increased SNS activity in BAT.

5. Increased SNS-mediated, insulin-dependent lipogenesis, i.e., synthesis of fats and TAGs, in BAT and elsewhere.

6. BAT proliferation; in an additive sense, single meals of 'cafeteria diet' in animals produce growth of BAT.[1072]

INFLUENCES OF FOOD CATEGORIES

Beyond the consideration of the effect of a typical meal on thermogenesis, we need to inspect the form of ingested calories. Sometimes, just altering composition of the diet can affect the amount of thermogenesis, even if caloric content is kept constant.[784] Consumption of moderate amounts of fat stimulates thermogenesis.[5,60,700,854,856,1086,1087,1317,1132] The consumption of a high fat diet, however, tends to overwhelm the thermogenic capacity. Hence, a high fat diet tends to reduce the ability of BAT to respond to insulin and inhibits FFA and TAG metabolism. Keeping the amount of consumed fat to a minimum reduces the stress to BAT and all other metabolic processes in the body.

In laboratory situations, the feeding of refined carbohydrates (especially glucose, fructose and sucrose) and fat usually stimulates the SNS, increases BMR, and promotes the growth of BAT.[169,173,430,467,1088,-1132,1295,1342,1344,1384,1386] It has been shown that the thermogenic action of sucrose does not involve protein catabolism.[514] Carbohydrates stimulate the SNS even when total caloric consumption is low.[289,1132] Chronic consumption of a high carbohydrate diet seems to increase sympathetic activity.[5,742] Much research has shown that carbohydrate consumption causes greater SNS stimulation than protein, or even fat.[5,688,1134,1253]

It appears that dietary protein has a modest thermogenic action, but it is also known that a protein-deficient diet can also activate BAT.[650,742,1317,1383,1395] It has been postulated that serotonergic pathways are involved,[1087] and that these pathways are activated by imbalances in certain amino acids caused by the protein insufficiency. In a recent Canadian study, researchers found that human subjects exhibited enhanced thermogenesis when fed highly palatable protein-rich foods (fish).[764] The palatability of the food accounted for an early 'cephalic' effect. But a later component appeared to be related to the nutrient itself. Compared to carbohydrate foods, the protein meal was more thermogenic. The researchers attributed NE-mediated thermogenesis to the cephalic phase, and the thermic effect of food (TEF) to the late phase.

The above study aside, substituting protein for carbohydrates on a calorie for calorie basis almost invariably results in reduced SNS activity.[512] The observation of enhance BAT activity under conditions of serious protein deficiency probably reflects an attempt by the body to produce heat sufficient to compensate for damage imparted to normal thermoregulatory processes by insufficiency. BAT activation under these circumstances cannot be viewed as a positive effect; rather it represents a damage-control mechanism.

Another benefit to be derived from a high carbohydrate diet is a significant impact on the regulation of eating. Carbohydrates tend to induce a feeling of satiation due to stimulation of vagal afferent neurons from the liver and duodenum. Also the metabolic control of carbohydrates is tighter than that of lipids. Carbohydrates, stored as glycogen in the liver and muscle, are brought into the metabolic arena and utilized much more efficiently that fats. Daily manipulation of glycogen stores may be contrasted to monthly turn over of lipid stores. The majority of consumed calories should therefore be provided by carbohydrates.

The U.S. Senate's Select Committee on Nutrition and Human Needs recommends that approximately 55% of total calories be derived from carbohydrate, 30% from fat, and 15% from protein. While 30% fat calories may seem high, 10% is too low to be practical, or palatable. Somewhere between 20-30% is probably the best. Recent research[1135] shows that fructose is better than sucrose in stimulating thermogenesis and carbohydrate oxidation; the authors recommend that fructose be used as a source of carbohydrate rather than sucrose in the diets of people with obesity and/or insulin resistance. A carbohydrate-rich diet may also help minimize cholesterol problems.[677]

Using carbohydrates as our model, let us observe some of the basic aspects of dietary manipulation of thermogenesis. An examination of the results of eating carbohydrates helps to illustrate the various components of thermogenesis. If needed, the reader should review the material in the chapter on Categories of Thermogenesis before continuing. The ingestion of energy in the form of carbohydrates is followed by an increase in energy expenditure equal to what is required for the digestion, assimilation and processing of the carbohydrate. This is the obligatory requirement. There follows an energy expenditure beyond that required for the obligatory requirements; this second, facultative, component is mediated by the sympathoadrenal system and can be reduced or blocked by the administration of drugs that block beta adrenergic receptors.[5,287,660] Carbohydrate-induced SNS activity is reflected by increased NE levels in brown adipose tissue.[60,1134] This activation of the SNS by carbohydrate is thought to be mediated by a central action of insulin on the hypothalamus; however, it has been pointed out that fructose ingestion also activates the SNS, and fructose does not increase insulin levels.[838,852] Some time after the initial responses to the carbohydrate, the adrenal medulla kicks in with the secretion of adrenaline.[60,66,67,256]

The SNS component of the response to carbohydrate ingestion is the thermogenic effect. The tissues targeted by this thermogenic action are BAT, liver, skeletal muscle and even white adipose tissue. The response is bi-phasic, i.e., there is an initial peak that coincides with the increased glucose uptake, and a later peak that coincides with peak NE concentration.

The early peak is, therefore, an obligatory component (TEF), and the later peak is a facultative component (DIT). A point of some controversy at the time of this writing is the importance of the role of the SNS in mediating thermogenesis in skeletal muscle;[60,337,347] some evidence suggests that adrenaline, not NE, may mediate skeletal muscle thermogenesis. According to this hypothesis, the biphasic response to carbohydrate ingestion dissociates into a sympathetic and an adrenal component. The sympathetic component stimulates thermogenesis in heart, liver, BAT and WAT, while the adrenal component stimulates thermogenesis primarily in the skeletal muscle.

While obligatory thermogenesis goes on without voluntary effort on our part, and in fact resists any kind of external manipulation, facultative thermogenesis responds easily to voluntary actions (exercise, eating, and consuming thermogenic for example); such action is mediated by suppressing or activating appropriate neural pathways. The situation can get quite complicated. For example, following the ingestion of a nice meal, a great deal of thermogenesis might occur: One, there is an overall obligatory component associated with the digestion and processing of the food (TEF), and two, there is superimposed a facultative component (DIT) occurring in BAT. Now, if the feeding were to take place in the cold, the thermogenic reactions could be further modified by CIST in the muscles and CINST in BAT. Furthermore, if one were to consume a thermogenic agent or sympathomimetic, the situation would become even more complex. All of these actions would be further complicated by the overall state of the organism (lean or obese, overfed or underfed, male or female, young or old, etc.) and by an incredible array of endocrine considerations involving interactions among the adrenals, pancreas, thyroid and pituitary glands.

INFLUENCES OF SPECIFIC NUTRIENTS

Many nutrients affect the metabolism of carbohydrates, fats and proteins; others affect some aspect of the thermogenic process directly. A few do both. In the following paragraphs, several nutrients will be briefly discussed, but only insofar as they impact thermogenesis. A comprehensive treatment of the properties and biochemistry of every nutrient is beyond the scope of this book.

Vitamins.

Several vitamins impact thermogenesis. The major thermogenic vitamin is probably vitamin C. Vitamin C helps maintain and/or restore tissue sensitivity to insulin. It also helps maintain normal function in the adrenal gland. B vitamins, especially B_6, niacin, and pantothenic acid are also very important in maintaining insulin responsiveness and adrenal health.[100,1182] A diet rich in raw fruits and vegetables will supply a good portion of these vitamins, but supplementation with C, B complex and beta-carotene will help insure healthy BAT functioning. Moderate amounts of thermogenic vitamins may be included in the primary thermogenic agent, but individuals may want and require additional support.

Minerals.

Zinc and magnesium aid the vitamins in maintaining tissue sensitivity to insulin, as do chromium and manganese which also help improve glucose tolerance. Zinc, chromium and selenium also facilitate some of the biochemical events directly involved in skeletal

291

muscle and/or BAT metabolism.[23,24,314,502,669] Swedish research has shown that the addition of potassium or magnesium to a meal increases the thermogenic response of obese patients to the meal.[612,613,1205] Recently phosphate supplements (2.7g/day) have been shown to increase thermogenesis in obese women on a calorie-restricted diet.[1405] The diet of most Americans is probably somewhat lacking in one or more of these thermogenic minerals. A simple mineral supplement will rectify the situation. Most of these are suitable for inclusion into a primary thermogenic agent.

Essential Fatty Acids (EFA).

EFA are known to play a crucial role in weight regulation. Deficient EFA increases appetite, promotes obesity, high serum cholesterol, and hypertension.[497,498,952] It also interferes with BAT metabolism and thermogenesis.[462] High serum EFA does the opposite.[56] It also increases BAT activity.[96] High EFA mimics DIT, and increases the concentration of UCP in BAT.[198,918,1160]

Dietary EFA is important because EFAs are precursors of the prostaglandins. PGE_1 synthesized from EFA performs the function of lipolysis, helping to break down triglycerides (TAGs) into FFA and glycerol.[201] PGE_1 stimulates adenylate cyclase, a key enzyme in BAT that facilitates lipolysis and the incorporation of FFA into mitochondrial metabolic pathways.[811] It also increases the sensitivity of NE-mediated events in BAT thermogenesis.[572] Finally, PGE_1 plays a critical role in insulin secretion and glucose tolerance. Diabetics have low levels of PGE_1, which helps regulate the release of insulin from pancreatic beta-cells.[488,857]

Sources of EFA are sometimes difficult to acquire. Evening primrose oil is the best. It contains significant quantities of cis-gamma-linoleic acid (cLA) and gamma-linolenic acid (GLA) which have been shown experimentally to cause weight loss.[422,1302] GLA is a direct precursor of PGE_1. Most other oils contain EFA that is several metabolic steps removed from PGE_1. The conversion of cLA to GLA can be easily disrupted by common dietary and environmental factors, such as trans-fatty acids derived from processed oils, saturated fats, alcohol, cholesterol, old age, free oxygen radicals, etc .

Next to evening primrose oil stand black currant seed oil and borage oil, both good sources of EFA. Cis-linoleic acid is the most important EFA. This acid, together with GLA, help raise the level of dihomogamma-linolenic acid (DGLA) which is converted into PGE_1 and arachidonic acid (AA). Sources of cis-linoleic acid are raw sunflower, pumpkin and sesame seeds, cold pressed oils of safflower, sunflower, soy and corn. Other good oils are those rich in linolenic acid: linseed and flaxseed.

It is important to maintain a balance between PGE_1 or DGLA and AA. AA is converted into PGE_2 which directly counteracts the effects of PGE_1 on adenylate cyclase and intracellular cAMP.[113] Under less than optimum conditions, the body sometimes elects to convert most DGLA into AA. To help insure the proper balance in this system, one should consider consuming supplemental B vitamins, vitamin C, and minerals such as zinc and magnesium.

Omega-3 fatty acids (eicosapentaenoic acid and docosahexaenoic acid) inhibit AA metabolism and are appropriate dietary thermogenic oils.[678,989] These oils are found in high concentration in fish oils and purslane oil. Oleic acid, found in olive oil also inhibits the formation of AA. The daily diet should contain up to 3 or 4 grams of EFA- or GLA-rich oil. Obviously, this much material cannot form a part of the primary thermogenic agent, but must be consumed separately.

As a final note, research now shows that dietary medium-chain triglycerides may be more effective in the stimulation of thermogenesis than long-chain triglycerides.[1080,1109]

Other Thermogenic Nutrients.

The restoration of insulin sensitivity is necessary for the expression of SNS-mediated thermogenesis. Chromium, niacin and manganese, along with GLA, fiber and vitamin C, help to do this. Each of these nutrients, in its own way, restores healthy glucose and lipid metabolism, improves the effectiveness of the glucose tolerance factor (GTF), and so forth.[24,359,721,1089,1301]

Also, because of the close relationship between insulin and growth hormone, and the necessity of maintaining the appropriate ratio between the two, dietary supplementation may be effectively targeted at enhancing GH levels. GH release has been shown to occur in response to the consumption of 1200mg or l-lysine combined with 1200mg l-arginine.

Cold exposure and certain dietary nutrients help stimulate the growth of BAT indirectly by increasing the effectiveness of a catecholaminergic neurotransmitter known as serotonin.[899] In fact, BAT growth may depend on the presence of serotonin.[900] Serotonin improves the effectiveness of the SNS, and plays an important role in the effects of PGE_1 on thermogenesis. Serotonin injections into the ventricles of rat brains increased oxygen consumption and thermogenesis, and decreased food intake.[769] Tryptophan and histidine are two nutrients that impact in a positive manner on serotonin synthesis and activity.[40] Histidine is a precursor of histamine, an important neurotransmitter in its own right, that probably helps control body temperature. During cold adaptation the level of histamine-containing mast cells in BAT increases significantly. Although we are not certain of the implications of this factor, the close functional relationship between histamine, PGE_1, serotonin and NE suggests that thermogenic metabolism is being affected in some way.[164,282,507,800] For one, histamine has been found to stimulate an increase in cAMP which would affect lipolysis in BAT.[227,671] Histidine is also important in the formation of PGE_1 and cAMP.[512,571] Foods high in histidine and/or histidine supplementation are recommended for improved control of thermogenesis through dietary means.[49,1154]

It is felt by some authorities that the transport of FFA into BAT mitochondria during DIT absolutely requires the action of the amino acid carnitine (this would be true of all cells in the body), but that the amount produced naturally by body (from lysine and methionine)[155] is woefully inadequate for the task. And since the synthesis also requires the presence of vitamin C, niacin and B_6, the odds quickly mount in favor of the forces that oppose the synthesis. In the diet, carnitine is found mainly in animal protein. As a supplement, carnitine is very expensive. While you will benefit from even a few milligrams of supplemental carnitine, studies show that it takes hundreds of milligrams to significantly impact metabolic events. Meanwhile, keep in mind that the incorporation of FFA into the mitochondria of cells is required before any large scale energy production can take place. Without it, you would die. The fact that you aren't dead yet argues in favor of the notion that you do have at least some natural carnitine in your cells. Finally, as noted in an earlier chapter in which carnitine-dependent and carnitine-independent pathways exist for FFA transport, the amount of carnitine measured in BAT cells closely approximates the amount required for maximum thermogenesis. Inordinate lethargy or chronic fatigue might signal a carnitine deficiency, but then again it might not. In favor of the role of carnitine is research that shows that inhibition of carnitine acyl transferase blocks mitochondrial respiration in BAT.[831]

Methionine, required for the synthesis of carnitine,[155] is also needed for the synthesis of albumin, the protein that acts as the primary carrier molecule for free fatty acids. Research has not examined what effect an albumin deficiency might have on BAT metabolism. However, since methionine is common to both carnitine and albumin, it, rather than carnitine, may be the supplement of choice to enhance thermogenesis. Limited research at APRL has indeed shown that some people who have not reacted to thermogenic stimulation, or who have plateaued before reaching their ideal body composition, begin to react for the first time or once more when methionine is added to their diet.

Anti-oxidants.

Persons employing exercise as a means for losing weight should be aware that exercise promotes the formation of free radicals. Free radicals are oxygen-based ions that interact with cell membranes in such a manner to weaken the cell. It is thought that most degenerative disease is the result of free radical damage. Although the body

295

has its own array of biochemicals for the inhibition of free radicals, these resources are quickly overwhelmed by the astonishing abundance of free-radical generating factors in today's world. For that reason, most authorities recommend the consumption of supplementary free-radical scavengers, or anti-oxidants. Free radical damage arising from exercise typically attacks the lipid components of cell membranes. This action can be significantly reversed by the consumption vitamin C, vitamin E, beta-carotene, zinc, selenium, reduced glutathione and a wide variety of flavonoid-rich plants.

GENETICS

One of the basic questions of thermogenesis is what role do the genes play in the etiology of obesity? Our working assumption is that obesity does have a genetic component. Research, mainly in animals, has shown that defects in the thermogenic pathway account for the majority of obesity-causing genetic disorders.[638,641,644] Stated another way, if an organism develops obesity due to some genetic fault, that fault probably involves a defect in thermogenesis. The goal of thermogenics is to correct or reverse this genetic fault, thereby creating lean people from obese people. Animals studies again suggest that this goal is feasible.

It has been said that the validation of the brown fat model of obesity requires one, that we demonstrate that defective thermogenesis indeed produces obesity in man, and two, that thermogenesis in man is primarily a function of BAT rather than skeletal muscle, for example. These requirements are not that easy to fulfill, but it can be argued that the eventual validation of the hypothesis should not prevent the immediate application of thermogenic theory to human health. In fact, at this point, validation of the thermogenic hypothesis is more a matter of academics than of practical worth, since empirical evidence is rapidly accumulating that demonstrates a clear and dramatic impact on body composition and fat loss following the consumption of thermogenic agents.

Generalizing from animal studies to humans is admittedly problematic and should not be engaged in haphazardly. In this case, however, while progress in basic research is necessarily slow, clinical work is providing a great deal of positive evidence for the theory that defective thermogenesis is the basis for much human obesity;[864,862] evidence for the brown fat hypothesis also continues to accumulate. The genetic basis for defective thermogenesis in humans may take decades to tease apart and conclusively demonstrate. Meanwhile, we can continue to operate on the reasonable assumption that what happens to rats and mice happens to humans also. So far, obese humans who consume the same nutrients as genetically obese animals exhibit a similar positive response towards leanness.

DEFICIENT THERMOGENESIS AND OBESITY: ANIMAL STUDIES

Throughout this book, we have repeatedly stated our assumption that obesity and body composition problems in humans have a genetic component that represents either a fault or a vestigial evolutionary process that has no functional place in the modern world. It is the view of most investigators in obesity research that this genetic disposition involves aspects of faulty thermogenesis in brown adipose tissue. While BAT thermogenesis probably accounts for no more than 10-15% of consumed calories even when operating correctly, scientists believe that the deposition of those calories as fat results in a large percentage of clinical obesity.

Theoretically, it appears that the only way that the energy and thermal equations can balance is if all consumed calories unavailable for the normal metabolic pathways (meeting metabolic needs and obligatory thermogenesis) must ultimately be stored as fat. Studies with a variety of experimentally/genetically obese animals uniformly show that the accumulation of fat is the result of defects in thermogenics.[537,810] More often than not, this process involves the suppression of thermogenesis in brown adipose tissue. In some animal models, the suppression of diet-induced BAT thermogenesis is totally absent. In others, there is a reduction in the energy expended for cold-induced non-shivering thermogenesis in BAT as well as a lowering of thermogenesis in other organs.[648] A third type of experimental obesity is characterized by gross overeating. An entire book could be written on the findings of animal-based genetic research in obesity. Since we do not have the space here, we present the following brief summaries of the characteristics of the major types of deficient thermogenesis created through experimental/genetic manipulation of laboratory rodents.

Animal Models of Obesity.

A. DIT Deficiencies.

1. PKC Rat. Produced by lesions of the hypothalamus that do not destroy DIT neural circuits. Exhibits excessive sensitivity to glucocorticoids that suppress DIT.[248,829,1057]

2. VMH Rat. Produced by surgical or chemical lesions of the ventromedial nucleus of the hypothalamus. Destruction of DIT, but not cold-induced non-shivering thermogenesis (CINST).[551,548,648,1091,-1314,1315]

3. fa/fa Rat (adult). Genetically obese. Lacks DIT; retains CINST, but is cold-intolerant.[14,150,453,566,783,787,788,1150,1291,1351] Decreased metabolic rate.[672] Inability to burn off calories as heat.[446,790] Exceptionally sensitive to suppression of hypothalamic control of DIT in BAT by glucocorticoids. Decreased /GDP binding.[90,1402] Presynaptic turnover of NE reduced.[791,792,794]

4. cp/cp Rat. Genetically obese. Similar signs as the fa/fa.

5. A^{vy} Mouse. Genetically obese. Slightly lower thermogenic state than normal (reduced GDP binding), reduced SNS activity, responds to dietary fat but not carbohydrate; responds to cold and NE.

6. Aging Rat. As rats (and humans?) get old they experience steadily decreasing DIT, and insulin resistance. Normal BAT stores; normal CINST. Few beta-adrenergic sites; reduced sensitivity of adenylate cyclase to NE.

7. Tube-fed Rat. Defective DIT results from lack of cephalic phase of DIT in BAT.

8. Capsaicin-Desensitized Rat. Atrophied BAT, low UCP, no DIT; not hyperphagic—only slight tendency to obesity; normal CINST.

9. SHR/N-cp Rat. Obese, diabetic; reduced BAT mitochondrial content.[832]

B. Deficient Thermoregulatory Thermogenesis.

1.   MSG Mouse. Treated with monosodium glutamate during neonatal period; torpor; reduced GDP binding and oxygen consumption in BAT.[810,913,1380,1264,1267] Condition reversed by combination of ephedrine and caffeine.[810]

2.   ob/ob Mouse. Genetically obese. Torpor; low SNS activity. Insulin resistant, no CINST, etc. Stimulation of SNS reverses the condition completely.[52,78,277,530,533,540,709,710,902,1266,1280,1257,1258,1280,1284,1402,1403]

3.   db/db Mouse. Genetically obese and diabetic. Similar to ob/ob mouse.

4.   fa/fa Rat (newborn). Thermoregulates at low body temperature all the time. BAT is inactive, huddling exaggerated.

5.   GTG Mouse. Lesions to the hypothalamus, induced by chemicals. Torpor like the MSG mouse except that it retains only the cephalic phase of DIT. DIT reappears later in life.[283,345,407,550,1404]

C. Deficient Control of Energy Intake.

1.   PVN Rat. Lesions to hypothalamus. Hyperphagic. PVN (paraventricular nucleus) controls food intake; altered metabolic efficiency. SNS normal; thermogenics normal.[405,863,1340]

2.   DIO Rats and Mice. (Diet-Induced Obesity) The "prodigal" genotype, involving a full spectrum of body compositions. At one end of the spectrum are animals characterized by gross obesity, with deficient BAT, small amounts of UCP, and so forth.[552,533,786,796,789,599,1096] At the other end of the spectrum are animals that do not readily become obese, in which BAT is very active, with doubled UCP. In between the extremes of DIOs are several strains of animals exhibiting varying degrees of propensity for obesity, usually on high fat diets.[1137,1014]

Genetic/surgical forms of obesity are characterized by an increase in metabolic efficiency. Increased efficiency means less wastage of calories as heat, and more calories stored as fat. This may be viewed as a condition characterized by reduced energy expenditure for thermogenesis that results from one genetic deficit or another, or through surgery. Genetic faults are of particular interest, because if it can be demonstrated that obesity is the result of an erroneous genetic expression, then solutions can be found to help counteract the hereditary problem. The typical characteristics of deficient DIT-induced obesity are as follows:

1.   Elevated metabolic efficiency

2.   Mild to moderate overeating

3.   Becoming obese even when eating the same diet as lean control animals

4.   Reduced sympathetic nervous system activity in BAT, secondary to lack of diet-induced activation of BAT thermogenesis.

The characteristics of deficient CINST-induced obesity are as follows:

1.   Thermoregulation at lower than normal body temperature: torpor

2.   Reduced BAT thermogenesis and obligatory thermogenesis in all organs as a consequence of slowing of metabolism at the lower body temperature

3.   Retention of ability to activate DIT in BAT

4.   Tendency to be cold intolerant, probably as another example of the excessive tendency to torpor.

Contrast the above deficiencies with the expression of fully functional brown adipose tissue in animal models. The cold-acclimated rat, for instance, is hypermetabolic and hyperphagic (eats a lot), eating twice as much as a rat kept at room temperature. Yet, in spite of the hyperphagia, the cold acclimated rat is very lean. BAT thermogenesis plays a major role in maintaining energy balance. Brown adipose tissue proliferates rapidly during cold-exposure, and cold-induced non-shivering thermogenesis increases correspondingly.

As another example, consider the normal cafeteria-fed rat. This animal is allowed to eat a variety of highly palatable foods in addition to, and following eating its normal laboratory chow. It is also hyperphagic and hypermetabolic. While the animal may gain weight, it gains much less than would be expected from its excess energy intake.[1066-1069] DIT is greatly increased.[542,1066,1067]

DEFICIENT THERMOGENESIS AND OBESITY: HUMAN STUDIES

Human counterparts to these animal models of human obesity exist and are currently the subject of intense study.[238,760,759,988,89] Since a state equivalent to torpor has not been observed in humans, it is unlikely that deficient CINST plays an important role in human obesity. That leaves deficient DIT obesity as the best model for human obesity. A defect in BAT thermogenesis occurs in all animal models of human obesity so fat studied.[528] Genetic differences in metabolism have been found by overfeeding studies in twin-pairs.[138] Thus, there definitely appears to be a genetic basis for obesity reflected in the presence of a gene that increases metabolic efficiency and results in greater energy store, i.e., fat.

The best conclusion from animal studies on deficient thermogenesis as an underlying cause of obesity appears to support the contention that obesity is caused by a lack of DIT facultative thermogenesis involving some underlying metabolic or genetic defect. This, in turn, suggests that a possible corrective action, based on a reversal of the metabolic or genetic fault, could successfully reverse obesity (research reviewed in chapter on Pharmacology). Remember that obesity here is defined as too much fat, not as too much weight. Even very thin, underweight, individuals can be obese if they have too much fat compared to their total weight, i.e., a low lean/fat ratio.

Of considerable importance in our discussion of deficient-BAT-induced obesity, are the various strains of DIO (Diet Induced Obesity) animals. Not much is known about the exact mechanisms that underlay these conditions, but the administration of thermogenic agents tends to reverse them. The existence of these strains suggests that there may be equivalent "strains" of obesity-inclined humans representing a broad spectrum of genetic faults, biochemical lesions, biochemical mistakes, etc. that predispose people toward obesity. How many of these are expressed through disruptions of BAT thermogenics? Only dozens of more years of research can answer that question with any degree of finality.

Meanwhile, it is in the best interest of overweight, over fat people that they take advantage of the opportunity to test out the viability of the thermogenic approach immediately. If, after a few months of consuming the thermogenic, a person does not experience an improvement in body composition, then perhaps his or her problem does not have a genetically-based error in BAT metabolism. We say "perhaps" because a failure to respond to the thermogenic may also be a function of a problem in adrenal, thyroid, pancreatic and CNS functioning.

Studies on normal, intact mice and rats in which thermogenesis is operating correctly present an entirely different picture of thermogenic action from that observed in genetically altered animals. In normals, the simple ingestion of food leads to rather significant events: Activation of thermogenesis in BAT; accumulation of brown fat tissue mass, increased sympathetic activity, and increased mitochondrial mass and GDP binding. This picture of healthy thermogenesis is the goal of thermogenic intervention.

Over the years, scientists have been discovering new ways to statistically manipulate body composition and weight data gathered on whole families, adoptees and twins, both monozygotic and dizygotic.[852] The following list summarizes these data.

1.   Obesity does run in families, both for environmental and genetic reasons. Thus obesity occurs because members of obese families share a common set of genes and a common environment. The unanswered question is how much does each component contribute to a given family member's body composition at any given point in that person's life.

2.   Family studies suggest that genes are responsible for about one-half of the phenotypic variation in obesity.

3.   Adoption studies show a clear genetic influence, since the body mass index (BMI) of adoptees much more closely resembles the biologic parent than the adoptive parent. Heritability for BMI of 20-40%.

4.   Genetic determination of BMI is fully expressed early in life.

5.   BMI of female adoptees is highly correlated to the mother, less to father. BMI of sons not significantly correlated to biologic parents.

6.   Twin studies offer strong support to a genetic component of obesity. Adopted twins reared apart, i.e., in different environments, show a greater similarity in body composition to one another than to the the other children of their respective adoptive families.

7.   Early expression of obesity appears to be the result of small effects of a number of genes, but later in life becomes the expression of larger effects due to a small set of genes. At least half of the genetic variance is due to the latter effect.

8.   Genetic effects of different sets of genes are expressed during childhood, between childhood and adulthood, and adulthood. However, about 50% of the genetic variation in childhood BMI is due to the same genes that affect adulthood BMI.

9.   About 40% of "adiposity rebound" - rapid growth in body fat that occurs somewhere between the ages of 4 and 7 - is due to genetic factors.

10. Evidence from both animal and human studies suggests the presence of single genetic loci or specific segments of the chromosome, that impact the development of obesity, i.e., a single recessive major gene that might account for 20-35% of the variance.

11. Alternatively, about 20-40% of the variance is thought to be due to polygenic effects, i.e., to the combined action of several different genes.

12. Environment may account for the remaining 23-60%. These environmental factors include such things as the following:

1. Dietary factors

    a. Nutrient type; relative proportions of fat, carbohydrate, and protein

    b. Palatability (how good the meals taste)

2. Physical Activity

    a. Sedentary lifestyle, watching too much television

    b. Active lifestyle, frequent exercise

3. Climate

    a. Overly warm climate, due to ambient temperature, clothing, etc., that tend to shut down thermogenic mechanisms

    b. Colder climate and habits that tend to create CINST.

> Bottom Line: ". . .genetic influences largely determine whether a person can become obese, but it is the environment that determines whether such a person does become obese, and the extent of that obesity."[852]

EPILOGUE

The science of thermogenics will continue to make new strides over the next two decades. Much of the information presented in this book will undoubtedly be superseded as research brings to light exciting and marvelous facts about the body's natural ability to deal with excess dietary fat.

We are convinced that genetic and environmentally-induced defects in thermogenesis are the root of much human misery. Having long ago rejected the kind of lifestyle that encourages healthy thermogenesis as well as healthy fat storage, we are faced with the necessity of dealing with a communal gene pool that has not, and may not ever, adapt to current patterns of living, including housing, diet and activity.

Calling upon the world's fantastic storehouse of natural elements as beautifully represented by the plant kingdom, we will continue to unlock to the secrets that nature has provided for the solutions to our evolving problems. Using good judgement and following the dictates of stellar research, development and production practices, we will be able to create both natural and synthetic compounds to aid us in achieving and maintaining health and vitality.

REFERENCES

1. Abbott, W.G.H., Howard, B.V., Christin, L., et.al. "Short-term energy balance: relationship with protein, carbohydrate and fat balances." **Am. J. Physiol.**, 255, E332-E337, 1988.
2. Abelenda, M. & Puerta, M.L. "Cold-induced thermogenesis in hypothyroid rats." **Pflugers Arch.**, 416(6), 663-666, 1990.
3. Abernethy, D.R., Todd, E.L. & Schwartz, J.B. "Caffeine disposition in obesity." **Br. J. Clin. Pharm.**, 20(1), 61-66, 1985.
4. Acheson, K.J. & Burger, A.G. **J. Clin. Endocr. Metab.**, 51, 84, 1980.
5. Acheson, K., Jequier, E. & Wahren, J. "Influence of beta-adrenergic blockade on glucose-induced thermogenesis in man." **J. Clin. Inv.**, 72, 893-902, 1983.
6. Acheson, K.J., Ravussin, E., Wahren, J., et.al. "Thermic effect of glugose in man. Obligatory and facultative thermogenesis." **J. Clin. Inv.**, 74, 1572, 1984.
7. Acheson, K.J., Ravussin, E., Schoeller, D.A., et.al. "Two-week stimulation or blockade of the sympathetic nervous system in man: influence on body weight, body composition, and twenty four-hour energy expenditure." **Metabolism**, 37, 91-98, 1988.
8. Acheson, K.J., Zahorska-Markiewicz, B., Pittet, P., et.al. "Caffeine and coffee: their influence on metabolic rate and substrate utilization in normal weight and obese individuals." **Am. J. Clin. Nutr.**, 33, 989-997, 1980.
9. Afzelius, B.A. "Brown adipose tissue: its gross anatomy, histology and cytology." In **Brown Adipose Tissue**, Lindberg, O., ed., NY, Elsevier, 1970, pp. 1-31.
10. Akinyanju, P. & Judkin, J. **Nature**, 214, 426-427, 1967.
11. Aktories, K., Schultz, G. & Jakobs, K.H. **FEBS Lett.**, 107, 100, 1979.
12. Aldridge, A., Aranda, J.V. & Neims, A.H. "Caffeine metabolism in the newborn." **Clin. Pharm. Ther.**, 25, 447, 1979.
13. Alexander, G. & Stevens, D. "Sympathetic innervation and the development of structure and function of brown adipose tissue: studies on lambs chemically sympathectomized in utero with 6-hydroxydopamine." **J. Develop. Physiol.**, 2, 119-137, 1980.
14. Allars, J., Holt, S.J. & York, D.A. "Energetic efficiency and brown adipose tissue uncoupling protein of obese Zucker rats fed high-carbohydrate and high-fat diets: the effects of adrenalectomy." **Int. J. Obes.**, 11, 591-601, 1987.
15. Amatruda, M. & Livingston, J. **Science**, 188, 264, 1975.
16. Aminian, K., Robert Ph., Seydoux, J., et.al. "Real time measurement of the contribution of the muscular activity to the metabolic rate in freely-moving rats." **Med. Biol. Eng. Comput.**, in press, 1993.
17. Amini-Sereshki, L. & Zarrindast, M. R. **Am. J. Physiol.**, 247, R154-R159, 1984.
18. Amir, S. "Activation of brown adipose tissue thermogenesis by chemical stimulation of the posterior hypothalamus." **Brain Res.**, 534(1-2), 303-308, 1990.
19. Amir, S. & De-Blasio, E. "Activation of brown adipose tissue thermogenesis by chemical stimulation of the hypothalamic supraoptic nucleus." **Brain Res.**, 563(1-2), 349-352, 1991.
20. Amir, S. & Schiavetto, A. "Injection of prostaglandin E2 into the anterior hypothalamic preoptic area activates brown adipose tissue thermogenesis in the rat." **Brain Res.**, 528(1), 138-142, 1990.
21. Ammaturo, V. & Monti, M. **Acta Med. Scand.**, 220, 181-184, 1986.
22. Andersen, T., Astrup, A. & Quaade, F. **Int. J. Obes.**, 16, 35-40, 1992.
23. Anderson, R.A., Polansky, M.M., Bryden, N.A., et.al. "Effect of exercise (running) on serum glucose, insulin, glucagon, and chromium excretion." **Diabetes**, 31, 212, 1982.
24. Anderson, R.A., Polansky, M.M., Bryden, N.A., et.al. "Chromium supplementation of human subjects: effects of glucose, insulin, and lipid variables." **Metabolism**, 32, 894, 1983.
25. Andersson, K.E. & Persson, C.G.A., eds. "Anti-asthma xanthines and adenosine." **Excerpta Medica**, Amsterdam, 3-484, 1985.
26. Ando, K. & Yanagita, T. **Pharm. Biochem. Behav.**, 41, 783, 1992.
27. Anson, J., Hinson, W.G., Pipkin, J., et.al. "Cell cycle and morphological changes induced 'in vitro' in spleen cells after exposure to phenylpropanolamine, caffeine and ephedrine singularly and in combination." **Govt Reports Announcements & Index**, 10, 1986.
28. Apfelbaum, M. In **Obesity in Perspective**, Bray, G.A., ed., Fogarty Conference, DHEY Publication, NIH, 75/708, 145, 1973.
29. Apfelbaum, M. "Metabolic effects of low and very low calorie diets." **Int. J. Obes.**, 17(suppl 1), S13-S16, 1993.
30. Appelt, G.D. "The safety of phenylpropanolamine." **J. Clin. Psychopharm.**, 3(5), 332-333, 1983.
31. Aquila, H., Link, T.A. & Klingenberg, M. **M. EMBO J.**, 4, 2369-2376, 1985.
32. Aranda, J.V., Gorman, W., Bergsteinsson, H., et.al. **J. Pediatr.**, 90, 467-472, 1977.
33. Aranda, J.V., Grondin, D. & Sasyniuk, B.I. **Pediatric Clinics of N. America**, 28, 113-133, 1981.
34. Arase, K., York, D.A. & Bray, G.A. "Corticosterone inhibition of the intracerebroventricular effect of 2-deoxy-D-glucose on brown adipose tissue thermogenesis." **Physiol. Behav.**, 40, 489-495, 1987.
35. Arase, K., York, D.A., Shimizu, H., Shargill, N., et.al. **Am. J. Physiol.**, 255, E255-E259, 1988.
36. Arch, J.S., Ainsworth, A.T. & Cawthorne, M.A. "Thermogenic and anorectic effects of ephedrine and congeners in mice and rats." **Life Sci.**, 30, 1817-1826, 1982.
37. Arch, J.R.S., Ainsworth, A.T., Cawthorne, M.A., et.al. "Atypical beta-adrenoreceptor on brown adipocytes as targets for anti-obesity drugs." **Nature**, 309, 163-165, 1984.
38. Arch, J.R.S., Ainsworth, A.T., Ellis, R.D.M., et.al. "Treatment of obesity with thermogenic beta-adrenoceptor agonists: studies on BRL 26830A in rodents." **Int. J. Obes.**, 8, 1-11, 1984.
39. Arieli, A. & Chinet, A. **Horm. Metab. Res.**, 17, 12-15, 1985.
40. Arimanana, L., Ashley, D.V.M., Furniss, D., et.al. **Progress in Tryptophan and Serotonin Research**, Schlossberger, H.B., Kochen, W., Linzen, B., et.al., eds., Berlin, DeGruyter & Co., 549-552, 1984.
41. Arnaud, M.J. "The pharmacology of caffeine." 273-313
42. Arnaud, M.J. & Welsch, C. In **Theophylline and other Methylxanthines**, Rietbrock, N., Woodcock, B.G. & Staib, A.H., eds., Friedr. Vieweg & Sohn, 135-148, 1982.
43. Arner, P. **Am. J. Clin. Nutr.**, 55, 228S, 1992.
44. Arner, P. "Adenosine, prostaglandins and phosphodiesterase as targets for obesity." **Int. J. Obes.**, 17(suppl 1), S57-S59, 1993.

309

45. Arner, P., Engfeldt, P. & Nowak, J. "In vivo observations on the lipolytic effect of noradrenaline during therapeutic fasting." **J. Clin. Endocr. Metab.**, 53, 1207-1212, 1981.
46. Arnold, J., LeBlanc, J., Cote, J., et.al. "Exercise suppression of thermoregulatory thermogenesis in warm- and cold-acclimated rats." **Can J. Physiol. Pharm.**, 64, 922-926, 1986.
47. Arnold, J. & Richard, D. **Am. J. Physiol.**, 252, R617-R623, 1987.
48. Arnold, J. & Richard, D. "Exercise during intermittent cold exposure prevents acclimation to cold in rats." **J. Physiol. (London)**, 390, 45-54, 1987.
49. Ashley, D.V.M. "Dietary control of brain 5-hydroxytryptamine synthesis: implications in the etiology of obesity." In **Nutrition: Neurotransmitter Function and Behavior**, Mauron, J., ed., Lewiston, NY, Hans Huber, 27-40, 1986.
50. Ashwell, M., Holt, S., Jennings, G., et.al. **FEBS Lett.**, 179, 233-237, 1985.
51. Ashwell, M., Rothwell, N.J., Stirling, D., et.al. **Proc. Nutr. Soc.**, 43, 148A, 1984.
52. Assimacopoulos, J.F., Giacobino, J.P., Seydoux, L., et.al. "Alterations of brown adipose tissue in genetically obese (ob/ob) mice. II. Studies of beta-adrenergic receptors and fatty acid degradation." **Endocrinology**, 110, 439-443, 1982.
53. Astrup, A. "Thermogenesis in human brown adipose tissue and skeletal muscle induced by sympathomimetic stimulation." **Acta Endocr.**, 112(suppl 278), 1-32, 1986.
54. Astrup, A.V. "Obesity and diabetes as side-effects of beta-blockers." **Ugeskr. Laeger.**, 152(40), 2905-2908, 1990.
55. Astrup, A.V., Andersen, T., Christensen, N.J., et.al. "Impaired glucose-induced thermogenesis and arterial norepinephrine response persist after weight reduction in obese humans." **Am. J. Clin. Nutr.**, 51(3), 331-337, 1990.
56. Astrup, A., Breum, L., Christensen, N.J., et.al. **Metabolism**, 41, 686, 1987.
57. Astrup, A., Breum, L., Toubro, S., et.al. **Int. J. Obes.**, 16, 269-277, 1992.
58. Astrup, A., Buemann, B., Christensen, N.J., et.al. **Metabolism**, 41, 686, 1992.
59. Astrup, A., Bulow, J., Christensen, N.J., et.al. "Ephedrine-induced thermogenesis in man: role for interscapular brown adipose tissue." **Clin. Sci.**, 64, 179-186, 1984.
60. Astrup, A., Bulow, J., Christensen, N.J., et.al. "Facultative thermogenesis induced by carbohydrate: a skeletal muscle component mediated by epinephrine." **Am. J. Physiol.**, 250, E226-E229, 1986.
61. Astrup, A., Bulow, J., Madsen, J., et.al. "Skin temperature and subcutaneous adipose blood flow in man." **Scand. J. Clin. Lab. Inv.**, 40, 135-138, 1980.
62. Astrup, A., Bulow, J., Madsen, J., et.al. "Contribution of BAT and skeletal muscle to thermogenesis induced by ephedrine in man." **Am. J. Physiol.**, 248(5 pt 1), E507-E515, 1985.
63. Astrup, A.V., Christensen, N.J., Simonsen, L., et.al. "Effects of nutrient intake on sympathoadrenal activity and thermogenic mechanisms." **J. Neurosci. Methods**, 34, 187-192, 1990.
64. Astrup, A., Lundsgaard, C., Madsen, J., et.al. "Enhanced thermogenic responsiveness during chronic ephedrine treatment in man." **Am. J. Clin. Nutr.**, 42(1), 83-94, 1985.
65. Astrup, A., Madsen, J., Holst, J.J., et.al. "The effect of chronic ephedrine treatment on substrate utilization, the sympathoadrenal activity, and energy expenditure during glucose-induced thermogenesis in man." **Metabolism**, 35(3), 260-265, 1986.
66. Astrup, A., Simonsen, L., Bulow, J., et.al. "Measurement of forearm oxygen consumption: role of heating the contralateral hand." **Am. J. Physiol.**, 255, E572-E578, 1988.
67. Astrup, A., Simonsen, L., Bulow, J., et.al. "Epinephrine mediates facultative carbohydrate-induced thermogenesis in human skeletal muscle." **Am. J. Physiol.**, 257, E340-E345, 1989.
68. Astrup, A., Simonsen, L., Bulow, J., et.al. "The contribution of skeletal muscle to carbohydrate-induced thermogenesis in man: the role of the sympathoadrenal system." In **Hormones, Thermogenesis and Obesity**, Lardy, H. & Stratman, F., eds., Elsevier, New York, NY, 187-196, 1989.
69. Astrup, A. & Toubro, S. "Thermogenic, metabolic, and cardiovascular responses to ephedrine and caffeine in man." **Int. J. Obes.**, 17(suppl 1), S41-S43, 1993.
70. Astrup, A.V., Toubro, S., Cannon, S., et.al. "Caffeine: a double-blind, placebo-controlled study of its thermogenic, metabolic, and cardiovascular effects in healthy volunteers." **Am. J. Clin. Nutr.**, 51(5), 759-767, 1990.
71. Astrup, A., Toubro, S., Cannon, S., et.al. **Cur. Ther. Res.**, 48, 1087, 1990.
72. Astrup, A., Toubro, S., Cannon, S., et.al. "Thermogenic, metabolic and cardiovascular effects of a beta-agonist, ephedrine. A double blind placebo-controlled study in humans." **Cur. Ther. Res.**, in press, 1991.
73. Astrup, A., Toubro, S., Cannon, S., et.al. "Thermogenic synergism between ephedrine and caffeine in healthy volunteers: a double-blind, placebo-controlled study." **Metabolism**, 40, 323-329, 1991.
74. Astrup, A., Toubro, S., Christensen, N.J., et.al. "Pharmacology of thermogenic drugs." **Am. J. Clin. Nutr.**, 55(suppl 1), 246S-248S, 1992.
75. Avogaro, P., Capri, C., Pais, M., et.al. "Plasma and urine cortisol behaaviour and fat mobilization in man after coffee ingestion." **Isr. J. Med. Sci.**, 9, 114-119, 1973.
76. Axelrod, J. "Studies on sympathomimetic amines. I. The biotransformation and physiological disposition of l-ephedrine and l-norephedrine." **J. Pharm. Exp. Ther.**, 109, 62-73, 1953.
77. Axelrod, L. **Diabetes**, 40, 1223, 1991.
78. Bailey, C.J., Thornburn, C.C. & Flatt, P.R. "Effects of ephedrine and atenolol on the development of obesity and diabetes in ob/ob mice." **Gen. Pharm.**, 17(2), 243-246, 1986.
79. Balint, G.A. "Effect of capsaicin on the hexobarbital necrosis of rats." **Kiseroetes Orvostudomany**, 24, 101-103, 1972.
80. Banet, M., Hensel, H. & Liebermann, H. "The central control of shivering and non-shivering thermogenesis in the rat." **J. Physiol.**, 383, 569-584, 1978.
81. Barnard, T., **J. Ultrastruct. Res.**, 29, 311, 1969.
82. Barnard, T. & Skala, J. "The development of brown adipose tissue." In **Brown Adipose Tissue**, Lindberg, O., ed., NY, Elsevier, 1970, pp. 33-72.
83. Barnard, T., Mory, G. & Nechad, M. "Biogenic amines and the trophic response of brown adipose tissue." In **Biogenic Amines in Development**, Parvez, H. & Parvez, S., eds., Holland, Elsevier, 391-439, 1980.
84. Bartness, T.J., Billington, C.J., Levine, A.S., et.al. **N.E. Am. J. Physiol.**, 251, R1109-R1117, 1987.
85. Bartness, T.J., Billington, C.J., Levine, A.S., et.al. **Am. J. Physio.**, 251, R1118-R1125, 1986.
86. Battig, K. **Rev. Environ, Health**, 9, 53, 1991.
87. Battig, K. "Acute and chronic cardiovascular and behavioural effects of caffeine, aspirin and ephedrine." **Int. J. Obes.**, 17(suppl 1), S61-S64, 1993.

88. Battig, K. & Welzl, H. In **Caffeine, Coffee, and Health**, Garattini, S., ed., NY Raven Press, in press, 1993.

89. Bazelmans, J., Nestel, P.J., O'Dea, K., et.al. "Blunted norephephrine responsiveness to changing energy states in obese subjects." **Metabolism**, 34, 154-160,1985.

90. Bazin, R., Eteve, D. and Lavau, M. "Evidence for decreased GDP binding to brown-adipose-tissue mitochondria of obese Zucker (fa/fa) rats in the very first days of life." **Biochem. J.**, 221, 241-245, 1984.

91. Beavers, W.R. & Covino, B.G. "Effects of glycine during cold exposure in man." **J. Appl. Physiol.**, 14, 390-392, 1959.

92. Beavo, A., Rogers, N.L., Crofford, O.B., et.al. **Molec. Pharm.**, 6, 597-603, 1970.

93. Becker, A.B., Simons, K.J., Gillespie, C.A., et.al. "The bronchodilator effects and pharmkinetics of caffeine in asthma." **N. Eng. J. Med.**, 310, 743, 1984.

94. Beckett, A.H., Tucker, G.T. & Moffat, A.C. "Routine detection and identification in urine of stimulants and other drugs, some of which may be used to modify performance in sport." **J. Pharm. Pharmacol.**, 19, 273-294, 1967.

95. Becker, A.B., Simons, K.J., Gillespie, C.A., et.al. "The bronchodilator effects and pharmkinetics of caffeine in asthma." **N. Eng. J. Med.**, 310, 743-746, 1984.

96. Becker, W., Bruce, A. & Larsson, B. "Autoradiographic studies with albumin bound carbon-14 labeled linoleic acid in normal and essential fatty acid deficient rat." **Ann. Nutr. Metab.**, 27, 415, 1983.

97. Beley, A., Beley, P., Rochette, L., et.al. "Evolution in vivo of the synthesis rate of catecholamines in various peripheral organs of the rat during cold exposure." **Pflugers Arch.**, 366, 259-264, 1976.

98. Belfrage, P., Fredrikson, G., Olsson, H. & Staelfors, P. "Control of adipose tissue lipolysis by phosphorylation/ dephosphorylation of hormone-sensitive lipase." In **The Adipocyte and Obesity: Cellular and Molecular Mechanisms**, Angel, A., Hollenberg, C.H. & Roncari, D.A., eds., Raven Press, New York, 1983, pp. 217-223.

100. Belko, A.Z., Obarzanek, E., Roach, R., et.al. "Effects of aerobic exercise and weight loss on riboflavin requirements of moderately obese, marginally deficient young women." **Am. J. Clin. Nutr.**, 40, 553, 1984.

101. Bellet, S., Feinberg, L.J., Sandberg, H., et.al. **J. Pharm. Exp. Ther.**, 159, 250-254, 1969.

102. Bellet, S., Kershbaum, A. & Aspe, J. **Arch. Intern. Med.**, 116, 750-752, 1965.

103. Bellet, S., Kershbaum, A. & Fink, E.M. "Response of free fatty acids to coffee and caffeine." **Metabolism**, 17, 702-707, 1968.

104. Bellet, S., Kostis, J., Roman, L., et.al. "Effect of coffee ingestion on adrenocortical secretion in young men and dogs." **Metabolism**, 18, 1007-1012, 1969.

105. Bellet, S., Roman, L., DeCastro, O., et.al. "The effect of coffee ingestion on catecholamine release." **Metabolism**, 18, 288-291, 1969.

106. Benedict, F.G., Miles, W.R., Roth, P., et.al. Publication 280, Washington, D.C., Carnegie Inst., 1919.

107. Bengtsson, C. "Comparison between alprenolol and chlorthalidone as antihypertensive agents." **Acta Med. Scand.**, 191, 433-439, 1972.

108. Bennett, T., Macdonald, I.A. & Sainsbury, R. "The influence of acute starvation on the cardiovascular responses to lower body subatmospheric pressure or to standing in man." **Clin. Sci.**, 66, 141-146, 1984.

109. Benzi, R.H. & Girardier, L. **Pflugers Arch.**, 406, 37-4, 1986.

110. Benzi, R.H., Shipbata, M., Seydoux, J., et.al. **Pflugers Arch.**, 411, 593-599, 1988.

111. Bergh, U., Hartley, H., Landsberg, L., et.al. "Plasma norepinephrine concentration during submaximal and maximal exercise at lowered skin and core temperatures." **Acta Physiol. Scand.**, 106, 383-384, 1979.

112. Berglund, B. & Hemmingsson, P. **Int. J. Sports Med.**, 3, 234-236, 1982.

113. Bergstrom, S. & Carlson, L.A. "Inhibitory action of prostaglandin E on the mobilization of free fatty acids and glycerol from human adipose tissue in vitro." **Acta Physiol. Scand.**, 63, 195, 1987.

114. Berkowitz, B.A. & Spector, S. "Effect of caffeine and theophylline on peripheral catecholamines." **Eur. J. Pharm.**, 13, 193-196, 1971.

115. Berkowitz, B.A., Tarver, J.H. & Spector, S. "Release of norepinephrine in the central nervous system by theophylline and caffeine." **Eur. J. Pharm.**, 10, 64-71, 1970.

116. Bertin, R., Jallot, M. & Portet, R. "Perinatal regulation of brown fat lipid metabolism; effect of ambient temperatue." InSzelenyi, Z. & Szekely, M. eds. **Contributions to Thermal Physiology**, Pergamon Press, Akademaiai Kiado, Budapest, 1981, pp. 527-530.

117. Berthoud, H.R., Bereiter, D.A., Trimble, E.R., et.al. **Diabetologia**, 20, 393-401, 1981.

118. Bessard, T., Schutz, Y. & Jequier, E. "Energy expenditure and postprandial thermogenesis in obese women before and after weight loss." **Am. J. Clin. Nutr.**, 38, 680-694, 1983.

119. Bevan, J.A. "Action of lobeline and capsaicin on afferent endings in the pulmonary artery of the cat." **Circ. Res.**, 10, 792-797, 1962.

120. Bianchi, C.P. **J. Pharm. Exp. Ther.**, 138, 41-47, 1962.

121. Bianco, A.C. & Silva, J.E. **Am. J. Physiol.**, 253, E255-E263, 1987.

122. Bianco, A.C. & Silva, J.E. **Am. J. Physiol.**, 255, E496-E503, 1988.

123. Bianco, A.C. & Silva, J.E. **Endocrinology**, 120, 55-62, 1987.

124. Bianco, A.C. & Silva, J.E. **J. Clin. Inv.**, 79, 295-300, 1987.

125. Billington, C.J., Bartness, T.J., Briggs, J., et.al. **E. Am. J. Physiol.**, 252, R160-R165, 1987.

126. Billington, C.J., Briggs, J.E., Link, J.G., et.al. "Glucagon in physiological concentrations stimulates brown fat thermogenesis in vivo." **Am. J. Physiol.**, 261(2 pt 2), R501-507, 1991.

127. Bjorntorp, P. "The associations between obesity, adipose tissue distribution and disease." **Acta Med. Scand.**, 723, 121-134, 1988.

128. Bjorntorp, P. **Diabetes Care**, 14, 1132, 1991.

129. Bjorntorp, P., Bergman, H., Varnauskas, E., et.al. "Lipid mobilization in relation to body composition in man." **Metabolism**, 18, 840, 1969.

130. Blanchette-Mackie, E.J. & Scow, R.O. **J. Lipid Res.**, 24, 229-244, 1983.

131. Bogardus, C., Lillioja, S., Moh, D., et.al. "Evidence for reduced thermic effect of insulin and glucose infusions in PIMA Indians." **J. Clin. Inv.**, 75, 1264-1269, 1985.

132. Bogardus, C., Lillioja, S., Ravussin, E., et.al. "Familial dependence on the resting metabolic rate." **N. Eng. J. Med.**, 315, 96-100, 1986.

133. Bohr, D. & Goulet, P.L. "Role of electrolytes in contractile machinery of vascular smooth muscle." **Am. J. Cardiol.**, 8, 549-556, 1961.

134. Bonati, M., Latini, F., Galletti, J.F., et.al. **Clin. Pharm. Ther.**, 32, 98-106, 1982.

311

136. Borensztajn, J. & Kotlar, T.J. "Contribution of adipose tissue to the clearance of chylomicrons from plasma." In **The Adipocyte and Obesity: Cellular and Molecular Mechanisms**, Angel, A., Hollenberg, C.H. & Roncari, D.A., eds., Raven Press, New York, 1983, p. 149.

137. Borgman, R.F. "Dietary factors in essential hypertension." **Prog. Food Nutr. Sci.**, 9(1-2), 109-147, 1985.

138. Bouchard, C., Tremblay, A., Despres, J.P., et.al. **New. Eng. J. M.**, 322, 1477, 1990.

139. Bouillaud, F., Ricquier, D., Thibault, J., et.al. "Molecular approach to thermogenesis in brown tissue. cDNA cloning of mitochondrial uncoupling protein." **Proc. Natl. Acad. Sci.**, 82, 445-448, 1984.

140. Boulant, J.A. "Hypothalamic mechanisms in thermoregulation." **Fed. Proc.**, 40, 2843-2850, 1981.

141. Bracco, D., Ferrarra, J.M., Arnaud, M.J. "Effects of caffeine on thermogenesis and substrate oxidation in lean and obese women." **Int. J. Obes.**, 17(suppl 1), S80, 1993.

142. Brady, L.J., Knoeber, C.M., Hoppel, C.L., et.al. "Pharmacologic action of L-carnitine on hypertriglyceridemia in obese Zucker rats." **Metabolism**, 35, 555, 1986.

143. Bravo, E.L. "Phenylpropanolamine and other over-the-counter vasoactive compounds." **Hypertension**, 11(3 pt 2), 117-10, 1988.

144. Bray, G.A. "Effect of caloric restriction on energy expenditure in obese patients." **Lancet**, 2, 397-398, 1969.

145. Bray, G.A. "The myth of diet in the management of obesity." **Am. J. Clin. Nutr.**, 23, 141-148, 1970.

146. Bray, G.A. "Integration of energy intake and expenditure in animals and man: the autonomic and adrenal hypotheses." In **Clinics in Endocrinology and Metabolism: Obesity**, James, W.P.T., ed., Philadelphia, W.B. Saunders, 521-546, 1984.

147. Bray, G.A. "Obesity--a disease of nutrient or energy balance." **Nutr. Rev.**, 45, 33-43, 1987.

147a. Bray, G.A. "Peptides affect the intake of specific nutrients and the sympathetic nervous system." **Am. J. Clin. Nutr.**, 55, 265s-271s, 1992.

148. Bray, G.A. **Ann. Int. Med.**, 115(2), 152, 1991.

148a. Bray, G.A. "Metabolic factors in the control of energy stores." **Metabolism**, 24, 99-117, 1975.

148b. Bray, G.A. "Use and abuse of appetite-suppressant drugs in the treatment of obesity." **Ann. Intern. Med.**, 119(7 pt 2), 707-713, 1993.

149. Bray, G.A., Schwartz, M., Rozin, R., et.al. "Relationship between oxygen consumption and body composition of obese patients." **Metabolism**, 19, 418-429.

150. Bray, G.A. & York, D.A. "Hypothalamic and genetic obesity in experimental animals: an autonomic and endocrine hypothesis." **Physiol. Rev.**, 59, 719-809, 1979.

151. Bray, G.A., York, D.A. & Fisler, J.S. "Experimental obesity: a homeostatic failure due to defective nutrient stimulation of the sympathetic nervous system." **Vitamins and Hormones**, 45, 1-24, 1989.

152. Brezinova, V. **Br. J. Clin. Pharm.**, 1, 203-206, 1974.

153. Briggs, A.H. & Melvin, S. "Ion movements in isolated rabbit aortic strips." **Am. J. Physiol.**, 201, 365-368, 1961.

154. Broeder, C.E., Brenner, M., Hofman, Z., et.al. "The metabolic consequences of low and moderate intensity exercise with or without feeding in lean and borderline obese males." **Int. J. Obes.**, 15(2), 95-104, 1991.

155. Broquist, H. "Carnitine biosynthesis and function." **Fed. Proc.**, 41, 2840, 1982.

156. Bruce, B.K, King, B.M., Phelps, G.R., et.al. "Effects of adrenalectomy and corticosterone administration on hypothalamic obesity in rats." **Am. J. Physiol.**, 243, E152, 1982.

157. Brueck, K. "Nonshivering thermogenesis in brown adipose tissue in realtion to age, and their integration in the thermoregulatory system." In **Brown Adipose Tissue**, Lindberg, O., ed., NY, Elsevier, 1970, pp. 117-154.

158. Brueck, K. & Zeisbergrger, E. "Adaptive changes in thermoregulation and their neuropharmacological basis." **Pharm. Ther.**, 35, 163-215, 1987.

159. Brundin, T., Thorne, A. & Wahren, J. "Heat leakage across the abdominal wall and meal-induced thermogenesis in normal-weight and obese subjects." **Metabolism**, 41(1), 49-55, 1992.

160. Brunzell, J.B. "Disorders of lipid metabolism." In **Cecil Textbook of Medicine**, 17th ed., Wyungaarden, J.B. & Smith, L.H.Jr., eds., Philadelphia, W.B. Saunders, 1985.

161. Buck, S.H. & Burks, T.F. "Capsaicin: hot new pharmacological tool." **Trends in Pharm. Sciences**, 4, 84-87, 1983.

162. Buckley, M.G. & Rath, E.A. **Biochem. J.**, 243, 437-442, 1987.

163. Bucsies, A. & Lembeck, F. **Eur. J. Pharm.**, 71, 71, 1981.

164. Bugojski, J. & Zacny E. "The role of central histamine H1- and H2-receptors in hypothermia induced by histamine in the rat." **Agents Actions**, 11, 442, 1981.

165. Bukkens, S.G., McNeill, G., Morrison, D.C., et.al. **Int. J. Obes.**, 15(2), 147-154, 1991.

166. Bukowiecki, L. **Can. J. Biochem. Cell Biol.**, 62, 623-630, 1984.

167. Bukowiecki, L.J. "Regulation of energy expenditure in brown adipose tissue." **Int. J. Obes.**, 9(suppl 2), 31-41, 1985.

168. Bukowiecki, L.J. In **Brown Adipose Tissue**, Trayhurn, P. & Nicholls, D.G., eds., London, Arnold, 105-121, 1986.

169. Bukowiecki, L., Collet, A.J., Follea, N., et.al. "Brown adipose tissue hyperplasia: a fundamental mechanism of adaptation to cold and hyperphagia." **Am. J. Physiol.**, 242, E353-E359, 1982.

170. Bukowiecki, L., Follea, N., Lupien, J., et.al. **J. Biol. Chem.**, 256, 12840-12848, 1981.

171. Bukowiecki, L., Follea, N., Paradis, A., et.al. "Stereospecific stimulation of brown adipoyste respiration by catecholamines via beta-I-adrenoreceptors." **Am. J. Physiol.**, 238, E552-E563, 1980.

172. Bukowiecki, L., Follea, N., Vallieres, J., et.al. **J. Eur. J. Biochem.**, 92, 189-196, 1978.

173. Bukowiecki, L.J., Geloen, A. & Collet, A.J. **J. Am. J. Physiol.**, 250, C880-C887, 1986.

174. Bukowiecki, L.J., Jahjah, L. & Follea, N. "Ephedrine, a potential slimming drug, directly stimulates thermogenesis in brown adipocytes via beta-adrenoreceptors." **Int. J. Obes.**, 6, 343-350, 1982.

175. Bukowiecki, L.J., Lupien, J., Follea, N., et.al. "Mechanism of enhanced lipolysis in adipose tissue of exercise-trained rats." **Am. J. Physiol.**, 239, E422, 1980.

176. Bukowiecki, L.J., Lupien, J., Follea, N., et.al. "Effects of sucrose, caffeine, and cola beverages on obesity, cold resistance, and adipose tissue cellularity." **Am. J. Physiol.**, 244, R500-R507, 1983.

177. Bunker, M.L. & McWilliams, M. "Caffeine content of common beverages." **J. Am. Diet. Assn.**, 74, 28, 1979.

178. Burgi, U. & Burgi-Saville, M.E. **Am. J. Physiol.**, 251, E503-E508, 1986.

179. Burns, T.W. & Langley, P.E., **J. Lab. Clin. Med.**, 75, 983-997, 1970.

180. Burns, T.W., Langley, P.E. & Terry, B.E. **J. Clin. Inv.** 67, 467, 1981.

181. Burns, T.W., Langley, P.E., Terry, B.E. & Bylund, D.B. "Pharmacological characterization and regulation of adrenergic receptors in human adipocytes." In **The Adipocyte and Obesity: Cellular and Molecular Mechanisms**, Angel, A., Hollenberg, C.H. & Roncari, D.A., eds., Raven Press, New York, 1983, pp. 235-244.

184. Butcher, R.W., Ho, R.J., Meng, H.C., et.al. "Adenosine 3'5'-monophosphate in biological materials II. The measurement of adenosine 3'5'-monophosphate in tissues and the role of the cyclic nucleotide in the lipolytic response of fat to epinephrine." **J. Biol. Chem.**, 240, 4515-4523, 1965.

185. Butcher, R.W. & Sutherland, E.W. **J. Biol. Chem.**, 237, 1244-1250, 1962.

186. Butts, N.K. & Crowell, D. "Effect of caffeine ingestion on cardiorespiratory endurance in men and women." **Res. O. Exerc. Sport**, 56, 301-305, 1985.

187. Cabanac, M. "Temperature regulation." **Ann. Rev. Physiol.**, 37, 415-439, 1975.

188. Cadarette, B.S., Levine, L., Berube, C.L., et.al. "Effects of varied dosages of caffeine on endurance exercise to fatigue." In **International Series on Sports Sciences, Biochemistry of Exercise**, Knuttgen, H.G., Vogel, J.A. & Poortmans, J., eds., Champaign, IL, Human Kinetics, 871-876, 1983.

189. Cadrin, M., Tolszczuk, M., Guy, J., **J. Histochem. Cytochem.**, 33, 150-154, 1985.

190. Cahill, G.F. "Adipose tissue metabolism." In **Fat as a Tissue**, Rodahl, K., & Issekutz, B., eds. McGraw Hill Book Company, New York, 1964, pp. 169-183.

191. Calles-Escandon, J. & Horton, E.S. "The thermogenic role of exercise in the treatment of morbid obesity: a critical evaluation." **Am. J. Clin. Nutr.**, 55(suppl 2), 533S-537S, 1992.

192. Camann, W.R., Murray, R.S., Mushlin, P.S., et.al. **Anesthes. Analges.**, 70, 181, 1990.

193. Cannon, B., Hedin, A. & Nedergaard, J. "Exclusive occurrence of thermogenin antigen in brown adipose tissue." **FEBS Lett.**, 150, 129-132, 1982.

194. Cannon, B., Jacobsson, A., Carneheim, C., et.al. In **Living in the Cold**, Heller, H.C., Musacchia, X.J. & Wang, L.C.H., eds., Elsevier, NY, 71-81, 1986.

195. Cannon, B. & Nedergaard, J. "Biochemical aspects of acclimation to cold." **J. Therm. Biol.**, 8, 85-90, 1983.

196. Cannon, B. & Nedergaard, J. "The biochemistry of an inefficient tissue-brown adipose tissue." **Essays Biochem.**, 20, 110-164, 1985.

197. Cannon, B., Nedergaard, J., & Sundin, U. "Physiological uncoupling in brown fat mitochondria." Szelenyi, Z. & Szekely, M. eds. **Contributions to Thermal Physiology**, Pergamon Press, Akademaiai Kiado, Budapest, 1981, pp. 479-497.

198. Cannon, B., Nedergaard, J. & Sundin, U. "Thermogenesis, brown fat and thermogenin." In **Survival in Cold**, Musacchia, X.J. & Jansky, L., eds., Amsterdam, Elsevier/North Holland.

199. Cannon, W.B., Querido, A., Britton, S.W., et.al. "Studies on the conditions of activity in endocrine glands. XXI. The role of adrenal secretion in the chemical control of body temperature." **Am. J. Physiol.**, 79, 466-507, 1927.

200. Cappelletti, E.M., Trevisan, R. & Caniato, R. "External antirheumatic and antineuralgic herbal remedies in the traditional medicine of north-eastern Italy." **J. Ethnopharm.**, 6, 161-190, 1982.

201. Carlson, L.A. "Metabolic and cardiovascular effects in vitro of prostaglandins." In **Prostaglandins Proc.**, 2nd Nobel Symposium, Stockholm, Bergstrom, S. & Samuelsson, B., eds., Stockholm, Almqvist & Witsell, 123, 1967.

202. Carlson, L.D. "Nonshivering thermogenesis and its endocrine control." **Fed. Proc.**, 19, 25-30, 1960.

203. Carneheim, C., Nedergaard, J. & Cannon, B. "Beta-adrenergic stimulation of lipoprotein lipase in rat brown adipose tissue during acclimation to cold." **Am. J. Physiol.**, 246, E327-E333, 1984.

204. Carneheim, C., Nedergaard, J. & Cannon, B. **Am. J. Physiol.**, 254, E155-E161, 1988.

206. Carvalho, S.D., Kimura, E.T., Bianco, A.C., et.al. "Central role of brown adipose tissue thyroxine 5'-deiodinase on thyroid hormone-dependent thermogenic response to cold." **Endocrinology**, 128(4), 2149-2159, 1991.

207. Casal, D.C. & Leon, A.S. "Failure of caffeine to affect substrate utilisation during prolonged running." **Med. Sci. Sports Exerc.**, 14, 174-179, 1982.

208. Casteilla, L., Forest, C., Robelin, J. "Characterization of mitochondrial-uncoupling protein in bovine fetus and newborn calf." **Am. J. Physiol.**, E627-E636, 1987.

209. Cawthorne, M.A., Smith, S.A. & Young, P. In **Recent Advances in Obesity Research**, Berry, E.M., Blondheim, S.H., Eliahou, H.E., et.al., eds., London, Libbey, 5, 312-318, 1987.

210. Cesari, M.P., Pasquali, R., Casimirri, F., et.al. **Int. J. Obes.**, 13(suppl 1), 152, 1989.

211. Chad, K. & Quigley, B. **Eur. J. Appl. Physiol.**, 59, 48-54, 1989.

212. Champigny, O. & Ricquier, D. "Effects of fasting and refeeding on the level of uncoupling protein mRNA in rat brown adipose tissue: evidence for diet-induced and cold-induced responses." **J. Nutr.**, 120(12), 1730-1736, 1990.

213. Chapman, B.J., Forquahar, D.L., Galloway, M.L., et.al. **Int. J. Obes.**, 12, 119, 1988.

214. Chatzipanteli, K., Rudolph, S. & Axelrod, L. **Diabetes**, 41, 927, 1992.

215. Chen, A.C.N. & Chapman, C.R. **Exp. Brain Res.**, 39, 359, 1980.

216. Chen, K.K. & Schmidt, C.F. **J. Pharm. Exp. Ther.**, 24, 339, 1925.

217. Chen, K.K., Wu, C. & Henriksen, E. Relationship between the pharmacological action and chemical constitution and configuration of the optical isomers of ephedrine and related compounds." **J. Pharm. Exp. Ther.**, 36, 363-400, 1929.

218. Cheney, R.H. **J. Pharm. Exp. Ther.**, 53, 304-313, 1935.

219. Cherek, D.R., Steinberg, J.L. & Branchi, J.T. **Psych. Res.**, 8, 1374, 1983.

220. Choo, J.J., Horan, M.A., Little, R.A., et.al. **Metab. Exp. Clin.**, 39, 647, 1990.

221. Christin, L., Nacht, C.A., Vernet, O., et.al. "Insulin: its role in the thermic effect of glucose." **J. Clin. Inv.**, 77, 1747, 1986.

222. Chua, S.S & Benrimoj, S.I. "Non-prescription sympathomimetic agents and hypertionsion." **Med. Tox. Adv. Dr. Exp.**, 3(5), 387-417, 1988.

223. Chua, S.S., Benrimoj, S.I., Gordon, R.D., et.al. "A controlled clinical trial on the cardiovascular effects of single doses of pseudoephedrine in hypertensive patients." **Br. J. Clin. Pharm.**, 28(3), 369-372, 1989.

224. Chvasta, T.E. & Cook, A.R. **Gastroenterology**, 61, 838-843, 1971.

225. Clark, M.G., Rattigan, S. & Colquhoun, E.Q. "Hypertension in obesity may reflect a homeostatic thermogenic response." **Life Sci.**, 48(10), 939-947, 1991.

226. Clark, R.F. & Curry, S.C. "Pseudoephedrine dangers (letter)." **Pediatrics**, 85(3), 389-390, 1990.

227. Clement-Cormier, Y.C., Parrish, R.G., Petzold, G.L., et.al. "Characterization of a dopamine-sensitive adenylate cyclase in the rat caudate nucleus." **J. Neurochem.**, 25, 143, 1975.

228. Clubley, M., Bye, C.E., Henson, T.A., et.al. **J. Clin. Pharm.**, 7, 157-163, 1979.

229. Cole, J.O., Pope, H.G., Labrie, R., et.al. **Clin. Pharm. Ther.**, 24, 243-252, 1978.

230. Coleridge, H.M., Coleridge, J.C.G. & Kidd, C. "Role of the pulmonary arterial baroreceptors in the effects produced by capsaicin in the dog." **J. Physiol.**, 170, 272-285, 1964.

231. Collins, T.F.X., Welsh, J., Black, T., et.al. "A study of the teratogenic potential of caffeine given by oral intubation to rats." **Reg.**

Tox. Pharm., 1, 355, 1980.

232. Collins, T.F.X., Welsh, J., Black, T., et.al. "A study of the teratogenic potential of caffeine ingested in drinking water." **Food Chem. Tox.**, 6, 763, 1983.

233. Colquhoun, E.Q., Hettiarachchi, M., Ye, J.M., et.al. **Life Sci.**, 43, 1747-1754, 1988.

234. Colquhoun, E.Q. & Clark, M.G. **News in Physiol. Sci.**, 6, 256, 1991.

235. Colton, T., Gosselin, R.E. & Smith, R.P. "The tolerance of coffee drinkers to caffeine." **Clin. Pharm. Ther.**, 9, 31-39, 1967.

236. Connacher, A.A., Bennet, W.M. & Jung, R.T. "Clinical studies with the beta-adrenoceptor agonist BRL 26830A." **Am. J. Clin. Nutr.**, 55(suppl 1), 258S-261S, 1992.

237. Connacher, A.A., Jung, R.T. & Mitchell, P.E.G. "Weight loss in obese subjects on a restricted diet given BRL 26830A, a new atypical beta-adrenoceptor agonist." **Br. Med. J.**, 296, 1217-1220, 1988.

238. Connacher, A.A., Jung, R.T., Mitchell, P.E.G., et.al. **Int. J. Obes.**, 12, 267-276, 1988.

239. Connacher, A.A., Lakie, M., Powers, N., et.al. "Tremor and the anti-obesity drug BRL 26830A." **Br. J. Clin. Pharm.**, 30(4), 613-615, 1990.

240. Conners, C.K. **Int. J. Mental Hlth.**, 4, 132-143, 1975.

241. Conners, C.K. **J. Abnormal Child Psych.**, 7, 145-151, 1979.

242. Connolly, E., Nanberg, E. & Nedergaard, J. **J. Biol. Chem.**, 261, 14377-14385, 1986.

243. Connolly, E. & Nedergaard, J. **J. Biol. Chem.**, 263, 10574-10582, 1988.

244. Cooney, G.J., Astbury, L.D., Williams, P.F., et.al. **Diabetes**, 36, 152-158, 1987.

245. Cooney, G.J., Caterson, I.D. & Newsholme, F.A., et.al. The effect of insulin and noradrenaline on the uptake of 2-[1-14C]deoxyglucose in vivo by brown adipose tissue and other glucose utilizing tissues of the mouse." FEBS Lett., 188, 257-261, 1985.

246. Cooney, G., Curi, R., Mitchelson, A., et.al. **Biochem. Biophys. Res. Commun.**, 138, 687-692, 1986.

247. Cormareche-Leydier, M. "The effects of long warm and cold ambient exposures on food and water intake in the capsaicin treated rat." **Fed. Proc.**, 42(119), 1983.

248. Coscina, D.V., Chambers, J.W., Park, I., et.al. "Impaired diet-induced thermogenesis in brown adipose tissue from rats made obese with parasagittal hypothalamic knife cuts." **Brain Res. Bull.**, 14, 585-593, 1985.

249. Costill, D.L., Coyle, E., Dalsky, G., et.al. "Effects of elevated plasma FFA and insulin on muscle glycogen usage during exercise." **J. Appl. Physiol.**, 43, 695-699, 1977.

250. Costill, D.L., Dalsky, G.P. & Fink, W.J. "Effect of caffeine ingestion on metabolism and exercise performance." **Med. Sci. Sports Exer.**, 10, 155-158, 1978.

251. Cote, C., Thibault, M.C. & Vallieres, J. **Life Sci.**, 37, 695-701, 1985.

252. Cottle, M.K.W. & Cottle, W.H. "Adrenergic fibers in brown fat of cold-acclimated rats." **J. Histochem. Cytochem.**, 18, 116-119, 1970.

253. Cottle, W.H. "The innervation of brown adipose tissue." In **Brown Adipose Tissue**, Lindberg, O., ed., NY, Elsevier, 1970, pp. 155-178.

254. Cottle, W.H. & Carlson, L.D. **Proc. Soc. Exper. Biol. Med.**, 92, 845, 1956.

255. Cryer, P.E. "Physiology and pathophysiology of the human sympathoadrenal neuroendocrine system." **N. Eng. J. Med.**, 30, 436-444, 1980.

256. Cryer, P.E., Tse, T.F., Clutter, W.E., et.al. "Roles of glucagon and epinephrine in hypoglycemic and nonhypoglycemic glucose counterregulation in humans." **Am. J. Physiol.**, 247, E198-E205, 1984.

257. Cui, J., & Himms-Hagen, J. "Long-term decrease in body fat and in brown adipose tissue in capsaicin-desensitized rats." **Am. J. Physiol.**, 262(4 pt 2), R568-573, 1992.

258. Cui, J., & Himms-Hagen, J. "Rapid but transient atrophy of brown adipose tissue in capsaicin-desensitized rats." **Am. J. Physiol.**, 262(4 pt 2), R562-567, 1992.

259. Cui, J., Zaror-Behrens, G. & Himms-Hagen, J. "Capsaicin desensitization induces atrophy of brown adipose tissue in rats." **Am. J. Physiol.**, 259(2 pt 2), R324-332, 1990.

260. Cui, J., Zaror-Behrens, G. & Himms-Hagen, J. **FASEB J.**, 3, A346, 1989.

261. Cunningham, S., Leslie, P., Hopwood, D., et.al. **Clin. Sci.**, 69, 343-348, 1985.

262. Cunningham, S.A., & Nicholls, D.G. **Biochem. J.**, 245, 485-491, 1987.

263. Cunningham, S.A., Wiesinger, H. & Nicholls, D.G. **Eur. J. Biochem.**, 157, 415-420, 1986.

264. Curatolo, P.W. & Robertson, D. **Ann. Intern. Med.**, 98, 641-653, 1983.

265. Dahlstrom, M., Jansson, E., Nordevang, E., et.al. "Discrepancy between estimated energy intake and requirement in female dancers." **Clin. Physiol.**, 10(1), 11-25, 1990.

266. Dale, D., Saris, W.H.M. & Hoor, F.T. **Int. J. Obes.**, 14(4), 347-359, 1990.

267. Dallasso, H.M. & James, W.P.T. **Br. J. Nutr.**, 40, 542, 1984.

268. D'Allessio, D.A., Kavle, E.C., Mozzoli, M.A., et.al. **J. Clin. Inv.**, 81, 1781-1789, 1988.

269. Daly, J.W., Bruns, R.F. & Snyder, S.H. **Life Sci.**, 28, 2083-2097, 1981.

270. Daly, P.A., Drieger, D.R., Dulloo, A.G., et.al. "Ephedrine, caffeine and aspirin: safety and efficacy for treatment of human obesity." **Int. J. Obes.**, 17(suppl 1), S73-S78, 1993.

271. Daly, P.A. & Landsberg, L. "Hypertension in obesity and NIDDM. Role of insulin and sympathetic nervous system." **Diabetes Care**, 14(3), 240-248, 1991.

272. Dambisya, Y.M., Wong, C.L. & Chan, K. **Meth. Find. Exp. Clin. Pharm.**, 13, 239, 1991.

273. Danforth, E., Daniels, H.L., Ravussin, E., et.al. "Thermogenic responsiveness in Pima Indians." **Clin. Res.**, 29, 663A, 1981.

274. Danforth, E. & Sims, E.A.H. "Thermic effect of overfeeding: Role of thyroid hormones, catecholamines, and insulin resistance." In **The Adipocyte and Obesity: Cellular and Molecular Mechanisms**, Angel, A., Hollenberg, C.H. & Roncari, D.A., eds., Raven Press, New York, 1983, pp. 271-282.

275. Danforth, E. "Diet and obesity." **Am. J. Clin. Nutr.**, 41, 1132-1145, 1985.

276. Daubresse, J.C., Luyckx, A., Demey-Ponsart, E., et.al. "Effects of coffee and caffeine on carbohydrate metabolism, free fatty acid, insulin, growth hormone and cortisol plasma levels in man." **Acta Diabet. Lat.**, 10, 1069, 1973.

277. Dauncey, M.J. & Brown, D. "Role of activity-induced thermogenesis in twenty-four hour energy expenditure of lean and genetically obese (ob/ob) mice." **Q. J. Exp. Physiol.**, 72, 549-559, 1987.

278. Davis, I. "In vitro regulation of adipose tissue." **Nature**, 218, 349-352, 1968.

279. Davis, T.R.A. "Chamber cold acclimatization in man." **J. App. Physiol.**, 16, 1011-1015, 1961.

280. Dawber, T.R., Kannel, W.B. & Gordon, T. "Coffee and cardiovascular disease: observations from the Framingham study." **N. Eng. J. Med.**, 291, 871, 1974.

281. Dawkins, M.J.R. & Hull, D. "The production of heat by fat." **Sci. Am.**, 213, 62-67, 1965.

282. Debnath, P.K., Bhattacharya, S.K., Sanyal, A.K., et.al. "Prostaglandins: effect of prostaglandin E1 on brain, stomach and intestinal serotonin in rat." **Biochem. Pharm.**, 27, 130, 1978.

283. Debons, A.F., Siclori, E., Das, K.C., et.al. "Gold thioglucose-induced hypothalamic damage, hyperphagia, and obesity: dependence on the adrenal gland." **Endocrinology**, 110, 2024-2029, 1982.

284. Decombaz, J. "Inventory of presumed ergogenic dietary supplements." **Ther. Umsch.**, 44(1), 911-916, 1987.

285. Decombaz, J., Arnaud, M.H., Milan, H., et.al. "Energy metabolism of medium-chain triglycerides versus carbohydrates during exercise." **Eur. J. Appl. Physiol.**, 52, 9-14, 1983.

286. Defreitas, B. & Schwartz, G. **Am. J. Psych.**, 136, 1337-1338, 1979.

287. DeFronzo, R.A., Thorin, D., Felber, J.P. "Effect of beta and alpha adrenergic blockage on glucose-induced thermogenesis in man." **J. Clin. Inv.**, 73, 633-639, 1984.

288. DeGubareff, T. & Sleator, W. **J. Pharm. Exp. Ther.**, 148, 202-214, 1965.

289. DeHaven, J., Sherwin, R., Hendler, R., et.al. "Nitrogen and sodium balance and sympathetic-nervous-system activity in obese subjects treated with a low-calorie protein or mixed diet." **N. Eng. J. Med.**, 302, 477-482, 1980.

290. Delagrange, P., Koster-Van Hoffen, G.C., Mirmiran, M., et.al. "Effect of S 20242, a melatoninergic analogue on body temperature rhythms in aged rats." **Int. J. Obes.**, 17(suppl 1), S82, 1993.

291. Denton, R.M., McCormack, J.G. & Marshall, S.E. **Biochem. J.**, 217, 441-452, 1984.

292. Derry, D.M., Schoenbaum, E. & Steiner, G., **Can. J. Physiol. Pharmacol.**, 47, 57, 1969.

293. Depocas, F. "Calorigenesis from various organ systems in the whole animal." **Fed. Proc.**, 5, 19-24, 1960.

294. Desai, H.G., Venugopalan, K. & Antia, F.P. "The effect of capsaicin on the DNA content of gastric aspirate." **Ind. J. Med. Res.**, 64, 163-167, 1976.

295. Desautels, M. "Mitochondrial thermogenin content is unchanged during atrophy of BAT of fasting mice." **Am. J. Physiol.**, 249, E99-E106, 1985.

296. Desautels, M. & Dulos, R.A. **Am. J. Physiol.**, 255, E120-E128, 1988.

297. Desautels, M. & Himms-Hagen, J. "Roles of noradrenaline and protein syntheses in the cold-induced increase in purine nucleotide binding by rat brown adipose tissue mitochondria." **Can. J. Biochem.**, 57, 968-976, 1979.

298. Deshaies, Y., Arnold, J. & Richard, D.J. **Appl. Physiol.**, 65, 549-554, 1988.

299. Dews, P.B. In **Caffeine**, Dews, P.B., ed., Springer-Verlag, Berlin, 86-103, 1984.

300. Diamond, P., & LeBlanc, J. **Am. J. Physiol.**, 252, E719-E726, 1987.

301. Diamond, P., & LeBlanc, J. **Am. J. Physiol.**, 253, E521-E529, 1987.

302. Dickinson, A.L., Haymes, E.M., Sparks, K.E., et.al. "Effects of moderate caffeine ingestion on factors contributing to the quality of endurance performance." **Med. Sci. Sports Exerc.**, 16, 171, 1984.

304. Djazayery, A., Miller, D.S. & Stock, M.J. "Energy balances in obese mice." **Nutr. Metab.**, 23, 357, 1979.

305. Dobmeyer, D.J., Stine, R.A. & Leier, C.V. **N. Eng. J. Med.**, 308, 814-816, 1983.

306. Dominguez, M.J., Fernandez, M., Elliott, K., et.al. **Biochem. Biophys. Res. Commun.**, 138, 1390-1394, 1986.

307. Donnerer, J. & Lembeck, F. **Naunyn-Schmiedeberg's Arch. Pharm.**, 320, 54, 1982.

308. Dorfman, L.J. & Jarvik, M.E. **Clin. Pharm. Ther.**, 11, 869-872, 1970.

309. Doubt, T.J. "Physiology of exercise in the cold." **Sports Med.**, 11(6), 367-381, 1991.

310. Doubt, T.J. & Hsieh, S.S. **Med. Sci. Sports Exer.**, 23(4), 435, 1991.

311. Drahota, Z. "Fatty acid oxidation by brown adipose tissue mitochondria." In **Brown Adipose Tissue**, Lindberg, O., ed., NY, Elsevier, 1970, pp. 225-244.

312. Drahota, Z., Honova, E., Hahn, P. & Gazzotti, P., **FEBS Symposium, Mitochondria-Structure and Function**, L. Ernster & Z. Drahota, eds., Academic Press, New York, 1969, p. 153.

313. Dreisbach, R.H. & Pheiffer, C. **J. Lab. Clin. Med.**, 28, 1212-1219, 1943.

314. Dresendorfer, R. "Hypozincemia in runners." **Phys. and Sports Med.**, 84, 97, 1980.

315. Dugdale, A.E. & Payne, P.R. "Pattern of lean and fat deposition in adults." **Nature**, 266, 349-351, 1977.

316. Dulloo, A.G. Ph.D. Thesis, Univ. of London, 1982.

317. Dulloo, A.G. "Ephedrine, xanthines and prostaglandin-inhibitors: actions and interactions in the stimulation of thermogenesis." **Int. J. Obes.**, 17(suppl 1), S35-S40, 1993.

318. Dulloo, A.G., Geissler, C.A., Horton, T., et.al. **Am. J. Clin. Nutr.**, 49, 44-50, 1989.

319. Dulloo, A.G. & Girardier, L. "Adaptive changes in energy expenditure during refeeding following low-calorie intake: evidence for a specific metabolic component favoring fat storage." **Am. J. Clin. Nutr.**, 52, 415-420, 1990.

320. Dulloo, A.G. & Girardier, L. "24 hour energy expenditure several months after weight loss in the underfed rat: evidence for a chronic increase in whole-body metabolic efficiency." **Int. J. Obes.**, 16, in press, 1992.

321. Dulloo, A.G. & Miller, D.S. "Unimpaired thermogenic response to noradrenaline in genetic (ob/ob) and hypothalmic (MSG) obese mice." **Biosci. Rep.**, 4, 343-349, 1984.

322. Dulloo, A.G. & Miller, D.S. "Thermogenic drugs for the treatment of obesity: sympathetic stimulants in animal models." **Br. J. Nutr.**, 52, 179-196, 1984.

323. Dulloo, A.G. & Miller, D.S. "Energy balance following sympathetic denervation of brown adipose tissue." **Can. J. Physiol. Pharm.**, 62, 235-240, 1984.

324. Dulloo, A.G. & Miller, D.S. "Increased body fat due to elevated energetic efficiency following chronic administration of inhibitors of the sympathetic nervous system activity." **Metabolism**, 34, 1061-1065, 1985.

325. Dulloo, A.G. & MIller, D.S. "The Do-Do pill: potentiation of thermogenic effects of ephedrine by methylxanthines." **Proc. Nutr. Soc.**, 44, 16A, 1985.

326. Dulloo, A.G. & Miller, D.S. "The thermogenic properties of ephedrine/methylxanthine mixtures: animal studies." **Am. J. Clin. Nutr.**, 43(3), 388-394, 1986.

327. Dulloo, A.G. & Miller, D.S. "The thermogenic properties of ephedrine/methylxanthine mixtures: human studies." **Int. J. Obes.**, 10, 467-481, 1986.

328. Dulloo, A.G. & Miller, D.S. "Aspirin as a promoter of ephedrine-induced thermogenesis: potential use in the treatment of obesity." **Am. J. Clin. Nutr.**, 45, 564-569, 1987.

329. Dulloo, A.G. & Miller, D.S. "Prevention of genetic fa/fa obesity with an ephedrine-methylxanthines thermogenic mixture." **Am.**

J. Physiol., 252(3 pt 2), R507-R513, 1987.

330. Dulloo, A.G. & Miller, D.S. "Reversal of obesity in the genetically obese fa/fa Zucker rat with an ephedrine/methylxanthines thermogenic mixture." **J. Nutr.**, 117(2), 383-389, 1987.

331. Dulloo, A.G. & Miller, D.S. "Obesity--a disorder of the sympathetic nervous system." **Wrld. Rev. Nutr. Diet**, 50, 1-56, 1987.

332. Dulloo, A.G. & Miller, D.S. **Wrld. Rev. Nutr. Diet**, 50, 1-55, 1988.

333. Dulloo, A.G. & Miller, D.S. "Ephedrine, caffeine and aspirin: 'over-the-counter' drugs that interact to stimulate thermogenesis in the obese." **Nutr.**, 5(1), 7-9, 1989.

334. Dulloo, A.G., Seydoux, J. & Girardier, L. **Prog. Obes. Res.**, 135, 1990.

335. Dulloo, A.G., Seydoux, J. & Girardier, L. "Peripheral mechanisms of thermogenesis induced by ephedrine and caffeine in brown adipose tissue." **Int. J. Obes.**, 15(5), 317-326, 1991.

336. Dulloo, A.G., Seydoux, J. & Girardier, L. **Metabolism**, 41, in press, 1992.

337. Dulloo, A.G., Young, J.B. & Landsberg, L. "Sympathetic nervous system responses to cold exposure and diet in rat skeletal muscle." **Am. J. Physiol.**, 255, E180-E188, 1988.

338. Dureman, E.I. **Clin. Pharm. Ther.**, 3, 29-33, 1961.

339. Dyment, P.G. "The adolescent athlete and ergogenic aids." **J. Adolesc. Health Care**, 8(1), 68-73, 1987.

340. Ebner, S., Burnol, A.F., Ferre, P., et.al. **Eur. J. Biochem.**, 170, 469-474, 1987.

341. Eckerskorn, C. & Klingenberg, M. **FEBS Lett.**, 226, 166-170, 1987.

342. Eddy, N.B. & Downs, A.W. **J. Pharm. Exp. Ther.**, 33, 167-174, 1928.

343. Einhorn, D., Young, J.B. & Landsberg, L. "Hypotensive effect of fasting: possible involvement of the sympathetic nervous system and endogenous opiates." **Science**, 217, 727-729, 1982.

344. Elahi, D., Sclater, A., Waksmonski, C., et.al. **Clin. Res.**, 39, 355A, 1991.

345. Eley, J. & Himms-Hagen, J. "Brown adipose tissue of mice with goldthioglucose-induced obesity: abnormal circadian control." **Am. J. Physiol.**, in press, 1989.

346. Eley, J. & Himms-Hagen, J. "Attenuated response to cold of brown adipose tissue of mice made obese with gold thioglucose." **Can. J. Physiol. Pharm.**, 67, 116-121, 1989.

347. Elia, M., Folmer, F., Schlatmann, A., et.al. "Carbohydrate, fat, and protein metabolism in muscle and in whole body after mixed meal ingestion." **Metabolism**, 37, 542-551, 1988.

348. Elkins, R.N., Rapoport, J.L., Zahn, T.P., et.al. **Am. J. Psychiatry**, 138, 178-183, 1981.

349. Elks, M.L. & Manganiello, V.C. **Endocrinology**, 115, 1262, 1984.

350. Emorine, L.J., Marullo, S., Briend-Sutren, M.M., et.al. "Molecular characterization of the human beta-3 adrenergic receptor." **Science**, 245, 1118-1121, 1989.

351. Engfeldt, P., Arner, P. & Bolinder, J. **J. Clin. Endocr. Metab.**, 56, 501, 1983.

352. Engfeldt, P., Arner, P., Bolinder, J., et.al. **J. Clin. Endocr. Metab.**, 54, 625, 1982.

353. Engfeldt, P., Arner, P., Bolinder, J., et.al. **J. Clin. Endocr. Metab.**, 55, 983, 1982.

354. Engfeldt, P., Arner, P. & Ostman, J. **Metabolism**, 31, 910, 1982.

355. Ernster, V.L., Mason, L., Goodson, W.H.III, et.al. Effects of caffeine-free diet on benign breast disease: a randomized trial." **Surgery**, 91, 263, 1982.

356. Essig, D., Costill, D.C. & Van Handel, P. "Effect of caffeine ingestion on utilization of muscle glycogen and lipid during leg ergometer cycling." **Int. J. Sports Med.**, 1, 70-74, 1980.

357. Estler, C.J., Ammon, H.P.T. & Herzog, C. "Swimming capacity of mice after prolonged treatment with psychostimulants. I. Effect of caffeine on swimming performance and cold stress." **Psychopharm.**, 58, 161-166, 1978.

358. Evans, E. & Miller, D.S. "The effect of ephedrine on the oxygen consumption of fed and fasted subjects." (abstr) **Proc. Nutr. Soc.**, 36, 136A, 1977.

359. Everson, G.J. & Shrader, R.E. "Abnormal glucose tolerance in manganese deficient guinea pigs." **J. Nutr.**, 94, 89, 1968.

360. Fagher, B., Liedholm, H., Monti, M., et.al. "Thermogenesis in human skeletal muscle as measured by direct microcalorimetry and muscle contractile performance during beta-adrenoceptor blockade." **Clin. Sci.**, 70, 435-441, 1986.

361. Fain, J.N., Reed, N. & Saperstein, R. "The isolation and metabolism of brown fat cells." **J. Biol. Chem.**, 242, 1887-1894, 1967.

362. Fears, R. **Br. J. Nutr.**, 39, 363-374, 1978.

363. Feinberg, L.J., Sandberg, H., DeCastro, O., et.al. **Metabolism**, 17, 916-922, 1968.

364. Feist, D.D. & Feist, C.F. "Catecholamine-synthesizing enzymes in adrenals of seasonally acclimatized voles." **J. Appl. Physiol.**, 44, 59-62, 1978.

365. Feldman, D. "Evidence that brown adipose tissue is a glucocorticoid target organ." **Endocrinology**, 103, 2091-2097, 1978.

366. Felig, P. "Insulin is the mediator of feeding related thermogenesis: insulin resistance and/or deficiency results in a thermogenic defect which contributes to the pathogenesis of obesity." **Clin. Physiol.**, 4, 267, 1984.

367. Felig, P., Cunningham, P., Levitt, M., et.al. **Am. J. Physiol.**, 244, E45-E51, 1983.

368. Felt, J.M. "Lipid transport between adipose tissue and blood." In **Fat as a Tissue**, Rodahl, K., & Issekutz, B., eds. McGraw Hill Book Company, New York, 1964, pp. 95-109.

369. Ferre, P., Burnol, A.F., Leturque, A., et.al. "Glucose utilization in vivo and insulin sensitivity of rat brown adipose tissue in various physiological and pathological conditions." **Biochem. J.**, 233, 249-252, 1986.

370. File, S.E., Bond, A.J. & Lister, R.G. **J. Clin. Psychopharm.**, 2, 102-106, 1982.

371. Firestone, P., Poitras-Wright, H. & Douglas, V. **J. Learning Disability**, 11, 133-141, 1978.

372. Fisher, S.M., McMurray, R.G., Berry, M., et.al. "Influence of caffeine on exercise performance in habitual caffeine users." **Int. J. Sports Med.**, 7, 276-280, 1986.

373. Fisler, J.S., Yoshida, T. & Bray, G.A. "Catecholamine turnover in S5B/PL and Osborne-Mendel rats: response to a high-fat diet." **Am. J. Physiol.**, 247, R270-295, 1984.

374. Fitzgerald, M. "Capsaicin and sensory neurons-A review." **Pain**, 15, 109-130, 1983.

375. Flaim, K.E., Horwitz, B.A. & Horowitz, J.M. "Coupling of signals to brown fat: alpha- and beta-adrenergic responses in intact rats." **Am. J. Physiol.**, 232, R101-R109, 1977.

376. Flatmark, T. & Pedersen, J.I. **Biochem. Biophys. Acta**, 416, 53-103, 1975.

377. Flodin, N.W. "Atherosclerosis: an insulin-dependent disease." **J. Am. Coll. Nutr.**, 5, 417, 1986.

378. Flower, R.J., Moncada, S. & Vane, J.R. In **Goodman and Gilman's the Pharmacological Basis of Therapeutics**, Gilman, A.G., Goodman, L.S., Rall, T.W., et.al., eds., 7th ed. NY, Macmillan, 674, 1985.

379. Folk, G.E. **Textbook of Environmental Physiology**, Philadelphia, Lea and Febiger, 1974.

380. Foltz, E., Ivy, A.C. & Barborka, C.J. **Am. J. Physiol.**, 136, 79-86, 1942.
381. Foltz, E., Ivy, A.C. & Barborka, C.J. **J. Lab. Clin. Med.**, 28, 603-606, 1942.
382. Forbes, G.B., Kreipe, R.E. & Lipinski, B. **Hum. Nutr. Clin. Nutr.**, 36C, 485, 1982.
383. Forrest, W.H.,Jr., Bellville, J.W. & Brown, B.W. **Anesthesiology**, 36, 37-41, 1972.
384. Foster, D.O. "Quantitative role of brown adipose tissue in thermogenesis." In **Brown Adipose Tissue**, Trayhurn, P., Nicholls, D.G., eds., London, Arnold, 31-51, 1986.
385. Foster, D.O. & Depocas, F. "Evidence against noradrenergic regulation of vasodilation in rat brown adipose tissue." **Can. J. Physiol. Pharm.**, 588, 1418-1425, 1981.
386. Foster, D.O., Depocas, F. & Frydman, M.L. "Noradrenaline-induced calorigenesis in warm- and cold- acclimated rats: relations between concentration of noradrenaline in arterial plasma, blood flow to differently located masses of brown adipose tissue, and calorigenic response." **Can. J. Physiol. Pharm.**, 58, 915-924, 1980.
387. Foster, D.O. & Frydman, M.L. **Can. J. Physiol. Pharm.**, 57, 257-270, 1979.
388. Foster, D.O. & Frydman, M.L. "Nonshivering thermogenesis in the rat. II. Measurement of blood flow with microspheres point to brown adipose tissue as the dominant site of the calorigenesis induced by noradrenaline." **Can. J. Physiol. Pharm.**, 56, 110-122, 1978.
389. Foster, D.O. & Frydman, M.L. **Can. J. Physiol. Pharm.**, 58, 97-109, 1978.
390. Foster, D.O. & Frydman, M.L. "Tissue distribution of cold-induced thermogenesis in conscious warm- or cold-acclimated rats reevaluated from changes in tissue blood flow: the dominant role of brown adipose tissue in the replacement of shivering by nonshivering thermogenesis." **Can. J. Physiol. Pharm.**, 57, 257-270, 1979.
392. Fov, G. "Capsicum." **Med. Press**, 2, 191-192, 1886.
393. Franklin, B.A. & Rubenfire, M. "Losing weight through exercise." **JAMA**, 244, 377, 1980.
394. Fras, I. **Am. J. Psychiat.**, 131, 228-229, 1974.
395. Fredholm, B.B. "Are methylxanthine effects due to antagonism of endogenous adenosine?" **Trends Pharm. Sci.**, 115, 129-132, 1980.
396. Fredholm, B.B. **Acta Med. Scand.**, 217, 149, 1985.
397. Fredholm, B.B., Lunell, N.O., Persson, B., et.al. "Development of tolerance to the metabolic actions of beta2-adrenoceptor stimulating drugs." **Acta Obstet. Gynecol. Scand.**, 108(suppl), 53-59, 1982.
398. Fredholm, B.B. & Persson, C.G.A. **Eur. J. Pharm.**, 81, 673-676, 1982.
399. Freeman, P.H. & Wellman, P.J. **Brain Res. Bull.**, 18, 7-11, 1987.
400. French, R.R., Gore, M.G. & York, D.A. **Biochem. J.**, 251, 385-389, 1988.
401. Friedman, G.D., Siegelaub, A.B. & Seltzer, C.C. "Cigarettes, alcohol, coffee and peptic ulcer." **N. Eng. J. Med.**, 290, 469, 1974.
402. Fritz, I.B. & Yue, K.T.N., **J. Lipid. Res.**, 4, 279, 1963.
403. Fritz, I.B., in **Cellular Compartmentalization and Control of Fatty Acid Metabolism**, F.C. Gran, ed., p. 39, Universitetsforlaget, Oslo, 1968.
404. J. Bremer, in **Cellular Compartmentalization and Control of Fatty Acid Metabolism**, F.C. Gran, ed., p. 65, Universitetsforlaget, Oslo, 1968.
405. Fukushima, M., Tokunaga, K., Lupien, J., et.al. "Dynamic and static phases of obesity following lesions in PVN and VMH." **Am. J. Physiol.**, 253, R523-529, 1987.
406. Fuller, C.A., Horowitz, J.M. & Horwitz, B.A. "Spinal cord thermosensitivity and sorting of neural signals in cold-exposed rats." **J. Appl. Physiol.**, 42, 154-158, 1977.
407. Fuller, N.J., Stirling, D.M., Dunnett, S., et.al. "Decreased brown adipose tissue thermogenic activity following a reduction in brain serotonin by intraventricular p-chlorophenylalanine." **Biosci. Rep.**, 7, 121-127, 1987.
408. Furlong, F.W. **Can. Psychiatr. Assn. J.**, 20, 577-583, 1975.
409. Fyda, D.M., Cooper, K.E. & Veale, W.L. "Modulation of brown adipose tissue-mediated thermogenesis by lesions to the nucleus tractus solitarius in the rat." **Brain Res.**, 546(2), 203-210, 1991.
410. Fyda, D.M., Cooper, K.E. & Veale, W.L. "Nucleus tractus solitarii lesions alter the metabolic and hyperthermic response to central prostaglandin E1 in the rat." **J. Physiol. Lond.**, 442, 337-349, 1991.
412. Gale, C.C. "Neuroendocrine aspects of thermoregulation." **Ann. Rev. Physiol.**, 35, 391-430, 1973.
413. Galitzky, J., Vermorel, M., Lafontan, M., et.al. "Thermogenic and lipolytic effect of yohimbine in the dog." **Br. J. Pharm.**, 104(2), 514-518, 1991.
414. Gallen, I.W., Macdonald, I.A. & Allison, S.P. "Glucose metabolism and thermogenesis during glucose and insulin infusion in severely underweight patients." **Jpen. J. Parenter Enteral Nutr.**, 16(1), 5-10, 1992.
415. Galpin, K.S., Henderson, R.G., James, W.P.T., et.al. "GDP binding to brown adipose tissue mitochondria of mice treated chronically with corticosterone." **Biochem.**, 214, 265-268, 1983.
416. Galster, A.D., Clutter, W.E., Cryer, P.E., et.al. "Epinephrine plasma thresholds for lipolytic effects in man: measurements of fatty acid transport with [1-13C]palmitic acid." **J. Clin. Inv.**, 67, 1729-1738, 1981.
417. Gamse, R., Holzer, P. & Lembeck, F. "Decrease of substance P in primary sensory afferent neurons and impairment of neurogenic plasma extravasation by capsaicin." **Br. J. Pharm.**, 68, 207, 1980.
418. Gamse, R., Lackner, D., Gamse, G., et.al. "Effect of capsaicin pretreatment on capsaicin-evoked release of immunoreactive somatostatin and substance P from primary sensory neurons." **Naunyn-Schmiedeberg's Arch. Pharm.**, 316, 38-41, 1981.
419. Gamse, R., Leeman, S.E., Holzer, P., et.al. **Naunyn-Schmiedeberg's Arch. Pharm.**, 317, 140, 1981.
420. Gamse, R., Molnar, A. & Lembeck, F. "Substance release from spinal cord slices by capsaicin." **Life Sci.**, 24, 629, 1979.
421. Gamse, R., Wax, A., Zigmond, R.E., et.al. "Immunoreactive substance P in sympathetic ganglia: distribution and sensitivity towards capsaicin." **Neuroscience**, 6, 437, 1981.
422. Garcia, C. "Comparison of the effects of evening primrose oil and placebo in overweight patients with a family history of obesity." 2nd Int. Cong. EFAs, PGs, LTs, Abstr., London, UK, 44, 1985.
423. Garfinkel, B.D., Webster, C.D. & Sioman, L. **Am. J. Psychiat.**, 132, 723-729, 1975.
424. Garlick, P.J., Glugston, G.A. & Waterlow, J.C. **Am. J. Physiol.**, 258, E235, 1980.
425. Garriott, J.C., Simmons, L.M., Poklis, A., et.al. "Five cases of fatal overdose from caffeine-containing 'look-alive' drugs." **J. Anal. Tox.**, 9(3), 141-143, 1985.
426. Garrow, J.S. **Energy Balance and Obesity in Man**, 2nd ed., Amsterdam, Elsevier/North Holland, 1978.
427. Garrow, J.S. "The regulation of energy expenditure in man." In **Recent Advances in Obesity Research II**, Bray, G.A., ed., London, Newman Publishing, 200-210, 1978.

428. Geissler, C.A. "Effects of weight loss, ephedrine and aspirin on energy expenditure in obese women." **Int. J. Obes.**, 17(suppl 1), S45-S48, 1993.

429. Geissler, C.A., Miller, D.S., Shah, M. "The daily metabolic rate of the post-obese and the lean." **Am. J. Clin. Nutr.**, 45, 914-920, 1987.

430. Geloen, A., Collet, A.J., Guay, G., et.al. **J. Am. J. Physiol.**, 254, C175-C182, 1988.

431. Gerlach, E. & Becker, B.F., eds. **Topics and Perspectives in Adenosine Research**, Springer-Verlag, Berlin, 3-629, 1987.

432. Gibbins, J.M., Denton, R.M. & McCormack, J.G. **Biochem. J.**, 228, 751-755, 1985.

433. Gilbert, R.M., Marsham, J.A., Schwieder, M., et.al. "Caffeine content of beverages as consumed." **Can. Med. Assn. J.**, 114, 205-208, 1976.

434. Giles, D. & McLaren D. "Effects of caffeine and glucose ingestion on metabolic and respiratory functions during prolonged exercise." **J. Sports Sci.**, 2, 35-46, 1984.

435. Gilliland, K. & Andress, D. **Am. J. Psychiat.**, 138, 512-514, 1981.

436. Giovannini, P., Seydoux, J. & Girardier, L. **Pflugers Arch.**, 411, 273-277, 1988.

437. Girardier, L. "Current topics in brown adipose tissue research." in Szelenyi, Z. & Szekely, M. eds. **Contributions to Thermal Physiology**, Pergamon Press, Akademaiai Kiado, Budapest, 1981, pp. 469-474.

438. Girardier, L. "Brown fat: an energy dissipating tissue." In **Mammalian Thermogenesis**, Girardier, L. & Stock, M.J., eds., NY, Chapman & Hall, 50-98, 1983.

439. Girardier, L. "Control systems in the defense of body fat stores." **Int. J. Obes.**, 17(suppl 1), S3-S8, 1993.

440. Girardier, L. & Seydoux, J. "Neural control of brown adipose tissue." In **Brown Adipose Tissue**, Trayhurn, P. & Nicholls, D.G., eds., London, Arnold, 122-151, 1986.

441. Girardier, L. & Stock, M.J. "Mammalian thermogenesis: an introduction." In **Mammalian Thermogenesis**, Girardier, L. & Stock, M.J., eds., 1-7, 1983.

442. Glatzel, V.H. & Rettenmaier, G. "Kreislaufregulation und Ernahrung." **Arch. Kreislaufforschung**, 45(1-4), 195-213, 1965.

443. Glick, Z. "Meal intake and brown adipose tissue in the rat." **Nutr. Behav.**, 2, 65, 1984.

444. Glinsukon, T., Stitmunnaithum, V., Toskulkao, C., et.al. "Acute toxicity of capsaicin in several animal species." **Toxicon**, 18, 215-220, 1980.

445. Goebel, H., Ernst, M., Jeschke, J., et.al. **Pain**, 48, 187, 1992.

446. Godbole, V., York, D.A. & Bloxham, D.P. "Developmental changes in the fatty (fafa) rat: evidence for defective thermogenesis preceding the hyperlipogenesis and hyperinsulinaemia." **Diabetologia**, 15, 41-44, 1978.

447. Golay, A. "Blunted glucose-induced thermogenesis: a factor contributing to relapse of obesity." **Int. J. Obes.**, 17(suppl 1), S23-S27, 1993.

448. Golay, A., Jallut, D., Schutz, Y., et.al. "Evolution of glucose-induced-thermogenesis in obese subjects with and without diabetes: a six-year follow-up study." **Int. J. Obes.**, 15(9), 601-607, 1990.

449. Golay, A., Schutz, Y., Felber, J.P., et.al. **Int. J. Obes.**, 10, 107-116, 1986.

450. Golay, A., Schutz, Y., Felber, J.P. "Blunted glucose-induced-thermogenesis in overweight patients. A factor contributing to relapse of obesity." **Int. J. Obes.**, 13, 767-775, 1989.

451. Golay, A., Schutz, Y., Meyer, H.U., et.al. "Glucose-induced thermogenesis in non-diabetic" **Br. J. Nutr.**, 35, 281-292, 1982.

452. Golay, A., Schutz, Y., Meyer, H.U., et.al. "Glucose induced thermogenesis in non-diabetic and diabetic obese subjects." **Diabetes**, 31, 1023-1028, 1982.

453. Goldberg, J.C. & Morgan, B.L.G. "Brown adipose tissue in the developing Zucker rat." (abstr) **Fed. Proc.**, 41, 401, 1982.

454. Goldstein, A. **Naunyn-Schmiedeberg's Arch. Pharm.**, 248, 269-278, 1964.

455. Goldstein, A., Kaizer, S. & Warren, R. **J. Pharm. Exp. Ther.**, 150, 146-151, 1965.

456. Goldstein, A., Kaizer, S. & Whitby, O. **Clin. Pharm. Ther.**, 10, 489-497, 1969.

457. Goldstein, A. & Warren, R. **Biochem. Pharm.**, 11, 166-168, 1962.

458. Goldstein, A., Warren, R. & Kaizer, S. **J. Pharm. Exp. Ther.**, 149, 156-159, 1965.

459. Goodbody, A.E. & Trayhurn, P. "Functional changes in brown adipose tissue of diabetic-obese (db/db) mice." In Szelenyi, Z. & Szekely, M. eds. **Contributions to Thermal Physiology**, Pergamon Press, Akademaiai Kiado, Budapest, 1981, 515-517.

460. Goubern, M., Chapey, M.F. & Portet, R. "Time-course variations of effective proton conductance and GDP binding in brown adipose tissue mitochondria of rats during prolonged cold exposure." **Comp. Biochem. Physiol. B.**, 100(4), 727-732, 1991.

461. Goubern, M. & Portet, R. **Ann. Nutr. Metab.**, 30, 380-385, 1986.

462. Goubern, M., Yazbeck, J., Senault, C., et.al. "Non-shivering thermogenesis and brown adipose tissue activity in essential fatty acid deficient rats." **Arch. Int. Physiol. Biochim.**, 98(4), 193-199, 1990.

463. Gould, L., Venkataraman, K., Goswami, M., et.al. "The cardiac effects of coffee." **Angiology**, 24, 455-463, 1973.

464. Graham, D.M. "Caffeine--its density, dietary sources, intake and biological effects." **Nutr. Rev.**, 36, 97-102, 1978.

465. Graham, T.E., Sathasivam, P. & McNaughton, K.W. **J. Appl. Physiol.**, 70(5), 2052, 1991.

466. Granneman, J.G. **J. Pharm. Exp. Ther.**, 245, 1075-1080, 1988.

467. Granneman, J.G. & Campbell, R.G. **Metabolism**, 33, 257-261, 1984.

468. Granneman, J.G. & MacKenzie, R.G. **J. Pharm. Exp. Ther.**, 245, 1068-1074, 1988.

469. Granneman, J.G. & MacKenzie, R.G., Fluharty, S.J., et.al. **J. Pharm. Exp. Ther.**, 233, 163-167, 1985.

470. Grayson, J. & Mendel, D. **J. Physiol.**, 133, 334, 1956.

471. Greco-Perotto, R., Assimacopoulos-Jeannet, F. & Jeanrenaud, B. **Biochem. J.**, 247, 63-68, 1987.

472. Greco-Perotto, R., Zaninetti, D., Assimacopoulos-Jeannet, F., et.al. **J. Biol. Chem.**, 262, 7732-7736, 1987.

473. Greden, J.F. **The Science**, 19, 6-11, 1979.

474. Greden, J.F., Fontaine, P. & Lubetsky, M. **Am. J. Psychiat.**, 131, 1089-1092, 1974.

475. Green, A.J. **J. Clin. Biol. Chem.**, 262, 15707, 1987.

476. Green, A., Milligan, G. & Dobias, B.J. **J. Biol. Chem.**, 267, 3223, 1992.

477. Green, A., Swenson, S., Johnson, J.L., et.al. **Biochem. Biophys. Res. Commun.**, 163, 137, 1989.

478. Green, R.M. & Stiles, G.L. **J. Clin. Inv.**, 77, 222-227, 1986.

479. Greengard, P. **Fed. Proc.**, 38, 2208-2217, 1979.

480. Gregory, S.F., McNaughton, S.R., Tristram, S., et.al. "Caffeine ingestion prior to incremental cycling to exhaustion in recreational cyclist." **Int. J. Sports Med.**, 11, 188-193, 1990.

481. Gribskov, C.L., Henningfield, M.F., Swick, A.G., et.al. "Evidence for unmasking of rat brown adipose tissue mitochondrial GDP binding sites in response to acute cold exposure. Effects of BSA washing on GDP binding." **Biochem. J.**, 233, 743-747, 1986.

482. Griggio, M.A., Richard, D. & Leblanc, J. "The involvement of the sympathetic nervous system in meal-induced thermogenesis in mice." **Int. J. Obes.**, 15(11), 711-715, 1991.

483. Grollman, A. "The action of alcohol, caffeine and tobacco on the cardiac output of normal man." **J. Pharm. Exp. Ther.**, 39, 313-327, 1930.

484. Gross, H.A., Lake, C.R., Ebert, M.H., et.al. "Catecholamine metabolism in primary anorexia nervosa." **J. Clin. Endocr. Metab.**, 49, 805-809, 1979.

485. Grubb, B. & Folk, G.E. **J. Comp. Physiol.**, 110, 217-226, 1976.

487. Guidotti, A., Zivkovic, B., Pfeiffer, R., et.al. "Involvement of 3',5'-cyclic adenosine monophosphate in the increase of tyrosine hydroxylase activity elicited by cold exposure." **Naunyn- Schmiedeberg's Arch. Pharm.**, 278, 195-206, 1973.

488. Guigliano, D., Torella, R., Sgambato, S. "Effects of alpha- and beta-adrenergic inhibition and somatostatin on plasma glucose, free fatty acids, insulin, glucagon, and growth hormone responses to prostaglandin E1 in man." **J. Clin. Endocr. Metabol.**, 48, 302, 1979.

489. Gujral, S., Bhumbra, H. & Swaroop, M. **Nutr. Reps. Int.**, 17, 183, 1978.

490. Gumbiner, B., Thorburn, A.W. & Henry, R.R. "Reduced glucose-induced thermogenesis is present in noninsulin-dependent diabetes mellitus without obesity." **J. Clin. Endocr. Metab.**, 72(4), 801-807, 1991.

491. Gunn, T.R., Metrados, K., Riley, P., et.al. **J. Pediatr.**, 94, 106-109, 1979.

492. Hadzovic, S., Hukovic, S. & Stern, P. "A contribution to the pharmacology of capsaicin." **Veterinaria, Saraj.**, 13, 449-452, 1964.

493. Haffner, S.M., Knapp, J.A., Stern, M.P., et.al. "Coffee consumption, diet, and lipids." **Am. J. Epidemiol.**, 122(1), 1-12, 1985.

494. Hahn, P. & Skala, J. "The role of carnitine in brown adipose tissue of suckling rats." **Comp. Biochem. Physiol.**, 51B, 507, 1975.

495. Haldi, J., Bachmann, G., Ensor, C., et.al. **J. Nutr.**, 27, 287-293, 1944.

496. Hamilton, J.M., Mason, P.W., McElroy, J.F., et.al. "Dissociation of sympathetic and thermogenic activity in brown fat of Syrian hamsters." **Am. J. Physiol.**, 250, R389-R395, 1986.

497. Hansen, A.E. "Role of unsaturated dietary fat in infant nutrition." **Am. J. Public Hlth.**, 47, 1367, 1957.

498. Hansen, A.E., Haggard, M.E., Boelsche, A.N., et.al. "Essential fatty acids in human nutrition." **J. Nutr.**, 60, 565, 1958.

499. Hansen, E.S. & Knudsen, J. **Biosci. Rep.**, 6, 31-38, 1986.

500. Harada, M. "Pharmacological studies on herb peaony root.IV. Analysis of therapeutic effects of paeony- and licorice-containing frequent prescriptions in Chinese medicine and comparison with effects of experimental pharmacological tests." **Yakugaku Zasshi**, 89, 899-908, 1969.

501. Harada, M. & Nishimura, M. "Contribution of alkaloid fraction to pressor and hyperglycemic effect of crude ephedra extract in dogs." **J. Pharm. Dyn.**, 4(9), 691-699, 1981.

502. Harallamabie, G. "Serum zinc and athletes in training." **J. Int. Sports Med.**, 2, 135, 1981.

503. Harri, M., Dannenberg, T., Oksanen-Rossi, R., et.al. **J. Appl. Physiol.**, 57, 1489-1497, 1984.

504. Harrie, J.R. **JAMA**, 213, 628, 1970.

505. Harris, W.H., Foster, D.O. & Nadeau, B.E. "Evidence for a contribution by brown adipose tissue to the development of fever in the young rabbit." **Can. J. Physiol. Pharm.**, 63, 595-598, 1985.

506. Harvey, D.H.P. & Marsh, R.W. **Devl. Med. Child Neurol.**, 20, 81-86, 1978.

507. Haubrich, D.R., Perez-Cruet, J. & Reid, W.D. "Prostaglandin E1 causes sedation and increases 5-hydroxytryptamine turnover in rat brain." **Br. J. Pharm.**, 48, 80, 1973.

508. Hausberger, F.X. "Neurogenic factors affecting adipose tissue metabolism." In **Fat as a Tissue**, Rodahl, K., & Issekutz, B., eds. McGraw Hill Book Company, New York, 1964, pp. 239-249.

509. Heaton, J.M. "The distribution of brown adipose tissue in the human." **J. Anatomy**, 112, 35, 1972.

510. Hedqvist, P., Fredholm, P.B. & Olundh, S. **Circ. Res.**, 43, 592-598, 1978.

511. Heindel, J.J., Cushman, S.W. & Jeanrenaud, B. "Cell associated fatty acid levels and energy-requiring processes in mouse adipocytes." **Am. J. Physiol.**, 226, 16, 1974.

512. Heleniak, E.P. & Aston, B. "Prostaglandins, brown fat and weight loss." **Med. Hypoth.**, 28, 13-33, 1989.

513. Hellon, R.F. "Neurophysiology of temperature regulation: problems and perspectives." **Fed. Proc.**, 40, 2804-2807, 1981.

514. Hendler, R. & Bonde, A.A. "Effects of sucrose on resting metabolic rate, nitrogen balance, leucine turnover and oxidation during weight loss with low calorie diets." **Int. J. Obes.**, 14(11), 927-938, 1990.

515. Hennekens, C.H., Evans, D., Hutchinson, G.M., et.al. **N. Eng. J. Med.**, 318, 923, 1988.

516. Henningfield, M.F. & Swick, R.W. "Immunochemical detection and quantitation of brown adipose tissue uncoupling protein." **Biochem. Cell Biol.**, 65, 245-251, 1987.

517. Henny, C., Schutz, Y., Buckert, A., et.al. "Thermogenic effect of the new beta-adrenoreceptor agonist RO 16-8714 in healthy male volunteers." **Int. J. Obes.**, 11, 473-483, 1987.

518. Henry, C.J.K. "Relationship between basal metabolic rate (BMR) and fasting metabolic rate (FMR): some theoretical and practical implications." **Int. J. Obes.**, 17(suppl 1), S79, 1993.

519. Henry, C.J. & Emery, B. "Effect of spiced food on metabolic rate." **Hum. Nutr. Clin. Nutr.**, 40(2), 165-168, 1986.

520. Herridge, C.F. & A'Brook, M.F. **Br. Med. J.**, 2, 160, 1968.

521. Hettiarachchi, M., Colquhoun, E.Q., Ye, J.M., et.al. "Norephedrine (phenylpropanolamine) stimulates oxygen consumption and lactate production in the perfused rat hindlimb." **Int. J. Obes.**, 15(1), 37-43, 1991.

522. Heyden, S., Tyroler, H.A., Heiss, G., et.al. "Coffee consumption and mortality. Total mortality, stroke mortality, and coronary heart disease mortality." **Arch. Intern. Med.**, 138, 1472, 1978.

523. Himms-Hagen, J. "Cellular thermogenesis." **Ann. Rev. Physiol.**, 38, 315-351, 1976.

524. Himms-Hagen, J. "Obesity may be due to a malfunctioning of brown fat." **Can. Med. Assn. J.**, 121, 1361-1364, 1976.

525. Himms-Hagen, J. In: **Nutritional Factors: Modulating Effects on Metabolic Processes**, R.F. Beers, Jr. E IE.G. Bassett, ed., pp. 85-99, Raven Press, New York.

526. Himms-Hagen, J. "Thyroid hormones and thermogenesis." In **Mammalian Thermogenesis**, Girardier, L. & Stock, M.J., eds., London, Chapman & Hall, 141-177, 1983.

527. Himms-Hagen, J. "Brown adipose tissue thermogenesis in obese animals." **Nutr. R.**, 41, 261, 1983.

528. Himms-Hagen, J. "Role of thermogenesis in brown adipose tissue in the regulation of energy balance." In **The Adipocyte and Obesity: Cellular and Molecular Mechanisms**, Angel, A., Hollenberg, C.H. & Roncari, D.A., eds., Raven Press, New York, 1983, pp. 259-270.

529. Himms-Hagen, J. "Thermogenesis in brown adipose tissue as an energy buffer." **New Eng. J. Med.**, 311, 1549-1559, 1984.

530. Himms-Hagen, J. "Food restriction increases torpor and improves brown adipose tissue thermogenesis in ob/ob mice." **Am. J. Physiol.**, 248, E531-E539, 1985.
531. Himms-Hagen, J. "Brown adipose tissue metabolism and thermogenesis." **Ann. Rev. Nutr.**, 5, 69-94, 1985.
532. Himms-Hagen, J. "Brown adipose tissue and cold-acclimation." In **Brown Adipose Tissue**, Trayhurn, P. & Nicholls, D.G., eds., London, Arnold, 214-268, 1986.
533. Himms-Hagen, J. "Cold- versus diet-induced thermogenesis in brown adipose tissue: different strategies in different species." In **Living in the Cold**, Heller, H.C., Musacchia, X.J. & Wang, L.C.H., eds., NY, Elsevier, 93-100, 1986.
534. Himms-Hagen, J. "Defective thermogenesis in obese animals." **J. Obes. Wt. Reg.**, 6, 179-199, 1987.
536. Himms-Hagen, J. "Role of thermogenesis in the regulation of energy balance in relation to obesity." **Can. J. Physiol. Pharm.**, 67(4), 394-401, 1989.
537. Himms-Hagen, J. "Brown adipose tissue thermogenesis: role in thermoregulation, energy regulation and obesity." **Pharm. Ther.**, in press, 1989.
538. Himms-Hagen, J. "Brown adipose tissue thermogenesis and obesity." **Prog. Lipid Res.**, 28(2), 67-115, 1989.
539. Himms-Hagen, J. "Brown adipose tissue thermogenesis: interdisciplinary studies." **Faseb. J.**, 4(11), 2890-2898, 1990.
540. Himms-Hagen, J., Hogan, S. & Zaror-Behrens, G. "Increased brown adipose tissue thermogenesis in obese (ob/ob) mice fed on a palatable diet." **Am. J. Physiol.**, 250, E274-E281, 1986.
541. Himms-Hagen, J., Tokuyama, K., Eley, J., et.al. In **Hormones, Thermogenesis and Obesity**, Lardy, H.A. & Stratman, F., eds., Elsevier, NY, 1989.
542. Himms-Hagen, J., Triandafillou, J. & Gwilliam, C., **Am. J. Physiol.**, 241, E116-E120, 1981.
543. Hirsh, J. & Han, P.W. "Cellularity of rat adipose tissue: effects of growth, starvation and obesity." **J. Lipid Res.**, 10, 77-82, 1969.
544. Hittelman, K.J. & Lindberg, O. "Fatty acid uncoupling in brown fat mitochondria." In **Brown Adipose Tissue**, Lindberg, O., ed., NY, Elsevier, 1970, pp. 245-262.
545. Ho, T.F., Yip W.C.L., Tay, J.S.H. "Childhood obesity and glucose metabolism in obese Chinese children." **Int. J. Obes.**, 17(suppl 1), S79, 1993.
546. Hoepke, H. & Nikolaus, K. **Z. Mikroskop.-Anat. Forsch.**, 46, 1, 1939.
547. Hoffman, B.B. & Lefkowitz, R.J. In **The Pharmacological Basis of Therapeutics**, Gilman, A.G., Rall, T.W., Nies, A.S., et.al, eds., Pergamon Press, NY, 8th ed., 187, 1990.
548. Hogan, S., Coscina, D.V. & Himms-Hagen, J. "Brown adipose tissue of rats with obesity-inducing ventromedial hypothalamic lesions." **Am. J. Physiol.**, 243, E338-E344, 1982.
549. Hogan, S. & Himms-Hagen, J. "Abnormal brown adipose tissue in genetically obese mice (ob/ob): response to acclimation to cold." **Am. J. Physiol.**, 239, E301-E309, 1980.
550. Hogan, S. & Himms-Hagen, J. "Brown adipose tissue of mice with goldthioglucose-induced obesity: effect of cold and diet." **Am. J. Physiol.**, 244, E581-E588, 1983.
551. Hogan, S., Himms-Hagen, J. & Coscina, D.V. "Lack of diet-induced thermogenesis in brown adipose tissue of obese medial hypothalamic-lesioned rats." **Physiol. Behav.**, 35, 287-294, 1985.
552. Hogan, S., Sullivan, A.C. & Triscari, J. "Oxygen uptake and brown adipose tissue (BAT) of growing rats during development of dietary obesity." **Int. J. Obes.**, 9, A43, 1985.
553. Hogan, S., Sullivan, A.C. & Triscari, J. "Oxygen uptake (Vo2) and brown adipose tissue (BAT) of older rats resistant to development of dietary obesity." **Int. J. Obes.**, 9, A44, 1985.
554. Hogyes, A. "Beitrage zur physiologischen Wirkung der Bestandteile des Capsicum annuum." **Arch. Exp. Path. Pharm.**, 9, 117-130, 1878.
555. Hollands, M.A., Arch, J.R.S. & Cawthorne, M.A. **Am. J. Clin. Nutr.**, 34, 2291-2294, 1981.
556. Hollenberg, C.H., Roncari, D.A.K. & Djian, P. "Obesity and the fat cell: future prospects." In **The Adipocyte and Obesity: Cellular and Molecular Mechanisms**, Angel, A., Hollenberg, C.H. & Roncari, D.A., eds., Raven Press, New York, 1983, pp. 291-300.
557. Hollingworth, H.L. **Arch. Psychol.**, 22, 1-166, 1912.
558. Hollins, C. & Stone, T.W. **Br. J. Pharm.**, 69, 197-212, 1980.
559. Holloway, B.R., Howe, R., Rao, B.S., et.al. "ICI D7114: a novel selective adrenoceptor agonist of brown fat and thermogenesis." **Am. J. Clin. Nutr.**, 55(suppl 1), 262S-264S, 1992.
560. Holloway, F.A., Michaelis, R.C. & Huerta, P.L. "Caffeine-phenylethylamine combinations mimic the amphetamine discriminative cue." **Life Sci.**, 36(8), 723-730, 1985.
561. Holman, R.T. "Biological activities of and requirements for polyunsaturated acids." **Prog. Chem. Fats Lipids**, 9, 611, 1970.
562. Holt, S., Marchington, D. & York, D.A. "Defective sympathetic regulation of brown adipose tissue of Zucker obese rats." **Proc. Nutr. Soc.**, 43, 68A, 1984.
563. Holt, S.J., Wheal, H.V. & York, D.A. "Hypothalamic control of brown adipose tissue in Zucker lean and obese rats. Effect of electrical stimulation of the ventromedial nucleus and other hypothalamic centers." **Brain Res.**, 405, 227-233, 1987.
564. Holt, S.J., Wheal, H.V. & York, D.A. "Response of brown adipose tissue to electrical stimulation of hypothalamic centers in intact and adrenalectomized Zucker rats." **Neurosci. Lett.**, 84, 63-67, 1988.
565. Holt, S. & York, D.A. "The effect of adrenalectomy on GDP binding to brown adipose tissue mitochondria of obese animals." **Biochem. J.**, 208, 819-822, 1982.
566. Holt, S.J., York, D.A. & Fitzsimons, J.T.R. "The effects of corticosterone, cold exposure and overfeeding with sucrose on brown adipose tissue of obese Zucker rats (fa/fa)." **Biochem. J.**, 214, 215-223, 1983.
567. Holtzman, S.G. "Discriminative stimulus properties of caffeine in the rat: noradrenergic mediation." **J. Pharm. Exp. Ther.**, 239(3), 706-714, 1986.
568. Holzer, P., Jurna, I., Gamse, R., et.al. "Nociceptive thresholds after neonatal capsaicin treatment." **Eur. J. Pharm.**, 58, 511, 1979.
569. Holzer, P., Saria, A., Skofitsch, G., et.al. "Increase in tissue concentrations of histamine and 5-hydroxy-tryptamine following capsaicin treatment of newborn rats." **Life Sci.**, 29, 1099, 1981.
570. Horne, J.A., Percival, J.E. & Traynor, J.R. **EEG Clin. Neurophysiol.**, 49, 409, 1980.
571. Horrobin, D.F. **Prostaglandins: Physiology, Pharmacology and Clinical Significance**, Montreal, Eden Press, 1977.
572. Horrobin, D.F. "Essential fatty acids: a review." In **Clinical Uses of Essential Fatty Acids**, Horrobin, D.F., ed., Montreal, Eden Press, 3-36, 1982.
573. Horst, K., Wilson, R.J. & Smith R.G. "The effect of coffee and decaffeinated coffee on oxygen consumption, pulse rate and blood pressure." **J. Pharm. Exp. Ther.**, 58, 294-304, 1936.

574. Horton, R.W., LeFeuvre, R.A., Rothwell, N.J., et.al. "Opposing effects of activation of central GABA-a and GABA-b receptors on brown fat thermogenesis in the rat." **Neuropharm.**, 27, 363-366, 1988.

575. Horton, R., Rothwell, N.J. & Stock, M.J. "Chronic inhibition of GABA transaminase results in activation of thermogenesis and brown fat in the rat." **Gen. Pharm.**, 19, 403-405, 1988.

576. Horton, T.J. & Geissler, C.A. "Aspirin potentiates the effect of ephedrine on the thermogenic response to a meal in obese but not lean women." **Int. J. Obes.**, 15(5), 359-366, 1991.

577. Horwitz, B.A. "Cellular events underlying catecholamine-induced thermogenesis: cation transport in brown adipocytes." **Fed. Proc.**, 38, 2170-2176, 1979.

578. Horwitz, B.A. & Hamilton, J. **Comp. Biochem. Physiol.**, 78C, 99-104, 1984.

579. Hostmark, A.t., Spydevold, E., Lystad, A., et.al. **Nutr. Rep. Int.**, 35, 317-324, 1987.

580. "Hot peppers and substance P." **Lancet**, 1198, 1983.

581. Hsieh, A.C.L. & Carlson, L.D. "Role of the thyroid in metabolic response to low temperature." **Am. J. Physiol.**, 188, 40-44, 1957.

582. Hsieh, A.C.L. & Carlson, L.D. "Role of adrenaline and noradrenaline in chemical regulation of heat production." **Am. J. Physiol.**, 190, 243-246, 1957.

583. Hsieh, A.C.L., Carlson, L.D. & Gray, G. "Role of the sympathetic nervous system in the control of chemical regulation of heat production." **Am. J. Physiol.**, 190, 247-251, 1957.

584. Hughes, J.R., Oliveto, A.H., Helzer, J.E., et.al. **Am. J. Psychiatry**, 149, 33, 1992.

586. Hull, D. & Hardman, M.J. "Brown adipose tissue in newborn mammals. In **Brown Adipose Tissue**, Lindberg, O., ed., NY, Elsevier, 1970, pp. 97-116.

587. Huttunen, P., Hirvonen, J. & Kinnula, V. "The occurrence of brown adipose tissue in outdoor workers." **Eur. J. Appl. Physiol.**, 46, 339, 1981.

588. Ide, T. & Sugano, M. **Agric. Biol. Chem.**, 52, 511-518, 1988.

589. Ikezono, E., Takahashi, G. & Ikezono, T. "A trial of thermogenic agents for further weight reduction in obese patients with weights that have plateaued." **Int. J. Obes.**, 17(suppl 1), S80, 1993.

590. Ilyes, I. & Stock, M. J. "Effects of hypothyroidism and hyperthyroidism on thermogenic responses to selective and nonselective beta-adrenergic agonists in rats." **Acta Med. Hung.**, 47(3-4), 179-188, 1990.

591. IFT Expert Panel on Food Safety & Nutrition. "Evaluation of caffeine safety", a Scientific Status Summary, Chicago, IL, Inst. of Food Technologists.

592. Iriki, M., Riedel, W. & Simon, E. "Regional differentiation of sympathetic activity during hypothalamic heating and cooling in anesthetized rabbits." **Pflugers Arch.**, 328, 320-331, 1971.

593. Isaacs, G. "Permanent local anaesthesia and anhidrosis after clove oil spillage." **Lancet**, 882, 1983.

594. Isekutz, B., Miller, H.I. & Rokahl, K. "Lipid and carbohydrate metabolism during exercise." **Fed. Proc.**, 25, 1415-1420, 1966.

595. Ishigooka, J., Yoshida, Y. & Murasaki, M. **Prog. Neuro-Psychopharm. Biol. Psych.**, 15, 513, 1991.

596. Isidori, A., LoMonaco, A. & Cappa, M. "A study of growth hormone release in man after oral administration of amino acids." **Curr. Med. Res. Opin.**, 7, 473, 1981.

597. Isler, D., Hill, H.P. & Meier, M.K. **Biochem. J.**, 245, 789-793, 1987.

598. Ismail, M.N. "Variations in daily energy expenditure in animals and man." Ph.D. Thesis, Univ. of London, 121-143, 1983.

599. Ismail, M.N., Dulloo, A.G. & Miller, D.S. "Genetic and dietary influences on the levels of diet-induced thermogenesis and energy balance in adult mice." **Ann. Nutr. Metab.**, 30, 189-195, 1986.

600. Ismail, M.N., Tee, A.C. & Zaini, A. "24 hour energy expenditure and BMR in obese patients using a room respirometer." **Int. J. Obes.**, 17(suppl 1), S79, 1993.

601. Issekutz, B., Paul, P., Miller, H.I., et.al. "Oxidation of plasma FFA in lean and obese humans." **Metabolism**, 17, 62-73, 1968.

602. Isselbacher, K. "Lipid Absorption and transport across intestinal mucosa." In **Fat as a Tissue**, Rodahl, K., & Issekutz, B., eds. McGraw Hill Book Company, New York, 1964, pp. 7-21.

603. Ito, T., Tanuma, Y., Yamada, M., et.al. "Morphological studies on brown adipose tissue in the bat and in humans of various ages." **Arch. Histol. Cytol.**, 54(1), 1-39, 1991.

604. Ivy, J., Costill, D.L., Essig, D.A., et.al. "The relationship of blood lactate to the anaerobic threshold and hyperventilation." (abstr) **Med. Sci. Sports**, 11, 96-97, 1979.

605. Ivy, J., Costill, D.L., Fink, W., et.al. "Influence of caffeine and carbohydrate feedings on endurance performance." **Med. Sci. Sports Exerc.**, 10, 6-11, 1978.

606. Ivy, J.L., Costill, D.L., Fink, W.J., et.al. **Med. Sci. Sports**, 11, 6-11, 1979.

607. Ivy, J.L., Costill, D.L., Fink, W.J., et.al. "Contribution of medium and long chain triglycerides intake to energy metabolism during prolonged exercise." **Int. J. Sports Med.**, 1, 15-20, 1980.

608. Ivy, J.L., Costill, D.L., Van Handel, P.J., et.al. "Alteration in the lactate threshold with changes in substrate availability." **Int. J. Sports Med.**, 2, 139-142, 1981.

609. Iwai, M., Hell, N.S. & Shimazu, T. **Pflugers Arch.**, 410, 44-47, 1987.

610. Jacobsen, B.K. & Thelle, D.S. **Br. Med. J.**, 294, 4-5, 1987.

611. Jacobsson, A., Stadler, U., Glotzer, M.A., et.al. "Mitochondrial uncoupling protein from mouse brown fat. Molecular cloning, genetic mapping, and mRNA expression." **J. Biol. Chem.**, 260, 16250-16254, 1985.

61. Jaedig, S. "Increased energy expenditure during caloric restriction in obese women by adding minerals." **Int. J. Obes.**, 17(suppl 1), S81, 1993.

613. Jaedig, S. & Henningsen, N.C. "Increased metabolic rate in obese worem after ingestion of potassium, magnesium and phosphate-enriched orange juice or injection of ephedrine." **Int. J. Obes.**, 15(6), 429-436, 1991.

614. Jamal, Z. & Saggerson, E.D. **Biochem. J.**, 249, 415-421, 1988.

615. James, J.E. **Caffeine and Health**, London, Academic Press, 1991.

616. James, W.P.T. "Energy requirements and obesity." **Lancet**, 2, 386-389, 1983.

617. James, W.P.T., Dauncey, M.J., Jung, R.T., et.al. "Comparison of genetic models of obesity in animals with obesity in man." In **Amimal Models of Obesity**, Festing, M.F.W., ed., London, Macmillan, 221-235, 1979.

618. James, W.P.T., McNeill, G. & Ralph, A. **Am. J. Clin. Nutr.**, 51, 264, 1990.

619. Jancso, G. "Selective degeneration of chemosensitive primary sensory neurons induced by capsaicin." **Cell Tissue Res.**, 195, 145-152, 1978.

620. Jancso, G. & Kiraly, E. **Brain Res.**, 210, 83, 1981.

621. Jancso, N. "Desensitization with capsaicin and related acylamides as a tool for studying the function of pain receptors." Proceedings of the 3rd International Pharmacological Meeting, 1966, Pharmacology of Pain, 9, 33-55, Oxford and NY., Pergamon Press, 1968.

622. Jancso, N. "Role of the nerve terminals in the mechanism of inflammatory reactions." **Bull. Millard Fillmore Hosp. Buffalo, NY.**, 7, 53-77, 1960.

623. Jancso, N. "Speicherung. Stoffanreicherung im Retikuloendothel und in der Niere." Budapest: Akademiai Kiado, 1955.

624. Jancso, N. "Stimulation and desensitization of the heat-sensitive hypothalamic receptors by chemical agents." **IIIrd Hung. Conf. on Ther. and Pharm. Res. 1964**, Budapest: Akademiai Kiado, 1965.

625. Jancso, N. & Jancso-Gabor, A. "Die Wirkung des Capsaicins auf die hypothalamischen Thermoreceptoren." **Arch. Exp. Path. Pharm.**, 251, 136, 1965.

626. Jancso, N., Jancso-Gabor, A. & Szolcsanyi, J. "Direct evidence for neurogenic inflammation and its prevention by denervation and by pretreatment with capsaicin." **Br. J. Pharm. Chemother.**, 31, 138-151, 1967.

627. Jancso, N., Jancso-Gabor, A. & Szolcsanyi, J. "Effect of capsaicin on thermoregulation." **Acta Physiol. Hung.**, 29, 364, 1966.

628. Jancso, N., Jancso-Gabor, A. & Szolcsanyi, J. "The role of sensory nerve endings in neurogenic inflammation induced in human skin and in the eye and paw of the rat." **Br. J. Pharm. Chemother.**, 33, 32-41, 1968.

629. Jancso-Gabor, A. "Some data to the pharmacology of histamine releasers." Dissertation, Szeged, 1947.

630. Jancso-Gabor, A., Szolcsanyi, J. & Jancso, N. "Irreversible impairment of thermoregulation induced by capsaicin and similar pungent substances in rats and guinea-pigs." **J. Physiol.**, 206, 495-507, 1970.

631. Jancso-Gabor, A., Szolcsanyi, J. & Jancso, N. "Stimulation and desensitization of the hypothalamic heat-sensitive structures by capsaicin in rats." **J. Physiol.**, 208, 449-450, 1970.

632. Jankelson, O.M., Beaser, S.B., Howard, H.M., et.al. **Lancet**, 1, 527-529, 1967.

633. Janssens, W.J. & Vanhoutte, P.M. "Instantaneous changes of alpha-adrenoceptor affinity caused by moderate cooling in canine cutaneous veins." **Am. J. Physiol.**, 234, H330-H337, 1978.

634. Jarboe, C.H., Hurst, H.E., Rodgers, G.C.Jr., et.al. "Toxicokinetics of caffeine elimination in an infant." **J. Tox. Clin. Tox.**, 24(5), 415-428, 1986.

635. Jenkins, A.B., Storlien, L.H., Chisholm, D.J., et.al. **J. Clin. Inv.**, 82, 293-299, 1988.

636. Jennings, G., Richard, D. & Trayhurn, P. "Effect of caging singly or in groups of different sizes on the thermogenic activity of interscapular brown adipose tissue in mice." **Comp. Biochem. Physiol. Acta**, A85, 583-586, 1986.

637. Jequier, E. "Long term measurement of energy expenditure in man: direct or indirect calorimetry?" In **Recent Advances in Obesity Research III**, Bjorntorp, P., Cairella, M. & Howard, A.N., eds., London, J. Libbey, 130-135, 1981.

638. Jequier, E. "Does a thermogenic defect play a role in the pathogenesis of obesity?" **Clin. Physiol.**, 3, 1-7, 1983.

639. Jequier, E. "Thermogenic responses induced by nutrients in man: their importance in energy balance regulation." In **Nutritional Adequacy, Nutrient Availability and Needs**, Mauron, J., ed., Basel, Birkhauser Verlag, 26-44, 1983.

640. Jequier, E. "Energy expenditure in obesity." **J. Clin. Endocr. Metab.**, 13, 563-580, 1984.

641. Jequier, E. **J. Obes. Wt. Reg.**, 6, 225-233, 1987.

642. Jequier, E. In **Energy Metabolism: Tissue Determinants and Cellular Corollaries**, Kinney, J.M. & Tucker, H.N., eds., NY, Raven Press, 123, 1992.

643. Jequier, E. "Adaptations to low and high caloric intake in humans." **Int. J. Obes.**, 17(suppl 1), S9-S12, 1993.

644. Jequier, E. & Schutz, Y. **Diabetes/Metabolism Rev.**, 4, 583-593, 1988.

645. Jessell, T.M. Iversen L.L. & Cuello, A.C. "Capsaicin-induced depletion of substance P from primary sensory neurons." **Brain Res.**, 152, 183, 1978.

646. Jessen, C. "Thermal afferents in the control of body temperature." **Pharm. Ther.**, 28, 107-134, 1985.

647. Jessen, K., Rabol, A. & Winkler, K. "Total body and splanchnic thermogenesis in curarized man during a short exposure to cold." **Acta Anaesthesiol. Scand.**, 24, 339-344, 1980.

648. Jeszka, J., Grav, H.J., Holm, H., et.al. "Opposite effect of cold on energetic efficiency in normal and obese Wistar rats with hypothalamic lesions." **J. Nutr.**, 121(3), 386-394, 1991.

649. Jick, H., Miettinen, O.S. Neff, R.K., et.al. "Coffee and myocardial infarction." **N. Eng. J. Med.**, 289, 63, 1973.

650. Johnston, J.L. & Balachandran, A.V. **J. Nutr.**, 117, 2046-2053, 1987.

651. Johnson, D. & Drenick, E.J. "Therapeutic fasting in morbid obesity. Long term follow up." **Arch. Intern. Med.**, 137, 1381-1382.

652. Jones, D. & Egger, T. "Use of herbs containing natural source ephedrine alkaloids in weight loss programmes." **Int. J. Obes.**, 17(suppl 1), S81, 1993.

653. Jones, D.A., Howell, S., Roussos, C., et.al. **Clin. Sci.**, 63, 161-167, 1982.

654. Jones, R., Henschen, L., Mohell, N., et.al. **Biochim. Biophys. Acta**, 889, 366-373, 1986.

655. Jones, S.B. & Musacchia, X.J. "Norepinephrine turnover in heart and spleen of 7-, 22-, 34C-acclimated hamsters." **Am. J. Physiol.**, 230, 564-568, 1976.

656. Joo, F., Szolcsanyi, J. & Jancso-Gabor, A. "Mitochondrial alterations in the spinal ganglion cells of the rat accompanying the long-lasting sensory disturbance induced by capsaicin." **Life Sci. Oxford**, 8, 621-262, 1969.

657. Joy, R.J.T. "Responses of cold-acclimatized men to infused norepinephrine." **J. Appl. Physiol.**, 18, 1209-1212, 1963.

658. Juan, H., Lembeck, F., Seewan, S., et.al. "Nociceptor stimulation and PGE release by capsaicin." **Naunyn-Schmiedebergs Arch. Pharm.**, 312, 139-1980.

659. Jung, R.T., Shetty, P.S., Barrand, M., et.al. "Role of catecholamines in hypotensive response to dieting." **Br. Med. J.**, 1, 12-13, 1979.

660. Jung, R.T., Shetty, P.S. & James, W.P.T. "The effect of beta-adrenergic blockade on metabolic rate and peripheral thyroid metabolism in obesity." **Eur. J. Clin. Inv.**, 10, 179-182, 1980.

661. Jung, R.T., Shetty, P.S. & James, W.P.T. **Int. J. Obes.**, 4, 95, 1980.

662. Jung, R.T., Shetty, P.S. & James, W.P.T. "Heparin, free fatty acids and an increased metabolic demand for oxygen." **Postgraduate Med. J.**, in press, 1981.

663. Jung, R.T., Shetty, P.S., James, W.P.T., et.al. "Reduced thermogenesis in obesity." **Nature (London)**, 279, 322-323, 1979.

664. Jung, R.T., Shetty, P.S., James, W.P.T., et.al. "Caffeine: its effect on catecholamines and metabolism in lean and obese humans." **Clin. Sci.**, 60, 527-535, 1981.

665. Kaciuba-Uscilko, H. "The effect of previous thyroxine administration on the metabolic response to adrenaline in new-born pigs." **Biol. Neonate**, 19, 220-226, 1971.

666. Kalix, P. "The pharmacology of psychoactive alkaloids from ephedra and catha." **J. Ethnopharm.**, 32(1-3), 201-208, 1991.

667. Kamei, K., Murata, M., Ishii, K., et.al. "Detection of amphetamine and related amines in urine by gas chromatography and combined gas chromatography-mass spectrometry." **Chem. Pharm. Bull.**, 21, 1996-2003, 1973.
668. Kamimori, G.H., Somani, S.M., Knowlton, R.G., et.al. "The effects of obesity and exercise on the pharmacokinetics of caffeine in lean and obese volunteers." **Eur. J. Clin. Pharm.**, 31(5), 595-600, 1987.
669. Kanarek, R. & Aprille, J. "Amelioration of sucrose-induced obesity in rats fed diets with increased levels of zinc, chromium, selenium." **Fed. Proc.**, 45, 351, 1986.
670. Kang, B.S., Han D.S., Paik, K.S., et.al. "Calorigenic action of norepinephrine in the Korean women divers." **J. Appl. Physiol.**, 29, 6-9, 1970.
671. Kanof, P.D., Hegstrand, L.R. & Greengard, P. "Biochemical characterization of histamine-sensitive adenylate cyclase in mammalian brain." **Arch. Biochem. Biophys.**, 182, 321, 1977.
672. Kaplan, M.L. "Consumption of oxygen and early detection of fa/fa genotype in rats." **Metabolism**, 28, 1147-1151, 1979.
673. Kaplan, M.L. & Leveille, G.A. "Calorigenic response in obese and non-obese women." **Am. J. Nutr.**, 29, 1108-1113, 1976.
674. Kaplan, M.M. & Young, J.B. **Endocrinology**, 120, 886-893, 1987.
675. Kaplan, R.J., Daman, L., Rosenberg, E.W., et.al. **Arch. Dermat.**, 114, 60-62, 1978.
676. Karacan, I., Thornby, J.I., Anch, A.M., et.al. **J. Clin. Pharm. Ther.**, 20, 682-689, 1976.
677. Kariyama, S., Okizaki, S. & Yoshida, A. "Hypocholesterolemic effect of polysaccharides and polysaccharide-rich foodstuffs in cholesterol-fed rats." **J. Nutr.**, 97, 382-388, 1960.
678. Karmali, R.A. "Fatty acids: inhibition." **Am. J. Clin. Nutr.**, 45, 225, 1987.
679. Kates, A.L. & Himms-Hagen, J. "Defective cold-induced stimulation of thyroxine 5'deiodinase in brown adipose tissue of the genetically obese (ob/ob) mouse." **Biochem. Biophys. Res. Commun.**, 130, 188-193, 1985.
680. Kather, H., Bieger, W. & Aktories, K., **J. Clin. Inv.**, 76, 1559, 1985.
681. Kather, H., Wieland, E., Fischer, B., et.al. **Biochem. J.**, 231, 531, 1985.
682. Kather, H., Wieland, E., Scheurer, A., et.al. **J. Clin. Inv.**, 80, 566, 1987.
683. Katims, J.J., Annau, Z. & Snyder, S.H. **J. Pharm. Exp. Ther.**, 227, 167-173, 1983.
684. Kato, N. & Yoshida, A. **Agric. Biol. Chem.**, 43, 191-192, 1979.
685. Katz, J. "Energy balance and futile cycling." In **Assessment of Energy Metabolism in Health and Disease**, Kinney, J.M., ed., Columbus, OH, Ross Laboratories, 6-9, 1980.
686. Katzeff, H.L. & Danforth, E.Jr. "The thermogenic response to norepinephrine, food and exercise in lean man during overfeeding." **Clin. Res.**, 29, 663A, 1981.
687. Katzeff, H.L., O'Connell, M., Horton, E.S., et.al. **Metabolism**, 35, 166, 1986.
688. Kaufman, L.N., Young, J.B. & Landsberg, L. "Protein stimulates sympathetic nervous system (SNS) activity less than carbohydrate: evidence for nutrient specific SNS responses." **Clin. Res.**, 32, 478A, 1984.
689. Kaufman, L.N., Young, J.B. & Landsberg, L. **Metabolism**, 38, 91, 1989.
690. Kawada, T., Koh-Ichiro, H. & Iwai, K. "Effects of capsaicin on lipid metabolism in rats fed a high fat diet." **Am. Inst. Nutr.**, 1272-1278, 1986.
691. Kawada, T., Suzuki, T., Takahashi, M., et.al. "Gastrointestinal absorption and metabolism of capsaicin and dihydrocapsaicin in rats." **Tox. Appl. Pharm.**, 72, 449-456, 1984.
692. Kawada, T., Watanabe, T., Takaishi, T., et.al. "Capsaicin-induced beta-adrenergic action on energy metabolism in rats: influence of capsaicin on oxygen consumption, the respiratory quotient, and substrate utilization." **J. Nutr.**, 116, 1272, 1986.
693. Kelly, L. & Bielajew, C. "Ventromedial hypothalamic regulation of brown adipose tissue." **Neuroreport**, 2(1), 41-44, 1991.
694. Keller, U. "Drugs against obesity." **Ther. Umsch.**, 47(8), 658-663, 1990.
695. Kern, P.A., Svoboda, M.E., Eckel, R.H., et.al. **Diabetes**, 38, 710, 1989.
696. Keys, A., Brozek, J., Henschel, A., et.al. **Biology of Human Starvation"**, Minneapolis, University of Minnesota Press, 1950.
697. Khoo, J.c., Yamamotom M., et.al. "Hormone-senstive lipase system and insulin stimulation of protein phosphatase activities in 3T3-L1 adipocytes." In **The Adipocyte and Obesity: Cellular and Molecular Mechanisms**, Angel, A., Hollenberg, C.H. & Roncari, D.A., eds., Raven Press, New York, 1983, pp. 225-233.
699. Ki, P., Negulesco, J.A. & Murnane, M. "Decreased total serum, myocardial and aortic cholesterol levels following capsaicin treatment." **IRCS Med. Sci. Libr. Compend.**, 10, 446-447, 1982.
700. Kim, H.K. & Romsos, D.R. "Brown adipose tissue metabolism in ob/ob mice: effects of a high-fat diet and adrenalectomy." **Am. J. Physiol.**, 253, E149-E157, 1987.
701. King, B.M. "Glucocorticoids and hypothalamic obesity." **Neurosci. Biobehav. Rev.**, 12, 29-37, 1988.
702. Klaus, S., Casteilla, L., Hentz, E., et.al. "The mRNA of protein disulfide isomerase and its homologue the thyroid hormone binding protein is strongly expressed in adipose tissue." **Mol. Cell Endocr.**, 73(2-3), 105-110, 1990.
703. Klein, R.H. & Salzman, L.F. **Percept. Motor Skills**, 40, 126, 1975.
704. Klement, W., Medert, H.A. & Arndt, J.O. **Pain**, 48, 269, 1992.
705. Klingenberg, M. **Biochemistry**, 27, 781-791, 1988.
706. Klingenberg, M. & Winkler, E. **EMBO J.**, 4, 3087-3092, 1985.
707. Klingenberg, M. & Winkler, E. **Methods Enzymol.**, 127, 772-779, 1986.
708. Klitsch, T. & Siemen, D. "Inner mitochondrial membrane anion channel is present in brown adipocytes but is not identical with the uncoupling protein." **J. Membr. Biol.**, 122(1), 69-75, 1991.
709. Knehans, A.W. & Romsos, D.R. "Reduced sympathetic nervous system activity in brown adipose tissue of obese (ob/ob) mice." (abstr) **Fed. Proc.**, 40, 888, 1981.
710. Knehans, A.W. & Romsos, D.R. "Reduced norepinephrine turnover in brown adipose tissue of ob/ob mice." **Am. J. Physiol.**, 242, E253-E261, 1982.
711. Knehans, A.W. & Romsos, D.R. **Metabolism**, 33, 652-657, 1984.
712. Knight, B.L. & Skala, J.P. **Can. J. Biochem.**, 60, 734-740, 1982.
713. Knutti, R., Rothweiler, H. & Schlatter, C. "Effect of pregnancy on the pharmacokinetics of caffeine." **Eur. J. Clin. Pharm.**, 21(2), 121, 1981.
714. Kobal, G., Hummel, C., Nuernberg, B., et.al. **Agents Actions**, 29, 342, 1990.
715. Koller, W., Cone, S. & Herbster, G. **Neurology**, 37, 169-172, 1987.
716. Kolodeznikova, E.D., & Afanasiev, J.I. "Brown fat reactiion to extreme exposures." In Szelenyi, Z. & Szekely, M. eds. **Contributions to Thermal Physiology**, Pergamon Press, Akademaiai Kiado, Budapest, 1981, pp. 519-521.
717. Komaromi, I. "Effects of alpha-and beta-adrenergic blockers on the actions of noradrenaline on body temperature in the newborn

guinea-pig." **Experientia**, 33, 1083-1084, 1977.
718. Kopecky, J., Sigjurdson, L., Park, I.R.A., et.al. **Am. J. Physiol.**, 251, E1-E7, 1986.
719. Kornacker, M.S. & Ball, E.G., **J. Biol. Chem.**, 243, 1638, 1968.
720. Kosuge, S., Inagaki, Y. & Uehara, T. "Studies on the pungent principles of Capsicum.I. On the chemical constitution of the pungent principles.(1) On isolation of the pungent principles." **Nihon Nougeikagaku Kaishi**, 32, 578-581, 1958.
721. Kozlovsky, A.S., Moser, P.B., Reiser, S., et.al. "Effects of diets high in simple sugars on urinary chromium losses." **Metabolism**, 35, 515, 1986.
722. Kraegen, E.W., James, D.E., Storlein, L.H., et.al. **Diabetologia**, 29, 192-198, 1986.
723. Krall, J.F., Fernandez, E.I. & Connelly-Fittingoff, M. "Human aging: effect of the activation of lymphocyte cyclic AMP-dependent protein kinase by forskolin." **Proc. Soc. Exp. Biol. Med.**, 184, 396, 1987.
724. Kreider, M.B. & Buskirk, E.R. "Supplemental feeding and thermal comfort during sleep in the cold." **J. Appl. Physiol.**, 11, 339-343, 1957.
725. Krief, S., Bazin, R., Dupuy, F., et.al. **Am. J. Physiol.**, 254, E342-E348, 1988.
726. Krieger, D.R., Daly, P.A., Dulloo, A.G., et.al. "Ephedrine, caffeine and aspirin promote weight loss in obese subjects." **Trans. Assn. Am. Phys.**, 103, 307-312, 1990.
727. Krueger, J., Zulch, J. & Gandorfer, M.Z. **Ernahrungswissenschaft**, 18, 51-61, 1979.
728. Kucio, C., Jonderko, K. & Piskorska, D. "Does yohimbine act as a slimming drug?" **Isr. J. Med. Sci.**, 27(10), 550-556, 1991.
729. Laville, M., Cornu, C., Normand, S. et.al. "Decreased glucose-induced thermogenesis at the onset of obesity." **Am. J. Clin. Nutr.**, 57, 851-6, 1993
730. Kuroshima, A., Yahata, T. & Habara, Y. **J. Therm. Biol.**, 9, 81-86. 1984.
731. Kush, R.D., Young, J.B., Katzeff, H.L., et.al., **Metabolism**, 35, 1110-1120, 1986.
732. Kuznicki, J.T. & Turner, L.S. "Effects of caffeine on caffeine users and non-users." **Physiol. Behav.**, 37, 397-408, 1986.
733. Landsberg, L. **Q. J Med.**, 236, 1081-1090, 1986.
734. Landsberg, L. "Insulin resistance, energy balance and sympathetic nervous system activity." **Clin. Exp. Hypertens. A.**, 12(5), 817-830, 1990.
735. Landsberg, L. "The sympathoadrenal system, obesity and hypertension: an overview." **J. Neurosci. Methods**, 34(1-3), 179-186, 1990.
736. Landsberg, L., Greff, L., Gunn, S., et.al. "Adrenergic mechanisms in the metabolic adaptation to fasting and feeding: effects of phlorizin on diet-induced changes in sympathoadrenal activity in the rat." **Metabolism**, 29, 1128-1137, 1980.
737. Landsberg, L., Saville, M.E. & Young, J.B. "Sympathodrenal system and regulation of thermogenesis." **Am. J. Physiol.**, 247, E181-E189, 1984.
738. Landsberg, L. & Young, J.B. "Fasting, feeding and regulation of sympathetic nervous system." **N. Eng. J. Med.**, 298, 1295-1301, 1978.
739. Landsberg, L. & Young, J.B. "The role of the sympathetic nervous system and catecholamines in the regulation of energy metabolism." **Am. J. Clin. Nutr.**, 38, 1018, 1983.
740. Landsberg, L. & Young, J.B. "Autonomic regulation of thermogenesis." In **Mammalian Thermogenesis**, Girardier, L. & Stock, M.J., eds., London, Chapman & Hall, 99-140, 1983.
741. Landsberg, L. & Young, J.B. "Diet-induced changes in sympathetic nervous system activity." In **The Adipocyte and Obesity: Cellular and Molecular Mechanisms**, Angel, A., Hollenberg, C.H. & Roncari, D.A., eds., Raven Press, New York, 1983, pp. 283-290.
742. Landsberg, L. & Young, J.B. "The role of the sympathoadrenal system in modulating energy expenditure." **Clin. Endocr. Metab.**, 13(3), 475-499, 1984.
743. Landsberg, L. & Young, J.B. **Int. J. Obes.**, 9, 63, 1985.
744. Landsberg, L. & Young, J.B. "The influence of diet on the sympathetic nervous system." In **Neuroendocr. Perspectives**, Miller, E.E., MacLeod, R.M., & Frohman, L.A., eds., Amsterdam, Elsevier, 191, 1985.
745. Landsberg, L. & Young, J.B. "Sympathoadrenal activity and obesity: physiological rationale for the use of adrenergic thermogenic drugs." **Int. J. Obes.**, 17(suppl 1), S29-S34, 1993.
746. Lanoue, K.F., Koch, C.D. & Meditz, R.B. **J. Biol. Chem.**, 257, 13740-13748, 1982.
747. Lanoue, K.F., Strzelecki, T., Strzelecka, D., et.al. **J. Biol. Chem.**, 261, 298-304, 1986.
748. Lanzola, E., Tagliabue, A., Bozzi, G., et.al. "Obesity, diet and body temperature." **Ann. Nutr. Metab.**, 35(5), 274-283, 1991.
749. Lapidus, J.B., Tye, A. & Patil, P.N. "Steric aspects of adrenergic drugs VII. Certain pharmacological actions of d-(-)-pseudoephedrine." **J. Pharm. Sci.**, 56, 1125-1130, 1967.
750. Lardinois, C.K. "The role of omega 3 fatty acids on insulin secretion and insulin sensitivity." **Med. Hypoth.**, 24, 243, 1987.
751. Laska, E.M., Sunshine, A., Mueller, F., et.al. "Caffeine as an analgesic adjunct." **JAMA**, 251, 1711, 1984.
752. Laska, E.M., Sunshine, A., Zighelboim, I., et.al. "Effect of caffeine on acetaminophen analgesia." **Clin. Pharm. Ther.**, 33, 498, 1983.
753. Latini, R., Bonati, M., Marzi, E., et.al. **J. Pharm. Pharmacol.**, 32, 596-599, 1980.
754. Lean, M.E.J., Branch, W.J., James, W.P.T., et.al. "Measurement of rat brown-adipose-tissue mitochondrial uncoupling protein by radioimmunoassay-increased concentration after cold-acclimation." **Biosci. Rep.**, 3, 61-71, 1983.
755. Lean, M.E.J. & James, W.P.T. "Uncoupling protein in human brown adipose tissue mitochondria." **FEBS Lett.**, 163, 235-240, 1983.
756. Lean, M.E.J. & James, W.P.T. In **Brown Adipose Tissue**, Trayhurn, P. & Nicholls, D.G., eds., London, Arnold, 339-365, 1986.
757. Lean, M.E.J., James, W.P.T., Jennings, G., et.al. "Brown adipose tissue uncoupling protein content in infants, children and adult humans." **Clin. Sci.**, 71, 291-297, 1986.
758. Lean, M.E.J., James, W.P.T., Jennings, G., et.al. "Brown adipose tissue in patients with phaeochromocytoma." **Int. J. Obes.**, 10, 219-227, 1986.
759. Lean, M.E.J., Murgatroyd, P.R., Rothnie, I., et.al. **Clin. Endocr.**, 28, 665-674, 1988.
760. Lean, M.E.J. & Trayhurn, P.J. **Obes. Wt. Reg.**, 6, 234-253, 1987.
761. Lean, M.E.J., Trayhurn, P., Murgatroyd, P.R. In **Recent Advances in Obesity Research**, Berry, E.M., Blondheim, S.H., Eliahou, H.E., et.al., eds., London, Libbey, 5, 109-116, 1987.
762. LeBlanc, J. **Man in the Cold**, Springfield, IL, Thomas, 1975.
763. LeBlanc, J., Jobin, M., Cote, J., et.al. "Enhanced metabolic response to caffeine in exercise-trained human subjects." **J. Appl. Physiol.**, 59, 832-837, 1985.

764. LeBlanc, J., Diamond, P. & Nadeau, A. "Thermogenic and hormonal responses to palatable protein and carbohydrate rich food." **Horm. Metab. Res.**, 23(7), 336-340, 1991.
765. LeBlanc, J., Vallieres, J. & Vachon, C. "Beta-receptor sensitization by repeated injections of isoproterenol and by cold adaptation." **Am. J. Physiol.**, 222, 1043-1046, 1972.
766. LeBlanc, J. & Villemaire, A. "Thyroxine and noradrenaline sensitivity, cold resistance, and brown fat." **Am. J. Physiol.**, 218, 1742-1745, 1970.
767. LeDuc, J. "Catecholamine production and release in exposure and acclimation to cold." **Acta Physiol. Scand.**, 183, 1-101, 1961.
769. LeFeuvre, R.A., Aisenthal, L. & Rothwell, N.J. "Involvement of corticotrophin releasing factor (CRF) in the thermogenic and anorexic actions of serotonin (5-HT) and related compounds." **Brain Res.**, 555(2), 245-250, 1991.
770. LeFeuvre, R.A., Rothwell, N.J. & Stock, M.J. "Activation of brown fat thermogenesis in response to central injection of corticotropin releasing hormone in the rat." **Neuropharm.**, 26, 1217-1221, 1987.
771. Lefkowitz, R.J., Caron, M.G. & Stiles, G.L. "Mechanisms of membrane receptor regulation. Biochemical, physiological and clinical insights derived from studies of the adrenergic receptor." **N. Eng. J. Med.**, 310, 1570-1579, 1984.
772. Leibel, R.L. & Hirsch, J. "Dimished energy requirements in reduced-obese patients." **Metabolism**, 33, 164-170, 1984.
773. Leibowitz, S.F. "Brain monoamines and peptides: role in the control of eating behavior." **Fed. Proc.**, 45, 1396-1403, 1986.
774. Leigh, F.S.M., Kaufman, L.N. & Young, J.B. **Int. J. Obes.**, 16, 597, 1992.
775. Leinert, G.A. & Huber, H.P. **J. Psychol.**, 63, 269-274, 1966.
776. Leonard, J.L., Mellen, S.A. & Larsen, R. "Thyroxine 5'-deiodinase activity in brown adipose tissue." **Endocrinology**, 112, 1153-1155, 1983.
777. Lembeck, F. & Donnerer, J. **Naunyn-Schmiedeberg's Arch. Pharm.**, 322, 286, 1983.
778. Lembeck, F. & Gamse, R. In **Substance P in the Nervous System**, Porter, R. & O'Conner, M., eds., London, Pitman, 35-54, 1982.
779. LeMessurier, D.H. **J. Pharm. Exp. Ther.**, 57, 458-463, 1936.
780. Leonard, J.L., Mellen, S.A. & Larsen, P.R. **Endocrinology**, 112, 1153-1155, 1983.
781. Lever, J.D., Jung, R.T., Nnodim, J.O., et.al. **Anat. Rec.**, 215, 251-255, 1986.
782. Levi, L. "The effect of coffee on the function of the sympatho-adrenomedullary system in man." **Acta Med. Scand.**, 181, 431-438, 1967.
783. Levin, B.E., Comai, K., O'Brien, R.A., et.al. "Abnormal brown adipose composition and beta-adrenoreceptor binding in obese Zucker rats." **Am. J. Physiol.**, 243, E217-E224, 1982.
784. Levin, B.E., Finnegan, M.B., Marquet, E., et.al. "Effects of diet and obesity on brown adipose tissue metabolism." **Am. J. Physiol.**, 246, E418, 1984.
785. Levin, B.E., Finnegan, M.B., Marquet, E., et.al. "Defective brown adipose oxygen consumption in obese Zucker rats." **Am. J. Physiol.**, 247, E94-E99, 1984.
786. Levin, B.E., Finnegan, M., Triscari, J., et.al. "Brown adipose tissue and metabolic features of chronic diet-induced obesity." **Am. J. Physiol.**, 248, R717-R723, 1985.
787. Levin, B.E. & Sullivan, A.C. "Catecholamine synthesizing enzymes in various brain regions of the genetically obese Zucker rat." **Brain Res.**, 171, 560-566, 1979.
788. Levin, B.E. & Sullivan, A.C. "Catecholamine levels in discrete brain nuclei of seven month old genetically obese rats." **Pharm. Biochem. Behav.**, 11, 77-82, 1979.
789. Levin, B.E., Triscari, J. & Hogan, S., et.al. "Resistance to diet-induced obesity: food intake, pancreatic sympathetic tone, and insulin." **Am. J. Physiol.**, 252, R'7'-R478, 1987.
790. Levin, B.E., Triscari, J. & Sullivan, A.C. "Abnormal sympatho-adrenal function and plasma catecholamines in obese Zucker rats." **Pharm. Biochem. Behav.**, 13, 107 113, 1980.
791. Levin, B.E., Triscari, J. & Sullivan, A.C. "Defective catecholamine metabolism in peripheral organs of genetically obese Zucker rats." **Brain. Res.**, 224, 353-366, 1981.
792. Levin, B.E., Triscari, J. & Sullivan, A.C. "Sympathetic activity in thyroid-treated Zucker rats." **Am. J. Physiol.**, 243, R170-R178, 1982.
793. Levin, B.E., Triscari, J. & Sullivan, A.C. "Altered sympathetic activity during development of diet-induced obesity in rats." **Am. J. Physiol.**, 244, R347, 1983.
794. Levin, B.E., Triscari, J. & Sullivan, A.C. "Studies on the origins of abnormal sympathetic function in obese Zucker rats." **Am. J. Physiol.**, 245, E87-E93, 1983.
795. Levin, B.E., Triscari, J. & Sullivan, A.C. "Relationship between sympathetic activity and diet-induced obesity in two rat strains." **Am. J. Physiol.**, 245, R367-R371, 1983.
796. Levin, B.E., Triscari, J. & Sullivan, A.C. "Metabolic feature of diet-induced obesity without hyperphagia in young rats." **Am. J. Physiol.**, 251, R433-R440, 1983.
797. Levin, N., Shinsako, J. & Dallman, M.F. **Endocrinology**, 122, 694-701, 1988.
798. Levinson, W. & Dunn, P.M. "Nonassociation of caffeine and fibrocystic breast disease." **Arch. Intern. Med.**, 146, 1773, 1986.
799. Linn, S., Schoenbaum, S.C., Monson, R.R., et.al. "No association between coffee consumption and adverse outcomes of pregnancy." **N. Eng. J. Med.**, 306, 141, 1982.
800. Lomax, P. & Green, M.D. "Histaminergic neurons in the hypothalamic thermoregulatory pathways." **Fed. Proc.**, 40, 2741, 1981.
801. Loncar, D. "Convertible adipose tissue in mice." **Cell Tissue Res.**, 266(1), 149-161, 1991.
802. Londos, C., Cooper, D.M. & Wolff, J. **Proc. Natn. Acad. Sci. USA**, 77, 2551-2554, 1980.
803. Loennqvist, F. & Arner, P. **Biochem. Biophys. Res. Commun.**, 161, 654, 1989.
804. Loennqvist, F., Wennlund, A. & Arner, P. **Int. J. Obes.**, 13, 137, 1989.
805. Loennroth, P., Jansson, P.A., Fredholm, B.B., et.al. **Am. J. Physiol.**, 256, E250, 1989.
806. Loosmore, S. & Armstrong, D. "Do-Do abuse." **Br. J. Psychiatry**, 157, 278-281, 1990.
807. Lopes, J.M., Aubier, M., Jardim, J., et.al. "Effect of caffeine on skeletal muscle function before and after fatigue." **J. Appl. Physiol.**, 54, 1303-1305, 1983.
808. Lopez-Soriano, F.J. & Alemany, M. **Biochem. Int.**, 12, 471-478, 1986.
809. Lopez-Soriano, F.J. & Alemany, M. **Comp. Biochem. Physiol.**, 87B, 91-94, 1987.
810. Lowell, B.B., Napolitano, A., Usher, P., et.al. "Reduced adipsin expression in murine obesity: effect of age and treatment with the sympathomimetic-thermogenic drug mixture ephedrine and caffeine." **Endocrinology**, 126(3), 1514-1520, 1990.
811. Lowndes, R.H. & Mansel, R.E. "The effects of evening primrose oil administration on the serum lipids of normal and obese

human subjects." In **Clinical Uses of Essential Fatty Acids**, Horrobin, D.F., ed., Montreal, Eden Press, 37-52, 1982.

812.   Ma, S.W. & Foster, D.O. **Can. J. Physiol. Pharm.**, 64, 609-614, 1986.

813.   Ma, S.W. & Foster, D.O. "Brown adipose tissue, liver, and diet-induced thermogenesis in cafeteria diet-fed rats." **Can. J. Physiol. Pharm.**, 67(4), 376-381, 1989.

814.   Ma, S.W.Y., Nadeau, B.E. & Foster, D.O. "Evidence for liver as the major site of diet-induced thermogenesis of rats fed a 'cafeteria' diet." **Can. J. Physiol. Pharm.**, 65, 1802-1804.

815.   MacCormack, F.A. **Prevent. Med.**, 6, 104-119, 1977.

816.   MacLachlan, M., Connacher, A.A. & Jung, R.T. "Psychological aspects of dietary weight loss and medication with the atypical beta agonist BRL 26830A in obese subjects." **Int. J. Obes.**, 15(1), 27-35, 1991.

817.   Madsen, J., Bulow, J., Hartkop, A., et.al. "Thermogenic effect of ephedrine in skeletal muscle, heart and liver in the dog." (abstr) **Acta Physiol. Scand.**, 129A-150A, 1987.

818.   Maga, J.A. "Capsicum." **CRC Crit. Rev. Food Sci. Nutr.**, 6, 177-197, 1975.

819.   Mahdihassan, S. & Mehdi, F.S. "Soma of the Rigveda and an attempt to identify it." **Am. J. Chin. Med.**, 17(1-2), 1-8, 1989.

820.   Maickel, R.P., Matussek, N., Stern, D.N., et.al. "The sympathetic nervous system as a homeostatic mechanism. I. Absolute need for sympathetic nervous function in body termperature maintenance of cold-exposed rats." **J. Pharm. Exp. Ther.**, 157, 103-110, 1967.

821.   Maickel, R. & Snodgrass, W. **Tox. Appl., Pharm.**, 26, 218-230, 1973.

822.   Maiecka-Tendera, E. "Postheparin lipoprotein lipase activity in obese children treated with hypocaloric diet supplemented with ephedrine or theophylline." **Int. J. Obes.**, 17(suppl 1), S80, 1993.

823.   Maiecka-Tendera, E. "Effect of ephedrine and theophylline on weight loss in obese overfed rats." **Int. J. Obes.**, 17(suppl 1), S82, 1993.

824.   Makara, G.B., Gyorgy, L. & Molnar, J. "Circulatory and respiratory responses to capsaicin and histamine in rats pretreated with capsaicin." **Arch. Int. Pharmocodyn. Ther.**, 170, 39-45, 1967.

825.   Malchow-Moller, A., Larsen, S., Hey, H., et.al. "Ephedrine as an anorectic: the story of the 'Elsinore pill'." **Int. J. Obes.**, 5, 183-187, 1981.

826.   Mancini, M.C., Marsiaj, H.I., Hadoyama, M.M., et.al. **Int. J. Obes.**, 14(suppl 2), 141, 1990.

827.   Mansell, P.I., Fellows, I.W., MacDonald, I.A., et.al. "Defect in thermoregulation in malnutrition reversed by weight gain." **Q. J. Med.**, 76(280), 817-829, 1990.

828.   Marchington, D., Rothwell, N.J., Stock, M.J., et.al. "Energy balance, diet-induced thermogenesis and brown adipose tissue in lean and obese (fa/fa) Zucker rats after adrenalectomy." **J. Nutr.**, 113, 1395-1402, 1983.

829.   Marchington, D., Rothwell, N.J., Stock, M.J., et.al. "Thermogenesis and sympathetic activity in BAT of overfed rats after adrenalectomy." **Am. J. Physiol.**, 250, E362-E366, 1986.

830.   Marcus, C., Ehren, H., Bolme, P., et.al. **J. Clin. Inv.**, 82, 1793, 1988.

831.   Marette, A. & Bukowiecki, L.J. "Mechanism of norepinephrine stimulation of glucose transport in isolated rat brown adipocytes." **Int. J. Obes.**, 14(10), 857-867, 1990.

832.   Marette, A., Tulp, O.L. & Bukowiecki, L.J. "Mechanism linking insulin resistance to defective thermogenesis in brown adipose tissue of obese diabetic SHR/N-cp rats." **Int. J. Obes.**, 15(12), 823-831, 1991.

833.   Martin, W.R., Sloan, J.W., Sapira, J.D., et.al. **Clin. Pharm. Therap.**, 12, 245, 1971.

834.   Martins, R., Atgie, C., Gineste, L., et.al. "Increased GDP binding and thermogenic activity in brown adipose tissue mitochondria during arousal of the hibernating garden dormouse (Eliomys quercinus L.)." **Comp. Biochem. Physiol. A.**, 98(2), 311-316, 1991.

835.   Massoudi, M. & Miller, D.S. "Ephedrine, a thermogenic and potential slimming dr ,." **Proc. Soc. Nutr.**, 36, 135A, 1977.

836.   Massoudi, M. Evans, E. & Miller, D.S. "Thermogenic drugs for the treatment of c esity: screening using obese rats and mice." **Ann. Nutr. Metab.**, 27(1), 26-37, 1983.

837.   Mathew, A.G., Lewis, Y.S., Krishnamurphy, N., et.al. "Capsaicin." **The Flavour Industry**, 2, 691-695, 1971.

838.   Mathias, S., Garland, C., Barrett-Connor, E., et.al. "Coffee, plasma cholesterol, and lipoproteins. A population study in an adult community." **Am. J. Epidemiol.**, 121(4), 896-905, 1985.

839.   Matsushita, H., Kobayashi, K. & Kusumi, H. "Effect of cold exposure on lipoprotein lipase activity in brown adipose tissue." In Szelenyi, Z. & Szekely, M. eds. **Contributions to Thermal Physiology**, Pergamon Press, Akademaiai Kiado, Budapest, 1981, pp. 523-526.

840.   Mazzarelli, M., Jaspar, N., Zin, W.A. **J. Appl. Physiol.**, 60, 52-59, 1986.

841.   McCormack, J.G. & Denton, R.M. "Evidence that fatty acid synthesis in the interscapular brown adipose tissue of cold-adapted rats is increased in vivo by insulin by mechanisms involving parallel activation of pyruvate dehydrogenase and acetylcoenzyme A carboxylase." **Biochem. J.**, 166, 627-630, 1977.

843.   McCormack, K. & Brune, K. **Drugs**, 41, 533, 1991.

844.   McDonald, R.B., Day, C., Carlson, K., et.al. "Effect of age and gender on thermoregulation." **Am. J. Physiol.**, 257(4 pt 2), R700-704, 1989.

845.   McElroy, J.F. & Wade, G.N. "Short photoperiod stimulates brown adipose tissue growth and thermogenesis but not norepineph-rine turnover in Syrian hamsters." **Physiol. Behav.**, 37, 307-311, 1986.

846.   McGee, M.B. "Caffeine poisoning in a 19-year-old female." **J. Forensic Sci.**, 25(1), 29, 1980.

847.   McNaughton, K.W., Sathasivam, P., Vallerand, A.L., et.al. **J. Appl. Physiol.**, 68(5), 1889-1895, 1990.

848.   McNaughton, L.R. "Two levels of caffeine ingestion on blood lactate and free fatty acid responses during incremental exercise." **Res. Exerc. Sport**, 58, 255-259, 1987.

849.   McNeill, G., Bukkens, S.G.F., Morrison, D.C., et.al. **Proc. Nutr. Soc.**, 49(1), 14A, 1990.

850.   Means, J.H., Aub, J.C. & Dubois, E.F. **Arch. Intern. Med.**, 19, 832-839, 1917.

851.   Meijer, G.A., Janssen, G.M., Westerterp, K.R., et.al. "The effect of a 5-month endurance-training programme on physical activity: evidence for a sex-difference in the metabolic response to exercise." **Eur. J. Appl. Physiol.**, 62(1), 11-17, 1991.

852.   Mathias, C. J. "Effect of food intake on cardiovascular control in patients with imparied autonomic function." **J. Neuroscience Methods**, 34, 193-200, 1990.

852a. Meyer, J.M. & Stunkard, A. J. "Genetics and human obesity." **Obesity: Theory and Therapy**, 2nd Ed. A.J. Stunkard & T.A. Wadden, eds., Raven Press, Ltd. N.Y., 1993, pp. 137-149.

853.   Mercer, S.W. & Trayhurn, P. **Biochem. J.**, 212, 393-398, 1983.

854.   Mercer, S.W. & Trayhurn, P. **J. Nutr.**, 114, 1151-1158, 1984.

855.   Mercer, S.W. & Trayhurn, P. "Effects of ciglitazone on insulin resistance and thermogenic responsiveness to acute cold in brown

adipose tissue of genetically obese (ob/ob) mice." **FEBS Lett.**, 195, 12-16, 1986.

856. Mercer, S.W. & Trayhurn, P. "Effect of high fat diets on energy balance and thermogenesis in brown adipose tissue of lean and genetically obese (ob/ob) mice." **J. Nutr.**, 117, 2147-2153, 1987.

857. Metz, S., Fujimoto, W., Robertson, R.P. "Modulation of insulin secretion by cyclic AMP and prostaglandin E: the effects of theophylline, sodium salicylate and tolbutamide." **Metabolism**, 31, 1014, 1982.

858. Michaelis, R.C., Holloway, F.A., Bird, D.C., et.al. "Interactions between stimulants: effects on DRL performance and lethality in rats." **Pharm. Biochem. Behav.**, 27(2), 299-306, 1987.

859. Mihailova, L. "Morphological and functional changes of Bacillus anthracis under the action of capsaicin and piperine (1) communication. The action of capsaicin and piperine on the reproductive activity, morphological and cultural properties of Bacillus anthracis." **Bull. de L. 'Institut de Microbiologie**, 21, 277-289, 1970.

861. Millard, R.W. & Reite, O.B. "Peripheral vascular response to norepinephrine at temperatures from 2 to 40 C." **J. Appl. Physiol.**, 388, 26-30, 1975.

862. Miller, D.S. "Overfeeding in man." In **Obesity in Perspective**, Bray, G.A., ed., Washington, US Govt Printing Office, 137-143, 1973.

863. Miller, D.S. "Non-genetic models of obesity." In **Animal Models of Obesity**, Festing, M.F.W., ed., London, Macmillan, 131-140, 1979.

864. Miller, D.S. "Factors affecting energy expenditure." **Proc. Nutr. Soc.**, 41, 193-202, 1982.

865. Miller, D.S. "A controlled trial using ephedrine in the treatment of obesity." **Int. J. Obes.**, 10(2), 159-160, 1986.

866. Miller, D.S. & Mumford, P. "Gluttony I. An experimental study of overeating low- or high-protein diets." **Am. J. of Clin. Nutr.**, 20, 1212-1222, 1967.

867. Miller, D.S., Mumford, P.M. & Stock, M.J. "Gluttony II. thermogenesis in overeating man." **Am. J. Clin. Nutr.**, 20, 1223-1229, 1967.

868. Miller, D.S. & Parsonage, S. "Resistance to slimming. Adaptation or illusion?" **Lancet**, 2 773-775, 1975.

869. Miller, D.S., Stock, M.J. & Stuart, J.A. "The effects of carnitine and caffeine on the oxygen consumption of fed and fasted subjects." **Proc. Nutr. Soc.**, 33, 28A, 1974.

870. Mills, I., Barge, R.M., Silva, J.E., et.al. **Biochem. Biophys. Res. Commun.**, 143, 81-86, 1987.

871. Milner, R.F., Wang, I.C.H. & Trayhurn, P. "Brown fat thermogenesis during hibernation and arousal in Richardon's ground squirrel." **Am. J. Physiol.**, 256, R42-R48, 1989.

872. Minghelli, G., Schutz, Y., Charbonnier, A., et.al. **Am. J. Clin. Nutr.**, 51, 563, 1990.

873. Minghelli, G., Schutz, Y., Whitehead, R., et.al. **Am. J. Clin. Nutr.**, 53, 14, 1991.

874. Minokoshi, Y., Saito, M. & Shimazu, T. **Am. J. Physiol.**, 251, R1005-R1008, 1986.

875. Minton, J.P., Foecking, M.K., Webster, D.J.T., et.al. "Response of fibrocystic disease to caffeine withdrawal and correlation of cyclic nucleotides with breast disease." **Am. J. Obstet. Gynecol.**, 135, 157, 1979.

876. Minton, J.P., Foecking, M.K., Webster, D.J.T., et.al. "Caffeine, cyclic nucleotides and breast disease." **Surgery**, 86, 105, 1979.

877. Mitchell, M.D., Cleland, W.H. & Smith, M.E. **J. Clin. Endocr. Metab.**, 57, 771, 1983.

878. Mitchell, J.R., Carneheim, C.M.H., Jacobsson, A., et.al. In **Obesity in Europe '88**, Bjorntorp, P. & Rossner, S., eds., London, Libbey, in press, 1989.

879. Modan, M., Halkin, H., Almog, S., et.al. "Hyperinsuliemia: a link between hypertension, obesity and glucose intolerance." **J. Clin. Inv.**, 75, 809, 1985.

880. Mohell, N. **Acta Physiol. Scand.**, (suppl), 530, 1984.

881. Mohell, N., Connolly, E. & Nedergaard, J. **Am. J. Physiol.**, 253, C301-C308, 1987.

882. Mohell, N., Wallace, M. & Fain, J.N. **Mol. Pharm.**, 25, 64-69, 1984.

883. Molnar, D. "Effects of ephedrine and aminophylline on resting energy expenditure in obese adolescents." **Int. J. Obes.**, 17(suppl 1), S49-S52, 1993.

884. Molnar, J. "Die pharmakologischen Wirkungen des Capsaicins, des scharf schmeckenden Wirkstoffes im Paprika." **Arzneimittel-Forschung**, 15, 718-727, 1965.

885. Molnar, J. "Effect of capsaicin on the cat's nictitating membrane." **Acta Physiol. Hung.**, 30, 183-192, 1966.

886. Molnar, J. & Gyorgy, L. "Pulmonary hypertensive and other hacmodynamic effects of capsaicin in the cat." **Eur. J Pharm.**, 1, 86-92, 1967.

887. Monsereenusorn, Y. "The effects of capsaicin on intestinal sodium and fluid transport." **J. Pharmacobiodyn.**, 3, 631-635, 1980.

888. Monsereenusorn, Y. "Effect of Capsicum annum on blood glucose level." **Quart. J. of Crude Drug Res.**, 18, 1-7, 1980.

889. Monsereenusorn, Y. "In vitro intestinal absorption of capsaicin." **Tox. & Appl. Pharm.**, 53, 134-139, 1980.

890. Monsereenusorn, Y. "Subchronic toxicity studies of capsaicin and capsicum in rats." **Res. Comm. Chem. Path. Pharm.**, 41(1), 95-110, 1983.

891. Monsereenusorn, Y. "Thermo-pharmacology of capsaicin." **Thai J. Pharm.**, 3, 159-164, 1981.

892. Monsereenusorn, Y. & Glinsukon, T. "Inhibitory effect of capsaicin on intestinal glucose absorption in vitro." **Food & Cosmetic Tox.**, 16, 469-473, 1978.

893. Monsereenusorn, Y., Kongsamut, S. & Pezalla, P.D. "Capsaicin--a literature survey." **CRC Crit. Rev. Tox.**, 10, 231-239, 1982.

894. Monti, M., Edvinsson, L., Rankiev, E., et.al. **Acta Med. Scand.**, 220, 185-188, 1986.

895. Morabia, A., Slosman, D.O. & Pichard, C. "Relationships of weight-for-height indices with DXA measurements of bone, fat and lean masses, by body sites." **Int. J. Obes.**, 17(suppl 1), S79, 1993.

896. Morgan, J.B., York, D.A., Wasilewska, A., et.al. "A study of the thermic responses to a meal and to a sympathomimetic drug (ephedrine) in relation to energy balance in man." **Br. J. Nutr.**, 47, 21-32, 1982.

897. Morley, J.E. "Neuropeptide regulation of appetite and weight." **Endocr. Rev.**, 8, 256-287, 1987.

898. Mory, G., Bouillaud, F., Combes-Georges, M., et.al. "Noradrenaline controls the concentration of the uncoupling protein in brown adipose tissue." **FEBS Lett.**, 166, 393-397, 1984.

899. Mory, G. & Ricquier, D. "The trophic effect of serotonin on the brown adipose tissue of the rat and its mediation by the sympathetic nervous system." **Molecul. Physiol.**, 1, 113, 1981.

900. Mory, G., Ricquier, D. & Hemon, P. "Serotonine et histamine et reponse trophique du tissu adipeux brun." **J. Physiol.**, 75, 83A, 1979.

901. Mory, G., Ricquier, D., Nechad, M., et.al. "Impairment of trophic response of brown fat to cold in guanethidine-treated rats." **Am. J. Physiol.**, 242, C159-C165, 1982.

902. Moss, D., Ma, A. & Cameron, D.P. "Cafeteria feeding promotes diet-induced thermogenesis in monosodium glutamate-treated

mice." **Metab. Clin. Exp.**, 34, 1094-1099, 1985.

903. Mouroux, I., Bertin, R. & Portet, R. "Thermogenic capacity of the brown adipose tissue of developing rats; effects of rearing temperature." **J. Dev. Physiol.**, 14(6), 337-342, 1990.

904. Mroz, E. & Leeman, S.E. In **Methods of Hormone Radioimmunossay**, Jaffe, B.M. & Behrmann, M.R., eds., Academic Press, NY, 121-137, 1979.

905. Murciano, D., Auclair, M.H., Pariente, R., et.al. "A randomized, controlled trial of theophylline in patients with severe chronic obstructive pulmonary disease." **N. Eng. J. Med.**, 320, 1521-1525, 1989.

906. Myers, H.B. **J. Pharm. Exp. Ther.**, 23, 465-477, 1924.

907. Nagy, J.I. "Capsaicin: a chemical probe for sensory neuron mechanisms." **Handbook of Psychopharm.**, 15, 185-235, 1982.

908. Nagy, J.I. "Capsaicin's action on the nervous system." **Trends in Neurosci.**, 5, 362-365, 1982.

909. Nagy, J.I., Vincent, S.R., Staines, W.A., et.al. "Neurotoxic action of capsaicin on spinal substance P neurons." **Brain Res.**, 186, 435, 1980.

910. Naismith, D.J., Akinyanju, P.A., Szanto, S., et.al. **Nutr. Metab.**, 12, 144-151, 1970.

911. Naismith, D.J., Akinyanju, P.A. & Yudkin, J. **J. Nutr.**, 97, 375-381, 1969.

912. Nanberg, E. & Putney, J.,Jr. **FEBS Lett.**, 195, 319-322, 1986.

913. Napolitano, A., Lowell, B.B. & Flier, J.S. "Alterations in sympathetic nervous system activity do not regulate adipsin gene expression in mice." **Int. J. Obes.**, 15(3), 227-235, 1990.

914. Nechad, M. "Differentiation of the brown adipose tissue in the hamster: Evidence of a trophic role of the sympathetic innervation." In Szelenyi, Z. & Szekely, M. eds. **Contributions to Thermal Physiology**, Pergamon Press, Akademaiai Kiado, Budapest, 1981, pp. 503-505.

915. Nechad, M. "Structure and development of brown adipose tissue." In **Brown Adipose Tissue**, Trayhurn, P. & Nicholls, D.G., eds., London, Arnold, 1-30, 1986.

916. Nedergaard, J. & Cannon, B.C. "A possible metabolic effet of membrane depolarizatioin in brown adipose tissue." Szelenyi, Z. & Szekely, M. eds. **Contributions to Thermal Physiology**, Pergamon Press, Akademaiai Kiado, Budapest, 1981, pp. 475-477

917. Nedergaard, J. **Am. J. Physiol.**, 242, C250-C257, 1982.

918. Nedergaard, J., Becker, W. & Cannon, B. "Effects of dietary essential fatty acids on active thermogenin content in rat brown adipose tissue." **J. Nutr.**, 113, 1717, 1983.

919. Nedergaard, J., Jacobsson, A. & Cannon, B. **Hormones, Thermogenesis and Obesity** (Lardy, H. & Stratman, F., eds.), Elsevier, NY, 105-116, 1989.

920. Negulesco, J.A., Noel, S.A., Newman, H.A., et.al. "Effects of pure capsaicinoids (capsaicin and dihydrocapsaicin) on plasma lipid and lipoprotein concentrations of turkey poults." **Atherosclerosis**, 64(2-3), 85-90, 1987.

921. Neims, A.H. & von Borstel, R.W. In **Nutrition and the Brain**, Wurtman, R.J. & Wurtman, J.J., eds., Raven Press, NY, 1-30, 1983.

922. Nelson, H.S. "The effect of ephedrine on the response to epinephrine in normal men." **J. Allergy Clin. Immunol.**, 51, 191-198, 1973.

923. Nelson, H.S. Black, J.W., Branch, L.B., et.al. "Subsensitivity to epinephrine following the administration of epinephrine and ephedrine to normal individuals." **J. Allergy Clin. Immunol.**, 55, 299-309, 1975.

924. Neumann, R.O. **Arch. Hyg.**, 45, 1, 1902.

925. Newsholme, E.A. **Hormones, Thermogenesis and Obesity** (Lardy, H. & Stratman, F., eds.), Elsevier, NY, 47-58, 1989.

926. Newsholme, E.A. & Crabtree, B. "Substrate cycles in metabolic regulation and in heat generation." **Biochem. Soc. Symp.**, 41, 61-109, 1976.

927. Nicholls, D.G. "Cellular mechanisms in brown fat thermogenesis mitochondria." **Experientia**, 33, 1130-1131, 1977.

928. Nicholls, D.G. "Brown adipose tissue mitochondria." **Biochim. Biophys. Acata**, 549, 1-29, 1979.

929. Nicholls, D.G., Cunningham, S.A. & Rial, E. "The bioenergetic mechanisms of brown adipose tissue mitochondria." In **Brown Adipose Tissue**, Trayhurn, P. & Nicholls, D.G., eds., London, Arnold, 52-85, 1986.

930. Nicholls, D. & Locke, R. "Cellular mechanisms of heat dissipation." In **Mammalian Thermogenesis**, Girardier, L. & Stock, M.J., eds., NY, Chapman & Hall, 8-49, 1983.

931. Nicholls, D.G. & Locke, R.M. "Thermogenic mechanisms in brown fat." **Physiol. Rev.**, 64, 1-64, 1984.

932. Nieforth, K.A. & Cohen, M.L. In **Principles of Medicinal Chemistry**, Foye, W.O., ed., Lea & Febiger, Philadelphia, 282-285, 1974.

933. Nikaido, T., Ohmoto, T., Kuge, T., et.al. "The study on Chinese herbal medicinal prescription with enzyme inhibitory activity. III. The study of mao-to with adenosine 3',5'-cyclic monophosphate phosphodiesterase." **Yakugaku Zasshi**, 110(7), 504-508, 1990.

934. Nishikawa, T., Bruyere, H.J.Jr., Gilbert, E.F., et.al. "Potentiating effects of caffeine on the cardiovascular teratogenicity of ephedrine in chick embryos." **Tox. Lett.**, 29(1), 65-68, 1985.

935. Nishino, N. & Fujiwara, M. "Effect of capsaicin on the quinea-pig isolated atrium." **J. Pharm. Pharmacol.**, 21, 622-624, 1969.

936. Noble, R. "A controlled clinical trial of the cardiovascular and psychological effects of phenylpropanolamine and caffeine." **Drug Intell. Clin. Pharm.**, 22(4), 296-299, 1988.

937. Norman, D., Mukherjee, S., Symons, D., et.al. **J. Neurocytol.**, 17, 305-311, 1988.

938. Nopanitaya, W. "Long term effects of capsaicin on fat absorption and the growth of the rat." **Growth**, 37, 269-279, 1973.

939. Norgan, N.G., Durnin, J.V.G.A. **Am. J. Clin. Nutr.**, 33, 978, 1980.

940. Noronha, M., Raasmaja, A., Moolten, N., et.al. "Triiodothyronine causes rapid reversal of alpha 1/cyclic adenosine monophosphate synergism on brown adipocyte respiration and type II deiodinase activity." **Metabolism**, 40(12), 1327-1332, 1992.

941. Northam, W.J. & Jones, D.J. "Comparison of capsaicin and substance P induced cyclic AMP accumulation in spinal cord tissue slices." **Life Sci.**, 35, 293, 1984.

942. NTP Working Group "Toxicology and carcinogenesis studies of ephedrine sulfate in F344/N rats and B6C3F1 mice (feed studies)." **Natnl. Tox. Prog. Techn. Report Series**, 307, 1986.

943. Oberman, Z., Harell, A., Herzberg, M., et.al. "Changes in plasma cortisol, glucose and free fatty acids after caffeine ingestion in obese women." **Isr. J. Med. Sci.**, 11, 33-36, 1975.

944. Oberman, Z., Herzberg, M., Jaskolka, H., et.al. **Isr. J. Med. Sci.**, 11, 33-36, 1975.

945. Obregon, M.J., Mills, I., Silva, J.E., et.al. **Endocrinology**, 120, 1069, 1072, 1987.

946. O'Dea, K., Esler, M., Leonard, P., et.al. "Noradrenaline turnover during under- and over-eating in normal weight subjects." **Metabolism**, 31, 896-899, 1982.

947. Ohisalo, J.J., Ranta, S. & Huhtaniemi, T. **Metabolism**, 35, 143, 1986.

948. Oksbjerg, N. Ph.D. Thesis, Copenhagen, 1992.
949. Oliphant, L.W. "First observations of brown fat in birds." **Condor**, 88, 350-354, 1983.
950. Olivecrona, T. & Bengtsson, G. "Lipoprotein lipase." In **The Adipocyte and Obesity: Cellular and Molecular Mechanisms**, Angel, A., Hollenberg, C.H. & Roncari, D.A., eds., Raven Press, New York, 1983, pp. 117-126.
951. Oppenheimer, J.H., Schwartz, H.L., Lane, J.T., et.al. "Functional relationship of thyroid hormone-induced lipogenesis, lipolysis, and thermogenesis in the rat." **J. Clin. Inv.**, 87(1), 125-132, 1991.
952. Oster, P., Arab, L., Schellenberg, B., et.al. "Blood pressure and adipose tissue linoleic acid." **Res. Exp. Med.**, 175, 287, 1979.
953. Ostman, J. & Efendic, S., **Acta Med. Scand.**, 187, 471-476, 1970.
954. Owen, O.E., Kavle, E., Owen, R.S., et.al. **Am. J. Clin. Nutr.**, 44, 1-19, 1986.
955. Pachomov, N. "The effects of posterior and anterior hypothalamic lesion on the maintenance of body temperature in the rat." **J. Neuropath. Exp. Neurol.**, 21, 450-460, 1962.
956. Paffenbarger, R., Wing, A.L. & Hyde, R.T. "Chronic disease in former college students. XIII. Early precursors of peptic ulcer." **Am. J. Epidemiol.**, 100, 307, 1974.
957. Palmblad, J., Levi, L., Burger, A., et.al. "Effects of total energy withdrawal (fasting) on the levels of growth hormone, thyrotropin, cortisol, adrenaline, noradrenaline, T4, T3, and rT3 in healthy males." **Acta Med. Scand.**, 201, 15-22, 1977.
958. Pantelis, C., Hindler, C.G. & Taylor, J.C. "Use and abuse of khat (Catha edulis): a review of the distribution, pharmacology, side effects and a description of psychoses attributed to khat chewing." **Psychol. Med.**, 19(3), 657-668, 1989.
959. Pardoe, K.E., Gorecki, D.K.J. & Jones, D. "Ephedrine alkaloid patterns in herbal products based on ma huang (ephedra sinica)." **Int. J. Obes.**, 17(suppl 1), 1993.
960. Parizkova, J. & Roth, Z. **Hum. Bio.**, 44, 613, 1972.
961. Park, I.R.A. & Himms-Hagen, J. **Am. J. Physiol.**, 255, R874-R881, 1988.
962. Park, I.R.A. & Himms-Hagen, J. **FASEB J.**, 2, A1612, 1988.
963. Park, I.R.A., Himms-Hagen, J. & Coscina, D.V. "Lateral and medial hypothalamic lesions do not acutely affect brown adipose tissue." **Brain Res. Bull.**, 21, 805-811, 1988.
964. Park, I.R.A., Mount, D.B. & Himms-Hagen, J. **Am. J. Physiol.**, in press, 1989.
965. Parsons, W.D., Aranda, J.V. & Neims, A.H. **"Pediatr. Res.**, 10, 333, 1976.
966. Parsons, W.D. & Neims, A.H. "Effect of smoking on caffeine clearance." **Clin. Pharm. Ther.**, 24, 40, 1978.
967. Parsons, W.J., Ramkumar, V. & Stiles, G. **Am. Soc. Pharm. Exp. Ther.**, 33, 441, 1988.
968. Pasquali, R., Baraldi, G., Cesari, M.P., et.al. "A controlled trial using ephedrine in the treatment of obesity." **Int. J. Obes.**, 9(2), 93-98, 1985.
969. Pasquali, R. & Casimirri, F. "Clinical aspects of ephedrine in the treatment of obesity." **Int. J. Obes.**, 17(suppl 1), S65-S68, 1993.
970. Pasquali, R., Casimirri, F., Melchionda, N., et.al. "Chronic beta-receptor stimulation prevents nitrogen loss during semistarvation in obese subjects." **Int. J. Obes.**, 13(suppl 1 abstr 152), 1989.
971. Pasquali, R., Casimirri, F., Melchionda, N., et.al. **Clin. Sci.**, 82, 85-92, 1992.
972. Pasquali, R., Cesari, M.P., Besteghi, L., et.al. **Int. J. Obes.**, 11(suppl 3), 23, 1987.
973. Pasquali, R., Cesari, M.P., Melchionda, N., et.al. **Int. J. Obes.**, 9, 93, 1985.
974. Pasquali, R., Cesari, M.P., Melchionda, N., et.al. "Does ephedrine promote weight loss in low-energy-adapted obese women?" **Int. J. Obes.**, 11, 163-168, 1987.
975. Paterson, G. "The response to transmural stimulation of isolated arterial strips and its modification by drugs." **J. Pharm. Pharmacol.**, 17, 341-349, 1965.
976. Patil, P.N., Tye, A. & Lapidus, J.B. "A pharmacological study of the ephedrine isomers." **J. Pharm. Exp. Ther.**, 148, 158-168, 1965.
977. Patil, T.N. & Srinivasan, M. **Ind. J. Exp. Biol.**, 9, 167, 1971.
978. Paton, W.D.M. & Zaimis, E.J. **Pharm. Rev.**, 4, 219, 1952.
979. Patwardhan, R.V., Desmond, P.V., Johnson, R.F., et.al. **Clin. Pharm. Ther.**, 28, 398-403, 1980.
980. Pauling, L. **How to Live Longer and Feel Better**, NY, W.H. Freeman & Co., 1986.
981. Payne, P.R. & Dugdale, A.E. "A model for the prediction of energy balance and body weight." **Ann. Hum. Biol.**, 4(6), 525-535, 1977.
982. Peachey, T., French, R.R. & York, D.A. "Regulation of GDP binding and uncoupling-protein concentration in brown-adipose-tissue mitochondria." The effects of cold-acclimation, warm re-acclimation and noradrenaline." **Biochem. J.**, 249, 451-457, 1988.
983. Peckham, C.S., Stark, O., Simonite, V., et.al **Br. Med. J.**, 286, 1237, 1983.
984. Penicaud, L., Ferre, P., Terretaz, J., et.al. **Diabetes**, 36, 626-631, 1987.
985. Pentel, P. "Toxicity of over-the-counter stimulants." **JAMA**, 252(14), 1898-1903, 1984.
986. Perkins, R. & Williams, M.H. "Effect of caffeine upon maximal muscular endurance of females." **Med. Sci. Sports**, 7, 221-224, 1975.
987. Persson, C.G.A. **Thorax**, 40, 881-886, 1985.
988. Peterson, H.R., Rothschild, M., Weinberg, C.R., et.al. **New Eng. J. Med.**, 318, 1077-1083, 1988.
989. Phillipson, B.F., Rothrock, D.W. & Connor, W.E. "Reduction of plasma lipids, lipoproteins, and apoproteins by dietary fish oils in patients with hypertriglyceridemia." **N. Eng. J. Med.**, 312, 1210, 1985.
990. Pichard, C. "Protein calorie malnutrition." Privat Docent Theses, Hopital Cantonal Universitaire, Geneva, Switzerland, 1991.
991. Pierau, F. & Wurster, R.D. "Primary afferent input from cutaneous thermoreceptors." **Fed. Proc.**, 40, 2819-2824, 1981.
992. Pi-Sunyer, F.X. "Obesity: determinants and therapeutic initiatives." **Nutrition**, 7(4), 292-294, 1991.
993. Pittet, P., Chappuis, P., Acheson, K., et.al. "Thermic effect of glucose in obese subjects, studied by direct and indirect calorimetry." **Br. J. Nutr.**, 35, 281-292, 1976.
994. Poehlman, E.T., Despres, J.P., Bessette, H., et.al. **Med. Sci. Sports Exerc.**, 17, 689-694, 1985.
995. Porszasz, J. "Pharmaco-physiology of the vasomotor and respiratory reflex mechanishms." Dissertation, Szeged, 1966.
996. Porszasz, J., Gyorgy, L. & Porszasz-Gibiszer, K. "Cardiovascular and respiratory effects of capsaicin." **Acta Physiol. Acad. Sci. Hung.**, 8, 61-76, 1955.
997. Porszasz, J., Such, G. & Porszasz-Gibiszer, K. "Circulatory and respiratory chemoreflexes. I. Analysis of the site of action and receptor types of capsaicin." **Acta Physiol. Acad. Sci. Hung.**, 12, 189-205, 1957.
998. Porta, M., Jick, H. & Habakanga, J.K. **Ann. Allergy**, 57, 340, 1986.
999. Poulos, D.A. "Central processing of cutaneous temperature information." **Fed. Proc.**, 40, 2825-2829, 1981.

1000. Powers, S.K. & Dodd, S. "Caffeine and endurance performance." **Sports Med.**, 2, 165-174, 1985.
1001. Prentice, A.M., Davies, H.L., Black, A.E., et.al. "Unexpectedly low levels of energy expenditure in healthy women." **Lancet**, 1, 1419-1422, 1985.
1002. Preston, E., Triandafillou, J. & Haas, N. "Increased thermogenic capacity during tumor growth in the rat." **Nutr. Res.**, 8, 327-332, 1988.
1003. Prokop, H. **Nervenarzt**, 39, 71, 1968.
1004. Prusiner, S., Cannon, B. & Lindber, O. "Mechanisms controlling oxidative metabolism in brown adipose tissue." In **Brown Adipose Tissue**, Lindberg, O., ed., NY, Elsevier, 1970, pp. 283-318.
1005. Prusiner, S. & Poe, M. "Thermodynamic considerations of mammalian heat production." In **Brown Adipose Tissue**, Lindberg, O., ed., NY, Elsevier, 263-282, 1970.
1006. Raasmaja, A., Mohell, N. & Nedergaard, J. **Eur. J. Pharm.**, 106, 489-498, 1984.
1007. Raasmaja, A., & York, D.A. **Biochem. J.**, 249, 831-838, 1988.
1008. Radomski, M.W. & Boutlier, C. "Hormone response of normal and intermittent cold-preadapted humans to continuous cold." **J. Appl. Physiol.**, 53, 610-616, 1982.
1009. Rafael, J., Fesser, W. & Nicholls, D.G. **Am. J. Physiol.**, 250, C228-C235, 1986.
1010. Rall, T.W. In **The Pharmacological Basis of Therapeutics**, Gilman, A.G., Rall, T.W., Nies, A.S., et.al., eds., Pergamon Press, NY, 8th ed., 618, 1990.
1011. Rall, T.W. & Sutherland, E.W. **J. Biol. Chem.**, 232, 1065-1076, 1958.
1012. Rall, T.W. & West, T.C. "The potentiation of cardiac inotropic response to norepinephrine by theophylline." **Pharm. Exp. Ther.**, 139, 269-274, 1963.
1013. Ramadoss, C.S., Uyeda, K. & Johnson J.M. "Studies on the free fatty acid inactivation of phosphofructokinase." **J. Biol. Chem.**, 251, 98-107, 1976.
1014. Rampone, A.J. & Reynolds, P.J. "Food intake regulation by   diet-induced thermogenesis." **Med. Hypotheses**, 34(1), 7-12, 1991.
1015. Randle P.H., Newsholme, E.A. & Garland, P.B. "Regulation of glucose uptake by muscle 8. Effects of fatty acids, ketone bodies and pyruvate, and of alloxan diabetes and starvation, on the uptake and metabolic fate of glucose in the rat heart and diaphragm muscles." **Biochem. J.**, 93, 652-665, 1964.
1017. Rapoport, J.L., Jensvold, M., Elkins, R., et.al. **J. Nerv. Ment. Dis.**, 169, 726-732, 1981.
1018. Rappaport, E.B., Young, J.B., & Landsberg, L. "Effects of 2-deoxy-D-glucose on the cardiac sympathetic nerves and the adrenal medulla in the rat: further evidence for a dissociation of sympathetic nervous system and adrenal medullary responses." **Endocrinology**, 110, 650-656, 1982.
1019. Ratzmann, K.P., Riemer, D., Mannchen, E., et.al. **Endokrinologie**, 68, 319-326, 1976.
1020. Raven, P.B., Niki, I., Dahms, T.E., et.al. "Compensatory cardiovascular responses during an environmental cold stress, 5 degrees C." **J. Appl. Physiol.**, 29, 417-421, 1970.
1021. Ravussin, E., Acheson, K.J., Vernet, O., et.al. "Evidence that insulin resistance is responsible for the decreased thermic effect of glucose in human obesity." **J. Clin. Inv.**, 76, 1268-1273, 1985.
1022. Ravussin, E., Bogardus, C., Schwartz, R.S., et.al. "Thermic effect of infused glucose and insulin in man. Decreased response with increased insulin resistance in obesity and non insulin-dependent diabetes mellitus." **J. Clin. Inv.**, 72, 893-902, 1983.
1023. Ravussin, E., Burnand B., Schutz, Y., et.al. "Energy expenditure before and during energy restriction patients." **Am. J. Clin. Nutr.**, 41, 753-759, 1985.
1024. Ravussin, E., Lillioja, S., Knowler, W.C., et.al. **N. Eng. J. Med.**, 318, 467-472, 1988.
1025. Ravussin, E., Schutz, Y. & Acheson, J. "Short-term, mixed-diet overfeeding in man: no evidence for 'luxusconsumption'." **Am. J. Physiol.**, 249, E470-E477, 1985.
1026. Reckless, J.P.D., Gilbert, C.H. & Galton, D.J. "Alpha adrenergic receptor activity, cyclic AMP and lipolysis in adipose tissue of hypothyroid man and rat." **J. Endocr.**, 68, 419-430, 1976.
1027. Reed, C.E., Sims, J.H. & doPico, G.A. "Bronchodilator effect of ephedrine and theophylline after two to four weeks of repeated dosing." **Chest**, 73, 1019-1020, 1978.
1028. Reed, N. & Fain, J.N. "Hormonal regulation of the metabolism of free brown fat cells." In **Brown Adipose Tissue**, Lindberg, O., ed., NY, Elsevier, 1970, pp. 207-224.
1029. Reichard, C.C. & Elder, S.T. **Am. J. Psychiat.**, 134, 144-148, 1977.
1030. Reimann, H.A. **JAMA**, 202, 131-132, 1967.
1031. Reimann, H.A. **JAMA**, 202, 1105-1106, 1967.
1032. Reisin, E., Abel, R., Modan, M., et.al. **New Eng. J. Med.**, 298, 1-6, 1978.
1033. Rennie, M.J., Winder, W.W. & Holloszy, J.O. "A sparing effect of plasma free fatty acids on muscle liver glycogen content in exercising rats." **Biochem. J.**, 156, 647-655, 1976.
1034. Revelli, J.P., Muzzin, P., Paoloni, A., et.al. "Expression of the beta3-adrenergic receptor in human white adipose tissue." **Int. J. Obes.**, 17(suppl 1), S83, 1993.
1035. Reynolds, T.B., Paton, A., Freeman, M., et.al. **J. Clin. Inv.**, 32, 793, 1953.
1036. Richard, D., Arnold, J. & LeBlanc, J. **J. Appl. Physiol.**, 60, 1054-1059, 1986.
1037. Richard, D., Lachance, P. & Deshaies, Y. "Effects of exercise-rest cycles on energy balance in rats." **Am. J. Physiol.**, 256(4 pt 2), R886-891, 1989.
1038. Richard, D. & Rivest, S. "The role of exercise on thermogenesis and energy balance." **Can. J. Physiol. Pharm.**, 67, 1989.
1039. Richelsen, B. **Biochem. J.**, 247, 389, 1987.
1040. Richelsen, B. **Metabolism**, 37, 268, 1988.
1041. Richelsen, B., Eriksson, E.F. & Beck-Nilsen, H. **J. Clin. Endocr. Metab.**, 59, 7, 1984.
1042. Richelsen, B., Pedersen, O. & Sorensen, N. **J. Clin. Endocr. Metab.**, 62, 258, 1986.
1043. Richelsen, B., Pedersen, S.B., Moller-Pedersen, T., et.al. **Metabolism**, 40, 990, 1991.
1044. Richter, E.A., Christensen, N.J., Ploug, T., et.al. "Endurance training augments the stimulatory effect of epinephrine on oxygen consumption in perfused skeletal muscle." **Acta Physiol. Scand.**, 120, 613-615, 1984.
1045. Ricquier, D. "Brown adipose tissue: physiologic effector of thermogenesis." **Journ. Annu. Diabetol. Hotel Dieu.**, 27-31, 1990.
1046. Ricquier, D., Barlet, J.P., Garel, J.M., et.al. "An immunological study of the uncoupling protein of brown adipose tissue mitochondria." **Biochem. J.**, 210, 859-866, 1983.
1047. Ricquier, D. & Bouillaud, F. In **Brown Adipose Tissue**, Trayhurn, P. & Nicholls, D.G., eds., London, Arnold, 86-104, 1986.
1048. Ricquier, D. & Mory, G. "Factors affecting brown adipose tissue activity in animals and man." In **Clinics in Endocrinology and**

       Metabolism: Obesity, James, W.P.T., ed., Philadelphia, W.B. Saunders, 501-520, 1984.

1049. Ricquier, D., Nechad, M. & Mory, G. "Ultrastructural and biochemical characterization of human brown adipose tissue in pheochromocytoma." **J. Clin. Endocr. Metab.**, 54, 803-807, 1982.

1050. Ridley, R.G., Patel, H.V., Parfett, C.I.J., et.al. "Immunological detection of cDNA clones encoding the uncoupling protein of brown adipose tissue: evidence for an antigenic determinant within the C-terminal eleven amino acids." **Biosci. Rep.**, 6, 87-94, 1986.

1051. Ritchie, J.M. "Central nervous system stimulants (cont.), the xanthines." In **The Pharmacological Basis of Therapeutics**, Goodman, L.S. & Gilman, A., eds., NY, Macmillan, 1975.

1052. Roberts, S.B., Savage, S., et.al. **N. Eng. J. Med.**, 318, 461-466, 1988.

1053. Robertson, D., Froelich, J.C., Carr, R.K., et.al. "Effects of caffeine on plasma renin activity, catecholamines and blood pressure." **N. Eng. J. Med.**, 298, 181-186, 1978.

1054. Robertson, D., Wade, D., Workman, R., et.al. **J. Clin. Inv.**, 67, 1111-1117, 1981.

1056. Robinson, D.S., Parkin, S.M, Speake, B.K. & Little, J.A. "Hormonal control of rat adipose tissue lipoprotein lipase activity." In **The Adipocyte and Obesity: Cellular and Molecular Mechanisms**, Angel, A., Hollenberg, C.H. & Roncari, D.A., eds., Raven Press, New York, 1983, pp. 127-148.

1057. Romsos, D.R., Vander Tuig, J.G., Kerner, J., et.al. "Energy balance in rats with obesity-producing hypothalamic knife cuts: effects of adrenalectomy." **J. Nutr.**, 117, 1121-1128, 1987.

1058. Rosenberg, L., Mitchell, A.A., Shapiro, S., et.al. "Selected birth defects in relation to caffeine-containing beverages" **JAMA**, 247, 1429, 1982.

1059. Rosenbloom, D. & Sutton, J.R. "Drugs and exercise." **Med. Clin. N. Am.**, 69(1), 177-187, 1985.

1060. Rothwell, N.J. "Central activation of thermogenesis by prostaglandins: dependence on CRF." **Horm. Metab. Res.**, 22(12), 616-618, 1990.

1061. Rothwell, N.J. "Central effects of CRF on metabolism and energy balance." **Neurosci. Biobehav. Rev.**, 14(3), 263-271, 1990.

1062. Rothwell, N.J. "Neuroendocrine mechanisms in the thermogenic responses to diet, infection, and trauma." **Adv. Exp. Med. Biol.**, 274, 371-380, 1990.

1063. Rothwell, N.J., Hardwick, A., LeFeuvre, R.A., et.al. "Central actions of CRF on thermogenesis are mediated by pro-opiomelanocortin products." **Brain Res.**, 541(1), 89-92, 1991.

1064. Rothwell, N.J., Saville, M.E. & Stock, M.J. "Thermogenic responses to food, noradrenaline and triiodothyronine in lean and obese Zucker rats." **J. Physiol. (London)**, 519, 53P-54P, 1981.

1065. Rothwell, N.J., Saville, M.E. & Stock, M.J. "Sympathetic and thyroid influences on metabolic rate in fed, fasted, and refed rats." **Am. J. Physiol.**, 234, R339-R346, 1982.

1066. Rothwell, N.J. & Stock, M.J. "A role for brown adipose tissue in diet-induced thermogenesis." **Nature (London)**, 281, 31-35, 1979.

1067. Rothwell, N.J. & Stock, M.J. "Similarities between cold- and diet-induced thermogenesis in the rat." **Can. J. Physiol. Pharm.**, 58, 842-848, 1980.

1068. Rothwell, N.J. & Stock, M.J. "Regulation of energy balance." **Ann. Rev. Nutr.**, 1, 235-256, 1981.

1069. Rothwell, N.J. & Stock, M.J. "Influence of noradrenaline on blood flow to brown adipose tissue in rats exhibiting diet-induced thermogenesis." **Pflugers Arch.**, 389, 237-242, 1981.

1070. Rothwell, N.J. & Stock, M.J. "A role for insulin in the diet-induced thermogenesis of cafeteria-fed rats." **Metabolism**, 30, 673, 1981.

1071. Rothwell, N. J. & Stock, M. J. "Effects of cold and hypoxia on diet-induced thermogenesis." In Szelenyi, Z. & Szekely, M. eds. **Contributions to Thermal Physiology**, Pergamon Press, Akademaiai Kiado, Budapest, 1981, pp. 511-513.

1072. Rothwell, N.J. & Stock, M.J. **J. Nutr.**, 112, 426-435, 1982.

1073. Rothwell, N.J. & Stock, M.J. "Diet-induced thermogenesis." In **Mammalian Thermogenesis**, Girardier, L. & Stock, M.J., eds., London. Chapman & Hall, 208-233, 1983.

1074. Rothwell, N.J. & Stock, M.J. "Energy balance, thermogenesis and brown adipose tissue activity in tube-fed rats." **J. Nutr.**, 114, 1965-1970, 1984.

1075. Rothwell, N.J. & Stock, M.J. "Biological distribution and significance of brown adipose tissue." **Comp. Biochem. Physiol.**, 82A, 745-751, 1985.

1076. Rothwell, N.J. & Stock, M.J. "Influence of environmental temperature on energy balance, diet-induced thermogenesis and brown fat activity in 'cafeteria'-fed rats." **Br. J. Nutr.**, 56, 123-129, 1986.

1077. Rothwell, N.J. & Stock, M.J. "Brown adipose tissue and diet-induced thermogenesis." In **Brown Adipose Tissue**, Trayhurn, P. & Nicholls, D.G., eds., London, Arnold, 269-298, 1986.

1078. Rothwell, N.J. & Stock, M.J. **Int. J. Obes.**, 11, 319-324, 1987.

1079. Rothwell, N.J. & Stock, M.J. "Effect of environmental temperature on energy balance and thermogenesis in rats fed normal or low protein diets." **J. Nutr.**, 117, 833-837, 1987.

1080. Rothwell, N.J. & Stock, M.J. "Stimulation of thermogenesis and brown fat activity in rats fed medium chain triglyceride." **Metabolism**, 36, 128, 1987.

1081. Rothwell, N.J. & Stock, M.J. "Influence of adrenalectomy on age-related changes in energy balance, thermogenesis and brown fat activity in the rat." **Comp. Biochem. Physiol. Acta**, 89, 265-269, 1988.

1082. Rothwell, N.J. & Stock, M.J. **Diabetes/Metab. Rev.**, 4, 595-601, 1988.

1083. Rothwell, N.J. & Stock, M.J. "Insulin and thermogenesis." **Int. J. Obes.**, 12, 93-102, 1988.

1084. Rothwell, N.J., Stock, M.J. & Thexton, A.J. **J. Physiol. (London)**, 342, 15-21, 1983.

1085. Rothwell, N.M., Stock, M.J. & Tyzbir, R.S. "Mechanisms of thermogenesis induced by low protein diets." **Metabolism**, 32, 257-261, 1983.

1086. Rothwell, N.J., Stock, M.J. & Warwick, B.P. **Int. J. Obes.**, 7, 263-270, 1983.

1087. Rothwell, N.J., Stock, M.J. & Warwick, B.P. **Metabolism**, 34, 474-480, 1985.

1088. Rowe, J.W., Young, J.B., Minaker, K.L., et.al. "Effect of insulin and glucose infusions on sympathetic nervous system activity in normal man." **Diabetes**, 30, 219-225, 1981.

1089. Rubenstein, A.H., Levin, N.W. & Elliott, G.A. "Hypoglycemia induced by manganese." **Nature**, 194, 188, 1962.

1090. Saggerson, E.D., McAllister, T.W.J. & Baht, H.S. **Biochem. J.**, 251, 701-709, 1988.

1091. Sahakian, B.J., Trayhurn, P., Wallace, M., et.al. "Increased weight gain and reduced activity in brown adipose tissue produced by depletion of hypothalamic noradrenaline." **Neurosci. Lett.**, 39, 321-326, 1983.

1092. Saito, M., Minokoshi, Y. & Shimazu, T. **Life Sci.**, 41, 193-197, 1987.
1093. Sakaguchi, T. & Bray, G.A. **Brain Res. Bull.**, 18, 591-595, 1987.
1094. Sakaguchi, T. & Bray, G.A. **Brain Res. Bull.**, 21, 25-29, 1988.
1094a. Sakaguchi, T., Takashi, M. & Bray, G.A. "Diurnal changes in sympathetic activity; relation of food intake and to insulin injected in the ventromedial or suprachiasmatic nucleus." **J. Clin. Invest.**, 32, 282-286, 1988.
1095. Salimath, B.P. & Satyanarayana, M.N. "Inhibition of calcium and calmodulin-dependent phosphodiesterase activity in rats by capsaicin." **Biochem. Biophys. Res. Commun.**, 148(1), 292-299, 1987.
1096. Salmon, D.M.W. & Flatt, J.P. "Effect of dietary fat content on the incidence of obesity among ad libitum fed mice." **Int. J. Obes.**, 9, 443-449, 1985.
1097. Sambaiah, K. & Satyanarayana, M.N. "Hypocholesterolemic effect of red pepper & capsaicin." **Ind. J. Exp. Bio.**, 18, 898-899, 1980.
1098. Sambaiah, K. & Satyanarayana, M.N. "Influence of red pepper and capsaicin on body composition and lipogenesis in rats." **J. Biosci.**, 4, 425-430, 1982.
1099. Sambaiah, K. & Satyanarayana, M.N. "Lipotrope-like activity of red pepper." **J. Food Sci. Technol.**, 19, 30-31, 1982.
1100. Sambaiah, K., Satyanarayana, M.N. & Rao, M.V.L. "Effect of red pepper (chillies) and capsaicin on fat absorption and liver fat in rats." **Nutr. Reps. Int.**, 18(5), 521-529, 1978.
1101. Santiago, J.V., Clarke, W.L., Shah, S.D. "Epinephrine, norepinephrine, glucagon, and growth hormone release in association with physiological decrements in the plasma glucose concentration in normal and diabetic man." **J. Clin. Endocr. Metab.**, 51, 877-883, 1980.
1102. Saria, A., Lembeck, F. & Skofitsch, G. **J. Chromatogr.**, 208, 41, 1981.
1103. Saria, A., Skofitsch, G. & Lembeck **J. Pharm. Pharmacol.**, 34, 273, 1982.
1104. Saris, W.H.M. "The role of exercise in the dietary treatment of obesity." **Int. J. Obes.**, 17(suppl 1), S17-S21, 1993.
1105. Sakaguchi, T., Arase, K., Fesler, J.S., & Bray, G.A. "Effect of starvation and food intake on sympathetic activity." **Am. J. Physiol.**, 255, R284-R288, 1988.
1106. Sawyer, D.A., Julia, H.L. & Turin, A.C. 1982. "Caffeine and human behavior: arousal, anxiety, and performance effects." **J. Behav. Med.**, 5, 415, 1982.
1107. Sawyer, D.R., Conner, C.S. & Rumack, B.H. "Managing acute toxicity from nonprescription stimulants." **Clin. Pharm.**, 1(6), 529-533, 1982.
1108. Sbarbati, A., Baldassarri, A.M., Zancanaro, C., et.al. "In vivo morphometry and functional morphology of brown adipose tissue by magnetic resonance imaging." **Anat. Rec.**, 231(3), 293-297, 1991.
1109. Scalfi, L., Coltorti, A. & Contaldo, F. "Postprandial thermogenesis in lean and obese subjects after meals supplemented with medium-chain and long-chain triglycerides." **Am. J. Clin. Nutr.**, 53(5), 1130-1133, 1991.
1110. Scarpace, P.J., Baresi, L.A. & Morley, J.E. **Am. J. Physiol.**, 253, E629-E635, 1987.
1111. Scarpace, P.J., Matheny, M. & Borst, S.E. "Thermogenesis and mitochondrial GDP binding with age in response to the novel agonist CGP-12177A." **Am. J. Physiol.**, 262(2 pt 1), E185-190, 1992.
1112. Scatchard, G. "The attractions of proteins for small molecules and ions." **Ann. NY Acad. Sci.**, 51, 660-672, 1949.
1113. Schachtel, B.P., Fillingim, J.M., Lane, A.C., et.al. **Arch. Int. Med.**, 151, 733, 1991.
1114. Schairer, C., Brinton, L.A. & Hoover, R.N. "Methylxanthine and benign breast disease." **Am. J. Epidemiol.**, 124, 603, 1986.
1115. Scheidegger, K., O'Connell, M., Robbins, D.C., et.al. "Effects of chronic beta-receptor stimulation on sympathetic nervous system activity, energy expenditure, and thyroid hormones." **J. Clin. Endocr. Metab.**, 58, 895-903, 1984.
1116. Schimmel, R.J., Dzierzanowski, D., Elliott, M.E., et.al. **Biochem. J.**, 236, 757-764, 1986.
1117. Schimmel, R.J., Elliott, M.E. & Dehmel, V.C. **Mol. Pharm.**, 32, 26-33, 1987.
1118. Schimmel, R.J., Elliott, M.E. & McCarthy, L. In **Topics and Perspectives in Adenosine Research**, Gerlach, E. & Becker, B.F., eds., Springer-Verlag, Berlin Heidelberg, 261-274, 1987.
1119. Schimmel, R.J. & McCarthy, L. **Am. J. Physiol.**, 246, C301-C307, 1984.
1120. Schimmel, R.J. & McCarthy, L. **Am. J. Physiol.**, 250, C738-C743, 1986.
1121. Schimmel, R.J., McCarthy, L. & McMahon, K.K. **Am. J. Physiol.**, 244, C362-C368, 1983.
1122. Schmidt, A., Diamant, B., Bundgaard, A., et.al. "Ergogenic effect of inhaled beta 2-agonists in asthmatics." **Int. J. Sports Med.**, 9(5), 338-340, 1988.
1123. Schnackenberg, R.C. **Am. J. Psychiatr.**, 130, 796-798, 1973.
1124. Schneeberger, D., Tappy, L., Temler, E., et.al. "Effects of muscarinic blockade on insulin secretion and on glucose-induced thermogenesis in lean and obese human subjects." **Eur. J. Clin. Inv.**, 21(6), 608-615, 1991.
1125. Schneider-Picard, G., Carpentier, J.L. & Girardier, L. **J. Memb. Biol.**, 78, 85-89, 1984.
1126. Schoenbaum, E., Johnson, G.E., Sellers, E.A., et.al. "Adrenergic beta-receptors and non-shivering thermogenesis." **Nature**, 210, 426, 1966.
1127. Schoenbaum, E. Steiner, G. & Sellers, E.A. "Brown adipose tissue and norepinephrine." In **Brown Adipose Tissue**, Lindberg, O., ed., NY, Elsevier, 1970, pp. 179-196.
1128. Schutz, Y., Acheson, K., Jequier, E. **Int. J. Obes.**, 9(suppl 2), 111, 1985.
1129. Schutz, Y., Bessard, T. & Jequier, E. "Diet induced thermogenesis measured over a whole day in obese and nonobese women." **Am. J. Clin. Nutr.**, 40, 542-552, 1984.
1130. Schutz, Y., Golay, A., Felber, J.P., et.al. "Decreased glucose-induced-thermogenesis after weight loss in obese subjects: a predisposing factor for relapse of obesity?" **Am. J. Clin. Nutr.**, 39, 380-387, 1984.
1131. Schutz, Y., Ravussin, E., Diethelm, R., et.al. "Spontaneous physical activity measured by radar in obese and control subjects in a respiration chamber." **Int. J. Obes.**, 6, 23, 1982.
1132. Schwartz, J.H., Young, J.B. & Landsberg, L. "Effect of dietary fat on sympathetic nervous system activity in the rat." **J. Clin. Inv.**, 72, 361-370, 1983.
1133. Schwartz, R.S., Halter, J.B. & Bierman, E. "Reduced thermic effect of feeding in obesity: role of norepinephrine." **Metabolism**, 32, 114-117, 1983.
1134. Schwartz, R.S., Ravussin, E., Massari, M., et.al. "The thermic effect of carbohydrate versus fat feeding in man." **Metabolism**, 34, 285, 1985.
1135. Schwarz, J.M., Schutz, Y., Piolino, V., et.al. "Thermogenesis in obese women: effect of fructose vs. glucose added to a meal." **Am. J. Physiol.**, 262(4 pt 1), E394-401, 1992.
1136. Schwertz, M.T. & Marbach, G. **Arch. Sci. Physiol.**, 19, 425-479, 1965.

1137. Scotellaro, P.A., Ji, L.L, Gorski, J., et.al. "Body fat accretion: a rat model." **Med. Sci. Sports Exerc.**, 23(3), 275-279, 1991.
1138. Segal, K.R., Albu, J., Chun, A., et.al. "Independent effects of obesity and insulin resistance on postprandial thermogenesis in men." **J. Clin. Inv.**, 89(3), 824-833, 1992.
1139. Segal, K.R. & Gutin, B. "Thermic effects of food and exercise in lean and obese women." **Metabolism**, 32, 581-589, 1983.
1140. Segal, K.R, Gutin, B., Albu, J., et.al. "Thermic effects of food and exercise in lean and obese men of similar lean body mass." **Am. J. Physiol.**, 252, E110, 1987.
1141. Segal, K.R. Gutin, B., Nyman, A.M., et.al. "Thermic effect of food at rest, during exercise, and after exercise in lean and obese men of similar body weight." **J. Clin. Inv.**, 76, 1107, 1985.
1142. Segal, K.R., Presta, E. & Gutin, B. "Thermic effect of food during graded exercise in normal weight and obese man." **Am. J. Clin. Nutr.**, 40, 995-1000, 1984.
1143. Selbach, H. **Med. Klin.**, 68, 642-648, 1973.
1144. Selberg, O., Schlaak, S., Balks, H.J., et.al. "Thermogenic effect of adrenaline: interaction with insulin." **Eur. J. Appl. Physiol.**, 63(6), 417-423, 1991.
1145. Serra, F., Bonet, L. & Palou, A. **Arch. Int. Physiol. Biochim.**, 95, 263-268, 1987.
1146. Seydoux, J. Assimacopoulos-Jeannet, F., giacobino, J.P. & Girardier, L. "Reactivation of brown adipose tissue in obese hyperglycemic ob/ob mice by cold adaptation." In Szelenyi, Z. & Szekely, M. eds. **Contributions to Thermal Physiology**, Pergamon Press, Akademaiai Kiado, Budapest, 1981, pp. 507-509.
1148. Seydoux, J., Giacobino, J.P. & Girardier, L. **Endocrinology**, 25, 213-226, 1982.
1149. Seydoux, J. & Girardier, L. "Control of brown fat thermogenesis by the sympathetic nervous system." **Experientia**, 33, 1128-1130, 1977.
1150. Seydoux, J., Ricquier, D., Rohner-Jeanrenaud, F. "Decreased guanine nucleotide binding and reduced equivalent production by brown adipose tissue in hypothalamic obesity. Recovery after cold acclimation." **FEBS Lett.**, 146, 161-164, 1982.
1151. Seydoux, J., Trimble, E.R., Bouillaud, F., et.al. **FEBS Lett.**, 141-145, 1984.
1152. Shapiro, P. **Lancet**, 793, 1982.
1153. Shah, M., Miller, D.S. & Geissler, C.A. **Eur. J. Clin. Nutr.**, 42, 741-752, 1988.
1154. Sheiner, J.B., Morris, P. & Anderson, G.H. "Food intake suppression by histidine." **Pharm. Biochem. Behav.**, 23, 721, 1985.
1155. Sheldon, E.F. "The so-called hibernating gland in mammals: a form of adipose tissue." **Anat. Rec.**, 28, 331-347, 1924.
1156. Shetty, P.S., Jung, R.T. & James, W.P.T. "Effect of catecholamine replacement with levodopa on the metabolic response to semistarvation." **Lancet**, 1, 77-79, 1979.
1157. Shetty, P.S., Jung, R.T., James, W.P.T., et.al. "Postprandial thermogenesis in obesity." **Clin. Sci.**, 60, 519-525, 1981.
1158. Shibata, M., Benzi, R.H., Seydoux, J., et.al. **Brain Res.**, 436, 273-282, 1987.
1160. Shimomura, Y., Tamura, T. & Suzuki, M. "Less body fat accumulation in rats fed a safflower oil diet than in rats fed a beef tallow diet." **J. Nutr.**, 120(11), 1291-1296, 1990.
1161. Shirlow, M.J. & Mathers, C.D. **Int. J. Epidem.**, 14, 239-248, 1985.
1162. Shutz, Y., Golay, A., Felber, J.P., et.al. "Decreased glucose-induced thermogenesis after weight loss in obese subjects: a predisposing factor for relapse of obesity?" **Am. J. Clin. Nutr.**, 39, 380, 1984.
1163. Shutz, Y. & Jequier, E. In **Hormones, Thermogenesis and Obesity**, Lardy, H. & Stratman, F., eds., NY, Elsevier Science, 59, 1989.
1164. Silva, J.E. **Mol. Endocr.**, 2 706-713, 1988.
1165. Silva, J.E. & Larsen, P.R. "Adrenergic activation of triiodothyronine production in brown adipose tissue." **Nature**, 305, 712-713, 1983.
1166. Silva, J.E. & Larsen, P.R. **J. Clin. Inv.**, 76, 2296-2305, 1985.
1167. Silva, J.E. & Larsen, P.R. **Am. J. Physiol.**, 251, E639-E643, 1986.
1168. Silver, W. **Pediatrics**, 47, 635, 1971.
1169. Silverberg, A.B., Shah, S.D., Hammond, M.W., et.al. "Norepinephrine: hormone and neurotransmitter in man." **Am. J. Physiol.**, 234, E252-E256, 1978.
1170. Silverberg, D.S. "Non-pharmacological treatment of hypertension." **J. Hypertens. Suppl.**, 8(4), S21-26, 1990.
1171. Sims, E.A.H. "Experimental obesity, dietary-induced thermogenesis, and their clinical implications." **Clin. Endocr. Metab.**, 5, 377, 1976.
1172. Sims, E.A.H. "Energy balance in human beings: problems of plentitude." In **Vitamins and Hormones: Advances in Research and Applications**, Aurbach, G.D. & McCormick, D.B., eds., NY, Academic Press, 1-101, 1987.
1173. Sims, E.A.H. & Danforth, E. "Expenditure and storage of energy in man." **J. Clin. Inv.**, 79, 1019, 1987.
1174. Sims, E.A.H., Danforth, E.Jr, Bray, G.A., et.al. "Endocrine and metabolic effects of experimental obesity in man." **Rec. Prog. Horm. Res.**, 29, 457, 1973.
1175. Sjostrom, L. "A reveiw of weight maintenance and weight changes in relation to energy metabolism and body composition. In **Recent Advances in Obesity Research: IV**, London, John Libbeys, 82-95, 1985.
1176. Sjostrom, L., Schutz, Y., Gudinchet, F., et.al. "Epinephrine sensitivity with respect to metabolic rate and other variables in women." **Am. J. Physiol.**, 245, E431-E442, 1983.
1177. Skala, J.P. & Knight, B.L. **Biochim. Biophys. Acta**, 582, 122-131, 1979.
1178. Skala, J.P. & Shaikh, I.M. **Int. J. Biochem.**, 20, 15-22, 1988.
1179. Skala, J.P., Shaikh, I.M. & Cannon De Rodriguez, W. **Int. J. Biochem.**, 20, 7-13, 1988.
1180. Skofitsch, G., Donnerer, J. & Lembeck, F. "Comparison of nonivamide and capsaicin with regard to their pharmacokinetics and effects on sensory neurons." **Arzneim.-Forsch./Drug Res.**, 34(1 nr 2), 154-156, 1984.
1181. Smalley, R. L. "Changes in composition and metabolism during adipose tissue development." In **Brown Adipose Tissue**, Lindberg, O., ed., NY, Elsevier, 1970, pp. 33-72.
1182. Smith, C.M, Narrow, C.M., Kendrick, Z.V., et.al. "The effect of pantothenate deficiency in mice on their metabolic response to fast and exercise." **Metabolism**, 36, 115, 1987.
1183. Smith, J.G., Crounse, R.G. & Spence, D. "The effects of capsaicin on human skin, liver and epidermal lysosomes." **J. Inv. Derm.**, 54, 170-173, 1970.
1184. Smith, R.E. & Horwitz, B.A. "Brown fat and thermogenesis." **Physiol. Rev.**, 49, 330-425, 1969.
1185. Smith, S.A., Young, P. & Cawthorne, M.A. **Biochem. J.**, 237, 789-795, 1986.
1186. Snyder, S.H. In **Caffeine**, Dews, P.B., ed., Springer-Verlag, Berlin, 129-141, 1984.
1187. Snyder, S.H., Katims, J.J., Annau, A., et.al. **Proc. Natn. Acad. Sci. USA**, 78, 3260-3264, 1981.

1188. Soares, F.A. & Silveira, T.C. "Accumulation of brown adipose tissue in patients with Chagas heart disease." **Trans. R. Soc. rop. Med. Hyg.**, 85(5), 605-607, 1991.
1189. Sohar, E. & Sneh, E. "Follow up of obese patient 14 years after successful reducing diet." **Am. J. Clin. Nutr.**, 26, 845-848.
1190. Solanke, T.F. "The effect of red pepper (Capsicum frutescens) on gastric acid secretion." **J. Surg. Res.**, 15, 385-390, 1973.
1191. Sollman, I. & Pilcher, J.D. "The action of caffeine on the mammalian circulation." **J. Pharm. Exp. Ther.**, 3, 19-92, 1911.
1192. Solomon, S.J., Kurzer, M.S. & Calloway, D.H. "Menstrual cycle and basal metabolic rate in women." **Am. J. Clin. Nutr.**, 36,611-616, 1982.
1193. Sowers, J.R., Nyby, M., Stern, N., et.al. "Blood pressure and hormone changes associated with weight reduction in the obese." **Hypertension**, 4, 686-691, 1982.
1194. Spindel, E., Arnold, M., Cusack, B., et.al. **J. Pharm. Exp. Ther.**, 214, 58-62, 1980.
1195. Srinivasan, M., Aiyar, A.S., Kapur, O.P., et.al. **Ind. J. Exp. Biol.**, 2, 104, 1964.
1196. Srinivasan, M.R., Sambaiah, K., Satyanarayana, M.N., et.al. "Influences of red pepper and capsaicin on growth, blood constituents and nitrogen balance in rats." **Nutr. Reps. Int.**, 21, 455-467, 1980.
1197. Srinivasan, M.R. & Satyanarayana, M.N. "Effect of capsaicin on skeletal muscle lipoprotein lipase in rats fed high fat diet." **Ind. J. Exp. Biol.**, 27(10), 910-912, 1989.
1198. Stamler, J., Farinaro, E., Mojonniver, L.M., et.al. **JAMA**, 243, 1819-1823, 1980.
1199. Starr, I., Gamble, C.J. & Margolies, A. "A Clinical study of the action of 10 commonly used drugs on cardiac output, work and size: on respiration, on metabolic rate and on the electrocardiogram." **J. Clin. Inv.**, 16, 799-823, 1937.
1200. Steering Committee of the Physicians Health Study Research Group. **N. Eng. J. Med.**, 321, 129, 1989.
1201. Steffens, A.B., Van Der Gugten, J., Godeke, J., et.al. **Physiol. Behav.**, 37, 119-122, 1986.
1202. Steinberg, D., Nestel, P.J., Buskirk, E.R., et.al. "Calorigenic effect of norepinephrine correlated with plasma free fatty acid turnover and oxidation." **J. Clin. Inv.**, 43, 167-176, 1964.
1203. Steinberg, D., Vaughan, M., Nestel, P.J., et.al. **Biochem. Pharm.**, 12, 764, 1963.
1204. Steinberg, D., Pittman, R.C. Attie, A.D., et.al. "Uptake and degradation of low density lipoprotein by adipose tissue in vivo." In **The Adipocyte and Obesity: Cellular and Molecular Mechanisms**, Angel, A., Hollenberg, C.H. & Roncari, D.A., eds., Raven Press, New York, 1983, pp. 197-216.
1205. Stendig-Lindberg, G., Shapiro, Y., Epstein, Y., et.al. "Changes in serum magnesium concentration after strenuous exercise." **J. Am. Coll. Nutr.**, 6, 35, 1987.
1206. Stephenson, P.E. "Physiologic and psychotropic effects of caffeine on man." **J. Am. Diet. Assn.**, 71, 240-247, 1977.
1207. Stern, J.S., Brown, S., Stanhope, E., et.al. "Adrenalectomy reduced weight gain, adipose cell size and lipoprotein lipase activity in obese male Zucker rats (fa/fa)." **Fed. Proc.**, 42, 393, 1983.
1208. Stile, G.L. **Trends Pharm. Sci.**, 87, 486, 1986.
1209. Stock, M.J. "The role of brown adipose tissue in diet-induced thermogenesis." **Proc. Nutr. Soc.**, 48(2), 189-196, 1989.
1210. Stock, M. & Rothwell, N. **In: Nutritional Factors: Modulating Effects on Metabolic Processes**, R.F. Beers, Jr. & E.G. Bassett, eds., pp. 101-113, Raven Press, New Yor, 1981.
1211. Stock, M. & Rothwell, N. **Obesity and Leanness**, NY, John Wiley, 1982.
1212. Storlien, L.H., James, D.E., Burleigh, K.M., et.al. **Am. J. Physiol.**, 251, E576-E583, 1986.
1213. Strom, G. "Central nervous regulation of body temperature." In **Neurophysiology**, Washington, Am. Physiol. Soc., 1, 1173-1196, 1960.
1214. Stromme, S.B. & Hammel, H.T. "Effects of physical training on tolerance to cold in rats." **J. Appl. Physiol.**, 23, 815-824, 1967.
1215. Strubelt, O. In **Coffein und andere Methylxanthine**, Heim F. & Ammon, H.P.T., eds., F.K. Schattauer Verlag, Stuttgart, 121-128, 1969.
1216. Strubelt, O. & Siegers, C.P. "Zum Mechanismus der Kalorigenen wirkung von Theophyllin und Caffein." **Biochem. Pharm.**, 18, 1207-1212, 1969.
1217. Stunkard, A. & McLaren-Hume, M. "The results of treatment for obesity." **Arch. Intern. Med.**, 103, 79-85,, 1958.
1218. Sturanic, C., Papirus, S., Grossi, G., et.al. **Eur. J. Respir. Dis.**, 69, 557-563, 1986.
1219. Struthers, A.D., Reid, J.L., Whitesmith R., et.al. "The effects of cardioselective and non-selective beta-adrenoceptor blockade on the hypokalaemic and cardiovascular responses to adrenomedullary hormones in man." **Clin. Sci.**, 65, 143-147, 1983.
1220. Su, C. & Bevan, J.A. "The release of norepinephrine in arterial strips studied by the technique of superfusion and transmural stimulation." **J. Pharm. Exp. Ther.**, 172, 62-68, 1970.
1221. Subbarao, D., Chandrasekhara, N., Satyanarayana, M.N., et.al. **J. Nutr.**, 100, 1307, 1970.
1222. Sundin, U. "The influence of thyroxine on brown fat thermogenesis." Szelenyi, Z. & Szekely, M. eds. **Contributions to Thermal Physiology**, Pergamon Press, Akademaiai Kiado, Budapest, 1981, pp. 499-501.
1223. Sundin, U. & Fain, J.N. **Biochem. Pharm.**, 32, 3117-3120, 1983.
1224. Sundin, U., Mills, I. & Fain, J.N. **Metabolism**, 33, 1028-1033, 1984.
1225. Suter, E.R., **Experentia**, 25 p. 286, 1969.
1226. Suter, E.R., **Lab. Invest.**, 21, p. 246, 1969.
1227. Suter, E.R., **Lab. Invest.**, 21, p. 259, 1969.
1228. Suter, E.R., **J. Ultrastruct. Res.**, 26, 216, 1969.
1229. Sutherland, E.W. & Rall, T.W. **J. Biol. Chem.**, 232, 1077-1091, 1958.
1230. Suzuki, T. & Iwai, K. "Constituents of red pepper species: chemistry, biochemistry, pharmacology, and food science of the pungent principle of Capsicum species." In **The Alkaloids**, Brossi, A., ed., Academic Press, NY, 23, 227-299, 1984.
1231. Suzuki, T., Masukawa, Y. & Misawa, M. **Psychopharm.**, 102, 438, 1990.
1232. Svartengren, J., Svoboda, P. & Cannon, B. **Eur. J. Biochem.**, 128, 481-488, 1982.
1233. Svartengren, J., Svoboda, P., Drahota, Z., et.al. **Comp. Biochem.**, 78C, 159-170, 1984.
1234. Swaminathan, R., King, R.F.G.J., Holmfield, J., et.al. **Am. J. Clin. Nutr.**, 42, 177-181, 1983.
1235. Swanson, H.E. "Interrelations between thyroxin and adrenalin in the regulation of oxygen consumption in the albino rat." **Endocrinology**, 59, 217-225, 1956.
1236. (no entry)
1237. Szelenyi, Z. & Szekely, M. eds. **Contributions to Thermal Physiology**, Pergamon Press, Akademaiai Kiado, Budapest, 1981.
1238. Szillat, D. & Bukowiecki, L.J. **Am. J. Physiol.**, 245, E555-E559, 1983.
1239. Szolcsanyi, J. "Investigations on the metabolic rate in capsaicin desensitized rats." **Acta Physiol. Hung. Suppl.**, 32, 108, 1967.
1240. Szolcsanyi, J. & Jancso-Gabor, A. **Arzneim.-Forsch./Drug Res.**, 25, 1877, 1975.
1241. Szolcsanyi, J. & Jancso-Gabor, A. **Arzneim.-Forsch./Drug Res.**, 26, 33, 1976.

334

1242. Taira, K. & Shibasaki, S. **Biomed. Res.**, 5, 111-116, 1984.
1243. Tappy, l., Felber, J.P. & Jequier, E. "Energy and substrate metabolism in obesity and postobese state." **Diabetes Care**, 14(12), 1180-1188, 1991.
1244. Tauveron, I., Grizard, J., Thieblot, P., et.al. "Adaptation metabolique en hyperthyroidie. Implication de l'insuline." **Diabete Metab.**, 18(1 pt 2), 131-136, 1992.
1245. Tedesco, J.L. Flattery, K.V. & Sellers, E.A. "Effects of thyroid hormones and cold exposure on turnover of norepinephrine in cardiac and skeletal muscle." **Can. J. Physiol. Pharm.**, 55, 515-522, 1977.
1246. Tegowska, e. & Narebski, J. "The functional efficacy of BAT in cold and after noradrenaline injection, dependence on size of animal and on integumental isolation value." In Szelenyi, Z. & Szekely, M. eds. **Contributions to Thermal Physiology**, Pergamon Press, Akademaiai Kiado, Budapest, 1981, pp. 531-533.
1248. Tepperman, J. "Adipose tissue: Yang and Yin." In **Fat as a Tissue**, Rodahl, K., & Issekutz, B., eds. McGraw Hill Book Company, New York, 1964, pp. 394-409.
1249. Thelle, D.S. Arnensen, E. & Forde, O.H. "The Tromso heart study." **N. Eng. J. Med.**, 308, 1454, 1983.
1250. Ther, L., Muschaweck, R. & Hergott, J. **Arch. Exp. Path. Pharm.**, 231, 586-590, 1957.
1251. Theriault, E., Otsuka, M. & Jessell, T. "Capsaicin-evoked release of substance P from primary sensory neurons." **Brain Res.**, 87, 1, 1974.
1252. Therminarias, A., Chirpaz, M.F. & Tanche, M. "Catecholamines in dogs during cold adaptation by repeated immersion." **J. Appl. Physiol.**, 46, 662-668, 1979.
1253. Thiebaud, D., Acheson, K., Schutz, Y., et.al. "Stimulation of thermogenesis in man following combined glucose-long-chain triglyceride infusion." **Am. J. Clin. Nutr.**, 37, 603, 1983.
1254. Thoenen, H. "Induction of tyrosine hydroxylase in peripheral and central adrenergic neurons by cold-exposure of rats." **Nature**, 228, 861-862, 1970.
1255. Thompson, G.E. "Physiological effects of cold exposure." In **International Review of Physiology. Environmental Physiology II**, Robertshaw, D., ed., Baltimore, Univsity Park Press, 26-69, 1977.
1256. Thorne, A., Naslund, I. & Wahren, J. "Meal-induced thermogenesis in previously obese patients." **Clin. Physiol.**, 10(1), 99-109, 1990.
1257. Thurlby, P.L. & Trayhurn, P. "The role of thermoregulatory thermogenesis in the development of obesity in genetically obese (ob/ob) mice pairfed with lean siblings." **Br. J. Nutr.**, 42, 377-385, 1979.
1258. Thurlby, P.L. & Trayhurn, P. "Regional blood flow in genetically obese (ob/ob) mice. The importance of brown adipose tissue to the reduced energy expenditure on non-shivering thermogenesis." **Pflugers Arch.**, 385, 193-201, 1980.
1260. Toda, N. "Influence of cocaine and desipiamine on the contractile response of isolated rabbit pulmonary arteries and aortae to transmural stimulation." **J. Pharm. Exp. Ther.**, 179, 198-206, 1971.
1261. Toda, N., Usui, H. & More, J. "Contractile responses of spiral strips of some large blood vessels from rabbits to transmural stimulation, tyramine, dopamine and noradrenaline." **Jap. J. Pharm.**, 22, 59-69, 1972.
1262. Toda, N., Usui, H., Nishino, N., et.al. "Cardiovascular effects of capsaicin in dogs and rabbits." **J. Pharm. Exp. Ther.**, 181(3), 512-521, 1972.
1263. Toh, C.C., Lee, T.S. & Kiang, A.C. "The pharmacological actions of capsaicin and analogues." **Br. J. Pharm.**, 10, 175-182, 1955.
1264. Tokuyama, K. & Himms-Hagen, J. "Brown adipose tissue thermogenesis, torpor, and obesity of glutamate-treated mice." **Am. J. Physiol.**, 251, E407-E415, 1986.
1265. Tokuyama, K. & Himms-Hagen, J. "Increased sensitivity of the genetically obese mouse to corticosterone." **Am. J. Physiol.**, 252, E202-E208, 1987.
1266. Tokuyama, K. & Himms-Hagen, J. "Enhanced acute response to corticosterone in genetically obese (ob/ob) mice." **Am. J. Physiol.**, in press, 1989.
1267. Tokuyama, K. & Himms-Hagen, J. "Adrenalectomy prevents obesity in glutamate-treated mice." **Am. J. Physiol.**, in press, 1989.
1268. Toner, M.M., Kirkendall, D.T., Delio, D.J, et.al. "Metabolic and cardiovascular responses to exercise with caffeine." **Ergonomics**, 25, 1175-1183, 1982.
1269. Toubro, S., Astrup, A., Branebjerg, P.E., et.al. **Abstracts, 5th Nordic Nutr. Congress**, D16, 49, 1992.
1270. Toubro, S., Astrup, A.V., Breum, L., et.al. "Safety and efficacy of long-term treatment with ephedrine, caffeine and an ephedrine/caffeine mixture." **Int. J. Obes.**, 17(suppl 1), S69-S72, 1993.
1271. Trayhurn, P. **FEBS Lett.**, 104, 13-16, 1979.
1272. Trayhurn, P. **Biochim. Biophys. Acta**, 664, 549-560, 1981.
1273. Trayhurn, P. "Brown adipose tissue thermogenesis and the energetics of lactation in rodents." **Int. J. Obes.**, 9(suppl 2), 81-88, 1985.
1274. Trayhurn, P. "Brown adipose tissue and energy balance." In **Brown Adipose Tissue**, Trayhurn, P. & Nicholls, D.G., eds., London, Arnold, 299-338, 1986.
1275. Trayhurn, P. "The role of brown adipose tissue in the development of the obese-hyperglycemic syndrome in mice." In **Pathogenesis and New Approches to the Study of Noninsulin-dependent Diabetes Mellitus**, Chang, A.Y. & Diani, A.R., eds., NY, Alan R. Liss, 69-85, 1988.
1276. Trayhurn, P. "Thermogenesis and the energetics of pregnancy and lactation." **Can. J. Physiol. Pharm.**, 67, 1989.
1277. Trayhurn, P. "Energy expenditure and thermogenesis: animal studies on brown adipose tissue." **Int. J. Obes.**, 14(suppl 1), 17-29, 1990.
1278. Trayhurn, P. & Ashwell, M. "Control of white and brown adipose tissues by the autonomic nervous system." **Proc. Nutr. Soc.**, 46, 135-142, 1987.
1279. Trayhurn, P., Ashwell, M., Jennings, G., et.al. "Effect of warm or cold exposure on GDP binding and uncoupling protein in rat brown fat." **Am. J. Physiol.**, 252, E237-E243, 1987.
1280. Trayhurn, P. & James, W.P.T. "Thermoregulation and shivering thermogenesis in the genetically obese (ob/ob) mouse." **Pflugers Arch.**, 373, 189-193, 1978.
1281. Trayhurn, P. & James, W.P.T. "Thermogenesis and obesity." In **Mammalian Thermogenesis**, Trayhurn, P. & Stock, M.J., eds., London, Chapman & Hall, 234-258, 1983.
1282. Trayhurn, P. & Jennings, G. "Non-shivering thermogenesis and the thermogenic capacity of brown fat in fasted/refed mice." **Am. J. Physiol.**, 254, R11-R16, 1988.
1283. Trayhurn, P. & Jennings, G. **Biosci. Rep.**, 6, 805-810, 1986.

1284. Trayhurn, P., Jones, M.M., McGuckin, M.M., et.al. "Effects of overfeeding on energy balance and brown fat thermogenesis in obese (ob/ob) mice." **Nature (London)**, 295, 323-325, 1982.
1285. Trayhurn, P., Thurlby Pl. & James, W.P.T. "Thermogenic defect in preobese ob/ob mice." **Nature**, 266, 60-61, 1977.
1286. Trayhurn, P. & Wusteman, M.C. "Sympathetic activity in brown adipose tissue in lactating mice." **Am. J. Physiol.**, 253, E515-E520, 1987.
1287. Trayhurn, P. & Wusteman, M.C. "Apparent dissociation between sympathetic activity and brown adipose tissue thermogenesis during pregnancy and lactation in golden hamsters." **Can. J. Physiol. Pharm.**, 65, 2396-2399, 1987.
1288. Trembley, A., Sauve, L., et.al. **Int. J. Obes.**, 13, 357-366, 1989.
1289. Trendelenburg, U. "Supersensitivity and subsensitivity to sympathomimetic amines." **Pharm. Rev.**, 15, 225, 1963.
1290. Trendelenburg, U. "Mechanisms of supersensitivity and subsensitivity to sympathomimetic amines." **Pharm. Rev.**, 18, 629-640, 1966.
1291. Triandafillou, J. & Himms-Hagen, J. "Brown adipose tissue in genetically obese (fa/fa) rats: response to cold and diet." **Am. J. Physiol.**, 244, E145-E150, 1983.
1292. Troisi, R.J., Weiss, S.T., Parker, D.R., et.al. **Hypertension**, 17, 669, 1991.
1293. Truitt, E.B. In **Drill's Pharmacology in Medicine**, 4th ed. Di Palma, J.R., ed., McGraw-Hill Book Co., NY, 533-556, 1971.
1294. Trunet, P., Lhoste, F., Ansquer, J.C., et.al. "Decreased plasma epinephrine concentrations after glucose ingestion in humans." **Metabolism**, 33, 101-103, 1984.
1295. Tse, T.F., Clutter, W.E., Shah, S.D., et.al. "Neuroendocrine responses to glucose ingestion in man: specificity, temporal relationships, and quantitative aspects." **J. Clin. Inv.**, 72, 270-277, 1983.
1296. Tulp, O.L. & Buck, C.L. "Caffeine and ephedrine stimulated thermogenesis in LA-corpulent rats." **Comp. Biochem. Physiol. C. Comp. Pharm.**, 85(1), 17-19, 1986.
1297. Turnbull, A. "Tincture of capsicum as a remedy for chillblains and toothache." Med. Press, Dublin, 95-96, 1850.
1298. Tyrala, E.A. & Dodson, W.E. **Arch. Dis. Child.**, 54, 787-800, 1979.
1299. Uchida, Y. & Nomoto, T. "Intravenously infused adenosine increases the blood flow to brown adipose tissue in rats." **Eur. J. Pharm.**, 184(2-3), 223-231, 1990.
1300. Ungsurungsie, M., Suthienkul, O. & Paovalo, C. "Mutagenicity screening of popular Thai spices." **Food & Chem. Tox.**, 20, 527-530, 1982.
1301. Urberg, M. & Zenel, M.B. "Evidence for synergism between chromium and nicotinic acid in the control of glucose tolerance in elderly humans." **Metabolism**, 36, 896-899, 1987.
1302. Vaddadi, K.S. & Horrobin, D.F. "Weight loss produced by evening primrose oil administration in normal and schizophrenic individuals." **Psychol. Psychiat.**, 7, 52, 1979.
1303. Vagenakis, A.G., Burger, A., Portnay, G.I., et.al. **J. Clin. Endocr. Metab.**, 41, 191, 1975.
1304. Vallerand, A.L. "Effects of ephedrine/xanthines on thermogenesis and cold tolerance." **Int. J. Obes.**, 17(suppl 1), S53-S56, 1993.
1306. Vallerand, A.L. & Jacobs I. "Rates of energy substrate utilization during human cold exposure." **Eur. J. Appl. Physiol.**, 58, 872, 1989.
1307. Vallerand, A.L., Jacobs, I. & Kavanagh, M.F. "Mechanism of enhanced cold tolerance by an ephedrine-caffeine mixture in humans." **J. Appl. Physiol.**, 67(1), 438-444, 1989.
1308. Vallerand, A.L., Lupien, J. & Bukowiecki, L.J. "Cold exposure reverses the diabetogenic effects of high-fat feeding." **Diabetes**, 35, 329, 1986.
1309. Vallerand, A.L., Perusse, F. & Bukowiecki, L.J. **Am. J. Physiol.**, 253, E179-E186, 1987.
1310. Vallerand, A.L., Perusse, F. & Bukowiecki, L.J. **Diabetes**, in press, 1989.
1312. Van Calker, D., Muller, M. & Hamprecht, B. **J. Neurochem.**, 33, 999-1005, 1979.
1313. Vander Tuig, J.G., Kerner, J., Crist, K.A., et.al. "Impaired thermoregulation in cold-exposed rats with hypothalamic obesity." **Metab. Clin. Exp.**, 35, 960-966, 1986.
1314. Vander Tuig, J.G., Kerner, J. & Romsos, D.R. "Hypothalamic obesity, brown adipose tissue, and sympathoadrenal activity in rats." **Am. J. Physiol.**, 248, E607-E617, 1985.
1315. Vander Tuig, J.G., Knehans, A.W. & Romsos, D.R. "Lesions in the ventromedial hypothalamus reduce sympathetic nervous system activity in rats." (abstr) **Fed. Proc.**, 40, 887, 1981.
1316. Vander Tuig, J.G., Knehans, A.W. & Romsos, D.R. "Reduced sympathetic nervous activity in rats with ventromedial hypothalamic lesions." **Life Sci.**, 30, 913-920, 1982.
1317. Vander Tuig, J.G. & Romsos, D.R. **Metabolism**, 33, 26-33, 1984.
1318. Vaughan, M. "Effect of hormones on fat mobilization." In **Fat as a Tissue**, Rodahl, K., & Issekutz, B., eds. McGraw Hill Book Company, New York, 1964, pp. 203-214.
1319. Vikman, H.L., Ranta, S., Kiviluoto, T., et.al. **Acta Physiol. Scand.**, 142, 405, 1991.
1320. Villarroya, F., Felipe, A. & Mampel, T. **Biochim. Biophys. Acta**, 882, 187-191, 1986.
1321. Virus, R.M. & Gebhart, G.F. "Pharmacological actions of capsaicin: apparent involvement of substance P and serotonin." **Life Sci.**, 25, 1273, 1979.
1322. Virus, R.M., Knuepfer, M.M., McManus, D.Q., et.al. "Capsaicin treatment in adult male Wistar-Kyoto and spontaneously hypertensive rats: effects on nociception and cardiovascular regulation." **Eur. J. Pharm.**, 72, 209, 1981.
1323. Virus, R.M., McManus, D.Q. & Gebhart, G.F. "Capsaicin treatment in adult Wistar-Kyoto and spontaneously hypertensive rats: effects on substance P contents of peripheral and central nervous system tissues." **Eur. J. Pharm.**, 81, 67-73, 1982.
1324. Virus, R.R. & Gebhart, G.F. "Pharmacologic actions of capsaicin: apparent involvements of substance P and serotonin." **Life Sciences**, 25, 1273-1284, 1979.
1325. Walgren, M.C., Kaufman, L.N., Young, J.B., et.al **Metabolism**, 36, 585, 1987.
1326. Walker, H.C. & Romsos, D.R. "Glucocorticoids in the CNS regulate BAT metabolism and plasma insulin in ob/ob mice." **Am. J. Physiol.**, 262(1 pt 1), E110-117, 1992.
1327. Walker, R.B., Wood, D.M. & Akmal, M.M. "Hyperthermic and anorectic effects of oxazolidines derived from L-ephedrine in rats." **Life Sci.**, 47(7), 595-600, 1990.
1328. Walther, O.E., Iriki, M. & Simon, E. "Antagonistic changes of blood flow and sympathetic activity in different vascular beds following central thermal stimulation. II. Cutaneous and visceral sympathetic activity during spinal cord heating and cooling in anesthetized rabbits and cats." **Pflugers Arch.**, 319, 162-184, 1970.
1330. Wang, L.C.H. & Anholt, E.C. "Elicitation of supramaximal thermogenesis by aminophylline in the rat." **J. Appl. Physiol.**, 53, 16-20, 1982.

1331. Wang, L.C.H. & Lee, T.F. "Substrate preference in aminophylline-stimulated thermogenesis." **Life Sci.**, 36, 2539-2546, 1985.
1332. Wang, L.C. & Lee, T.F. "Enhancement of maximal thermogenesis by reducing endogenous adenosine activity in the rat." **J. Appl. Physiol.**, 68(2), 580-585, 1990.
1333. Wang, L.C.H., Man, S.F.P. & Belcastro, A.N. "Improving cold tolerance in men: effects of substrates and aminophylline." In **Homeostasis and Thermal Stress**, Cooper, K., Lomax, P., Schonbaum, E., et.al., eds., Basel, Karger, 22-26, 1986.
1334. Wang, L.C.H., Man, S.F.P. & Belcastro, A.N. "Metabolic and hormonal responses in theophylline-increased cold resistance in males." **J. Appl. Physiol.**, 63(2), 589-596, 1987.
1335. Ward, K.D., Garvey, A. & Bliss, R.E. **Pharm. Biochem. & Behav.**, 40, 937, 1991.
1336. Waugh, W.H. "Adrenergic stimulation of depolarized arterial muscle." **Circ. Res.**, 11, 264-276, 1962.
1337. Waugh, W.H. "Role of calcium in contractile excitation of vascular smooth muscle by epinephrine and potassium." **Circ. Res.**, 11, 927-940, 1962.
1338. Webb-Peploe, M.M. & Shepard, J.T. "Responses of the superficial limb veins of the dog to changes in termperature." **Circ. Res.**, 22, 737-746, 1968.
1339. Weigle, D.S., Sande, K.J., Iverius, P. "Weight loss leads to a marked decrease in nonresting energy expenditure in ambulatory human subjects." **Metabolism**, 37, 930-936, 1988.
1340. Weingarten, H.P., Chang, P. & McDonald, T.J. "Comparison of the metabolic and behavioral disturbances following paraventricular- and ventromedial-hypothalamic lesions." **Brain Res. Bull.**, 14, 551-559, 1985.
1341. Weiss, B. & Laties, V.G. "Enhancement of human performance by caffeine and the amphetamines." **Pharm. Rev.**, 14, 1-36, 1962.
1342. Welle, S. & Campbell, R.G. "Stimulation of thermogenesis by carbohydrate overfeeding. Evidence against sympathetic nervous system mediation." **J. Clin. Inv.**, 71, 916-925, 1983.
1343. Welle, S., Lilavivathana, U. & Campbell, R.G. "Increased plasma norepinephrine concentrations and metabolic rates following glucose ingestion in man." **Metabolism**, 29, 806-809, 1980.
1344. Welle, S., Lilavivathana, U. & Campbell, R.G. "Thermic effect of feeding in man: increased plasma norepinephrine levels following glucose but not protein or fat consumption." **Metabolism**, 30, 952-958, 1981.
1345. Welle, S.L., Seaton, T.B. & Campbell, R.G. "Some metabolic effects of overeating in man." **Am. J. Clin. Nutr.**, 44, 718, 1986.
1346. Welling, P.G., Lee, K.P., Patel, J.A., et.al. "Urinary excretion of ephedrine in man without pH control following oral administration of three commercial ephedrine sulfate preparations." **J. Pharm. Sci.**, 60, 1629-1634, 1971.
1347. Wellman, P.J. & Marmon, M.M. "Comparison of brown adipose tissue thermogenesis induced by congeners and isomers of phenylpropanolamine." **Life Sci.**, 37(11), 1023-1028, 1985.
1348. Wellman, P.J. & Marmon, M.M. "Synergism between caffeine and di-phenylpropanolamine on brown adipose tissue thermogenesis in the adult rat." **Pharm. Biochem. Behav.**, 22(5), 781-785, 1985.
1349. White, B.C., Lincoln, C.A., Pearce, N.W., et.al. **Science**, 209, 1547-1548, 1980.
1350. Whitesett, T.L., Manion, C.V. & Christensen, H.Dix **Am. J. Cardiol.**, 53, 918-922, 1984.
1351. Wickler, S.J., Horwitz, B.A. & Stern, J.S. "Regional blood flow in genetically obese rats during non-shivering thermogenesis." **Int. J. Obes.**, 6, 481-490, 1982.
1352. Widdowson, E.M. **J. Hyg. (London)**, 36, 269, 1936.
1353. Widdowson, E.M. **Proc. Nutr. Soc.**, 21, 121, 1962.
1354. Wiesinger, H., Klaus, S., Heldmaier, G., et.al. "Increased nonshivering thermogenesis, brown fat cytochrome-c oxidase activity, GDP binding, and uncoupling protein mRNA levels after short daily cold exposure of Phodopus sungorus." **Can. J. Physiol. Pharm.**, 68(2), 195-200, 1990.
1355. Wilcox, A.R. "The effects of caffeine and exercise on body weight, fat-pad weight, and fat-cell size." **Med. Sci. Sports Exer.**, 14, 317, 1982.
1356. Williams, J.H., Barnes, W.S. & Gadberry, W.L. "Influence of caffeine on force and EMG in rested and fatigued muscle." **Am. J. Phys. Med.**, 66(4), 169-183, 1987.
1357. Williams, R.J. **Nutrition Against Disease**, NY, Pitman Publ., 1971.
1358. Williamson, D.F., Madans, J., Anda, R.F., et.al. **N. Eng. J. Med.**, 324, 739, 1991.
1359. Wilson, M.A. "Treatment of obesity." **Am. J. Med. Sci.**, 299(1), 62-68, 1990.
1360. Wilson, S., Thurlby, P.L. & Arch, J.R.S. **Can. J. Physiol. Pharm.**, 65, 113-119, 1987.
1361. Winegrad, A.I., Goto, Y. & Lukens, F.D. "Adipose tissue in diabetes." In **Fat as a Tissue**, Rodahl, K., & Issekutz, B., eds. McGraw Hill Book Company, New York, 1964, pp. 344-361.
1362. Winsky, L. & Harvey, J.A. **J. Pharm. Exp. Ther.**, 241, 223-229, 1987.
1363. Winstead, D.K. **Am. J. Psychiatry**, 133, 1447-1450, 1976.
1364. Wolpaw, J.R. & Penry, J.K. **Electroencephalogr. Clin. Neurophysiol.**, 44, 568-574, 1978.
1365. Woodcock, A.A., Gross, E.R., Gellert, A., et.al. **N. Eng. J. Med.**, 305, 1611-1616, 1981.
1366. Woodward, J.A. & Saggerson, E.D. **Biochem. J.**, 238, 395-403, 1986.
1368. Wu, S.Y., Stern, J.S., Fisher, D.A., et.al. **Am. J. Physiol.**, 252, E63-E67, 1987.
1369. Yaksh, T.L., Farb, D.H., Leeman, S.E., et.al. "Intrathecal capsaicin depletes substance P in the rat spinal cord and produces prolonged thermal analgesia." **Science**, 206, 481, 1979.
1370. Yamasaki, K., Fujita, K., Sakamoto, M., et.al. "Separation and quantitative analysis of Ephedra alkaloids by gas chromatography and its application to evaluation of some, Ephedra species collected around Himalaya." **Chem. Pharm. Bull.**, 22, 2898-2902, 1974.
1371. Yen, T.T., McKee, M.M. & Bemis, K.G. "Ephedrine reduces weight of viable yellow obese mice (A/a)." **Life Sci.**, 28, 119-128, 1980.
1372. Yokogoshi, H., Mochizuki, S., Takahata, M., et.al. **Nutr. Rep. Int.**, 28, 805-814, 1983.
1373. York, D.A. "Neural activity in hypothalamic and genetic obesity." **Proc. Nutr. Soc.**, 46, 105-117, 1987.
1374. York, D.A., Holt, S.J. & Marchington, D. **Int. J. Obes.**, 9(suppl 1), 65-78, 1984.
1375. York, D.A., Holt, S.J. & Marchington, D. "Regulation of brown adipose tissue thermogenesis by corticosterone in obese fa/fa rats." **Int. J. Obes.**, 9(suppl 2), 89-95, 1985.
1376. Yoshida, T. & Bray, G.A. "Catecholamine turnover in rats with ventromedial hypothalamic lesions." **Am. J. Physiol.**, 246, R558-R565, 1984.
1377. Yoshida, T., Yoshioka, K., Hiraoka, N., et.al. "Effect of nicotine on norepinephrine turnover and thermogenesis in brown adipose tissue and metabolic rate in MSG obese mice." **J. Nutr. Sci. Vit. Tokyo**, 36(2), 123-130, 1990.
1378. Yoshida, T., Yoshioka, K., Wakabayashi, Y., et.al. "Effects of capsaicin and isothiocyanate on thermogenesis of interscapular

brown adipose tissue in rats." **J. Nutr. Sci. Vit. Tokyo**, 34(6), 587-594, 1988.

1379. Yoshioka, K., Yoshida, T., Kamanaru, K., et.al. "Caffeine activates brown adipose tissue thermogenesis and metabolic rate in mice." **J. Nutr. Sci. Vit. Tokyo**, 36(2), 173-178, 1990.

1380. Yoshioka, K., Yoshida, T. & Kondo, M. "Reduced brown adipose tissue thermogenesis and metabolic rate in pre-obese mice treated with monosodium-L-glutamate." **Endocr. Jap.**, 38(1), 75-79, 1991.

1381. Young, J.B. **Life Sci.**, 43, 193-200, 1988.

1382. Young, J.B., Einhorn, D. & Landsberg, L. "Decreased sympathetic (SNS) activity in interscapular brown adipose tissue (BAT) of streptozotocin-treated rats." **Diabetes**, 32(suppl 1), 26A, 1983.

1383. Young, J.B., Kaufman, L.N., Saville, M.E., et.al. **Am. J. Physiol.**, 248, R627-R637, 1985.

1384. Young, J.B. & Landsberg, L. "Stimulation of the sympathetic nervous system during sucrose feeding." **Nature**, 269, 615-617, 1977.

1385. Young, J.B. & Landsberg, L. "Suppression of sympathetic nervous system during fasting." **Science**, 196, 1473-1475, 1977.

1386. Young, J.B. & Landsberg, L. "Effect of diet and cold exposure on norepinephrine turnover in pancreas and liver." **Am. J. Physiol.**, 236, E524-E533, 1979.

1387. Young, J.B. & Landsberg, L. "Sympathoadrenal activity in fasting pregnant rats: dissociation of adrenal medullary and sympathetic nervous system responses." **J. Clin. Inv.**, 64, 109-116, 1979.

1388. Young, J.B. & Landsberg, L. "Impaired suppression of sympathetic activity during fasting in the gold thioglucose-treated mouse." **J. Clin. Inv.**, 65, 1086-1094, 1980.

1389. Young, J.B. & Landsberg, L. "Effect on concomitant fasting and cold exposure on sympathoadrenal activity in rats." **Am. J. Physiol.**, 240, E314-E319, 1981.

1390. Young, J.B. & Landsberg, L. "Diminished sympathetic nervous system activity in genetically obese (ob/ob) mouse." **Am. J. Physiol.**, 245, E148-E154, 1983.

1392. Young, J.B., Mullen, D. & Landsberg, L. "Caloric restriction lowers blood pressure in the spontaneously hypertensive rat." **Metabolism**, 27, 1711-1714, 1978.

1393. Young, J.B., Ross, R.M. & Landsberg, L. "Dissociation of sympathetic nervous system and adrenal medullary responses." **Am. J. Physiol.**, 247, E35-E40, 1984.

1394. Young, J.B., Rowe, J.W., Pallotta, J.A., et.al. "Enhanced plasma norepinephrine response to upright posture and glucose administration in elderly human subjects." **Metabolism**, 29, 532-539, 1980.

1395. Young, J.B., Saville, M.E. & Landsberg, L. "Increased sympathetic (SNS) activity (norepinephrine turnover) in rats fed a low protein diet: evidence against a role for dietary tyrosine." **Clin. Res.**, 31, 466A, 1983.

1396. Young, J.B., Saville, E. & Rothwell, N.J., et.al. "Effect of diet and cold exposure on norepinephrine turnover in brown adipose tissue of the rat." **J. Clin. Inv.**, 69, 1061-1071, 1982.

1397. Young, P., Cawthorne, M.A., Levy, A.L., et.al. **FEBS Lett.**, 176, 16-20, 1984.

1398. Young, P., Cawthorne, M.A. & Smith, S.A. **Biochem. Biophys. Res. Comm.**, 130, 241-248, 1985.

1399. Young, P. & Wilson, S. "Prolonged beta-receptor stimulation increases the amount of GDP-binding protein in brown adipose tissue mitochondria." **Life Sci.**, 34, 1111-1117, 1984.

1400. Zahorska-Markiewicz, B. **Acta Physiol. Pol.**, 31, 17-20, 1980.

1401. Zahorska-Markiewicz, B., Waluga, M. & Klin, M. "The effect of yohimbine on autonomic nervous function in obesity." **Int. J. Obes.**, 17(suppl 1), S80, 1993.

1402. Zanko, M.T., Sullivan, A.C. & O'Brien, R.A. "Defective purine nucleotide binding in brown adipose tissue mitochondria of genetically obese rats." (abstr) **Fed. Proc.**, 41, 714, 1982.

1403. Zaror-Behrens, G. & Himms-Hagen, J. "Cold-stimulated sympathetic activity in brown adipose tissue of obese (ob/ob) mice." **Am. J. Physiol.**, 244, E361-E366, 1983.

1404. Zaror-Behrens, G. & Himms-Hagen, J. "Diet and noradrenaline turnover in brown adipose tissue of goldthioglucose-obese mice." **Int. J. Obes.**, in press, 1988.

1405. Ziemba, A.W., Nazar, K., Kaciuba-Usciiko, H., et.al. "Thermogenic effect of phosphate supplementation in obese dieting women." **Int. J. Obes.**, 17(suppl 1), S82, 1993.

1406. Zirm, K.L., **Z. Naturforsch.**, 11b, 530, 1956.

1407. Zylan, K.D. & Carlisle, H.J. "Paradoxical effects of exogenous norepinephrine on cold-induced thermogenesis in the rat." **Pharm. Biochem. Behav.**, 39(1), 21-24, 1991.

1408. CRM reports on caffeine. **Pharmaceut. J.**, 19, 430, 1979.

1409. Brem, A.S., et.al., "Toxicity from ea ingestin in an infant: a computer simulation analysis." Pediat. Res. (Balt.), 11, 414, 1977.

1410. Turner, J.E. & Cravey, R.H. "A fatal ingstion of caffeine." **Clinical Toxicology**, 10, 341, 1977.

1411. Greden, J.F. & Domino, L. "Headache." **New Engl. J. Med.**, 303, 221, 1980.

1412. Shen, W.W. & Souza, T.C.D. "Cola-induced pyschotic organic brain syndrome." **Rocky Mtn. Med. J.**, 76, 312, 1979.

1413. Shorofsky, M.A. & Lamm, N. "Caffeine-withdrawal headache and fasting." **N.Y. St. J. Med.**, 77, 217, 1977.

1414. Farkas, C.S. "Caffeine intake and potential effect on health of a segment of Northern Canadian Indigeous People." **Int. J. Addiction**, 14, 27, 1979.

1415. Victor, C.S., et.al., "Somatic manifestations of caffeinism." **J. Clin. Psychiat.**, 42, 185, 1981.

1416. James, J.E. & Stirling, K.P. "Caffeine: a survey of some of the known and suspected deleterious effects of habitual use." **Br. J. Addict.**, 78, 251, 1983.

1417. Raebel, M.A. & Black, J. "The caffeine controversy: what are the facts?" **Hosp. Pharm.**, 19, 257, 1984.

1418. Tyraba, E.E. & Dodson, W.E. "Caffeine secretion into breast milk." **Arch. Dis. Childh.**, 54, 787, 1979.

1419. Minton, J.P., et.al., "Caffeine, cyclic nucleotides and breast diseae." **Surgery**, 86, 105, 1979.

1420. Boyle, C.A., et.al., "Caffeine consumption and fibrocystic breast disease: a case-control epidemiologic study." **J. Natl. Canc. Inst.**, 72, 1015, 1984.

1421. Heyden, S. & Fodor, J.G. "Coffee consumption and fibrocystic breasts: an unlikely association." **Can. J. Surg.**, 29, 208, 1986.

1422. Wilcox, A., Weinberg, C. & Band, D. "Caffeinated beverages and decreased fertility." **Lancet**, 2, 1453, 1988.

1423. Anonymous. "Caffeine may not be good for sperm." **New Scientist**, July 20, 183, 1978.

1424. Friedman, L., et. al., "Testicular atrophy and impaired spermatogenesis in rats fed high levels of the methylxanthines, caffeine, theobromine or theophylline." **J. Environm. Pathol. Toxicol.**, 2, 687, 1979.

1425. Wethersbee, P.S., et.al., "Caffeine and pregnancy - a retrospective survey." **Postgrad. Med.**, 62, 64, 1977.

1426. Weathersbee, P.S., et.al., "Caffeine and pregnancy." **Postgrad. med. J.**, 62, 64, 1977.

1427. Jacobsen, M.F., et.al., "Coffee and birth defects." **Lancet,** I, 1415, 1981.
1428. Borlee, I., et.al., "Facteur de risque pendant la grosesse?" **Louvain Med.**, 97, 279, 1978.
1429. Furmanova, M. & Guzewska, J. "Wplyw kofeiny na biolgoie jadk komorkowego." **Farm. Pol.**, 32, 1027, 1976.
1430. Morris, M.B. & Weinstein, L. "Caffeine and the fetus: Is trouble brewing?" **Amer. J. Obstet. Gynec.**, 140, 607, 1981.
1431. Worthington-Roberts, B. & Wiggle, A. "Caffeine and pregnancy outcome." **J. Obstet. Gynecol. Neonat. Nursing,** 12, 21, 1983.
1432. Mann, J.I. & Thorogood, M. "Coffee-drinking and myocardial infarction." **Lancet**, 2, 1215, 1975.
1433. Josephson, G.W. & Stine, R.J. "Caffeine intoxication: a case of paroxysmal atrial tachycardia." **J. Amer. Coll. Emergency Phycns.**, 5, 776, 1976.
1434. Hemminki, E. & Pesonen, T. "Regional coffee consumption and mortality from ischemic heart disease in Finland." **Acta Med. Scand.**, 201, 127, 1977.
1435. Handel, P.J., et.al. "Physiological responses to cola ingestion." **Res. Quart.**, 48, 436, 1977.
1436. Gould, L., et.al., "Electrophysiological properties of coffee in man." **J. Clin. Pharmacology**, 1, 46, 1979.
1437. Conrad, K.A. & Blanchard, J. & Trang, J.M. "Cardiovascular effects of caffeine in elderly men." **J. Am. Geriatar. Soc.**, 30, 267, 1982.
1438. Dobmeyer, D.J., et.al., "Arrhythmogenic effects of caffeine in man as determined by programmed electrical stimulation." **Clin. Res.**, 30, 182A, 1982.
1439. MacMahon, B., et.al., "Coffee and cancer of the pancreas." **New Engl. J. Med.**, 304, 630, 1981.
1440. Bray, G.A. "Calorgenic effect of human growth hormone." **J. Clin. Edocrin. and Metab.,** 29, 119, 1969.
1141. Woods, S. C., Decke, E. & Vasseli, J.R. "Metabolic hormones and regulation of body weight." **Pysch. Rev.**, 81, 26, 1974.

339

Index

A

abdomen 55, 189
acetic acid 19, 39, 252
ACTH 3, 135, 137, 151, 176, 180, 263, 267. *See also* adrenocorticotropic hormone
adapt to the cold 84
adapted to the cold 117
adenosine 119, 129, 146
   and lipolysis 138-139
   increases blood flow 146
   inhibition of 200-201, 229
   and NE 129, 187
adenylate cyclase 119, 123, 124, 125, 126, 130, 131, 132, 133, 146, 161, 173, 176, -
   189, 200, 202, 222, 233, 260, 292, 293
adipocyte 100
adrenaline 3, 52, 85, 97, 120-123, 134-137, 141-148, 159, 175-179, 183, 188-194, 228, 239,
   259, 261-266, 290
   and caffeine 200
   and carbohydrate 290
   and cayenne 259
   and exercise 183
   and hypertension 159
   and lipolysis 134-137
   and the SAS 141-148
   mechanism of action 175-179
   stores 52
   thermogenic action 203-205
adrenals 22, 50-51, 175-178
exhaustion 51, 178
function of 175-178
herbal support 261-266
rebound 50-51
adrenergic agents 211
Adrenergic Receptors 190
Adrenergic receptors 123
adrenergic receptors 85, 123, 131, 132, 133, 145, 146, 147, 173, 188, 190, 192, -
   198, 200, 201, 211, 229, 289
adrenocorticotropic hormone 3
aerobic exercise 44, 87, 183
age 14, 15, 33, 40, 54, 72, 74, 92, 124, 197, 200, 216, 282, 293
agonists 183, 185, 186, 191, 192, 198, 233
albumin 135, 136, 295
alcohol 160, 212, 280, 293
allergic reactions 23
alpha adrenergic 123, 181, 183, 193-194, 198
ambient temperature 72, 77, 78, 83, 108
American Phytotherapy Research Laboratory 7, 19, 35, 36, 37, 39, 41, 43, 46, 47, -

B

E